INTRODUCTORY
FOOD CHEMISTRY

some other AVI books

Food Science and Technology

BASIC FOOD CHEMISTRY   Cloth and Soft Cover   *Lee*
CEREAL SCIENCE   *Matz*
CHOCOLATE, COCOA AND CONFECTIONERY: SCIENCE AND
   TECHNOLOGY   *Minifie*
ELEMENTARY FOOD SCIENCE   Cloth and Soft Cover   *Nickerson and
   Ronsivalli*
ENCYCLOPEDIA OF FOOD ENGINEERING   *Hall, Farrall and Rippen*
ENCYCLOPEDIA OF FOOD TECHNOLOGY   *Johnson and Peterson*
FABRICATED FOODS   *Inglett*
FOOD ANALYSIS LABORATORY EXPERIMENTS   *Meloan and Pomeranz*
FOOD ANALYSIS: THEORY AND PRACTICE   *Pomeranz and Meloan*
FOOD AND THE CONSUMER   Soft Cover   *Kramer*
FOOD CHEMISTRY   *Aurand and Woods*
FOOD CHEMISTRY   Soft Cover   *Meyer*
FOOD ENGINEERING SYSTEMS, VOL. 1, OPERATIONS   Cloth and Soft
   Cover   *Farrall*
FOOD FLAVORINGS, 2ND EDITION   *Merory*
FOOD FOR THOUGHT   Soft Cover   *Labuza*
FOOD OILS AND THEIR USES   *Weiss*
FOOD PROCESS ENGINEERING   Cloth and Soft Cover   *Heldman*
FOOD PRODUCTS FORMULARY, VOL. 1   *Komarik, Tressler and Long*   VOL.
   2   *Tressler and Sultan*   VOL. 3   *Tressler and Woodroof*
FUNDAMENTALS OF DAIRY CHEMISTRY, 2ND EDITION   *Webb, Johnson and
   Alford*
MICROWAVE HEATING, 2ND EDITION   *Copson*
POULTRY PRODUCTS TECHNOLOGY, 2ND EDITION   *Mountney*
RHEOLOGY AND TEXTURE IN FOOD QUALITY   Cloth and Soft
   Cover   *deMan, Voisey, Rasper, and Stanley*
SNACK FOOD TECHNOLOGY   *Matz*
STORY OF FOOD   Soft Cover   *Garard*
SUGAR CHEMISTRY   *Shallenberger and Birch*
WATER IN FOODS   *Matz*

Nutrition and Biochemistry

CARBOHYDRATES AND THEIR ROLES   *Schultz, Cain and Wrolstad*
CHEMISTRY AND PHYSIOLOGY OF FLAVORS   *Schultz, Day and Libbey*
DIETARY NUTRIENT GUIDE   Soft Cover   *Pennington*
FOOD ENZYMES   *Schultz*
LIPIDS AND THEIR OXIDATION   *Schultz, Day and Sinnhuber*
NUTRITIONAL EVALUATION OF FOOD PROCESSING, 2ND EDITION   Cloth and Soft
   Cover   *Harris and Karmas*
PROTEINS AND THEIR REACTIONS   *Schultz and Anglemier*
PROTEINS AS HUMAN FOOD   *Lawrie*

INTRODUCTORY
FOOD CHEMISTRY

by IRA D. GARARD, Ph.D.

Professor of Chemistry, Emeritus
Rutgers—The State University
New Brunswick, New Jersey

THE AVI PUBLISHING COMPANY, INC.
WESTPORT, CONNECTICUT

Preface

This is a textbook for those college students who have a professional interest in the chemistry of food. It will also be of use to chemists and home economists who are working in food science, as an introduction to unfamiliar parts of the field.

To be able to understand all the chemistry in the book, the student will need a knowledge of organic chemistry and quantitative analysis. Some knowledge of biochemistry and microbiology will be helpful.

The extent of the field of food chemistry is determined by the requirements of nutrition, food processing and the law, and it has expanded enormously in the past half century. Vitamins, amino acids, additives, the increasing transfer of food preparation for consumption from the kitchen to industry and the increase in legal regulation are the main causes of the increase. Our knowledge of the composition of natural foods has also been increased by the development of chromatography and spectroscopy.

Some knowledge of the sources of the natural foods and the processing that they are subjected to, even the history of the food or the process, is essential to an understanding of many of the problems encountered. Lack of space in a textbook requires a limited treatment of these topics; it is hoped that this book treats them adequately for its purpose.

The selection of topics from the vast field of food chemistry is a problem, for a treatment of all of them is impossible. The author has included a thorough treatment of the chemistry of the main food components and a briefer treatment of many topics that may confront the food chemist. The references appended to each chapter will lead to a more thorough grasp of these subjects.

The author is indebted to several scientific journals, monographs, industrial companies, government agencies and personal communications for information outside his own personal experience.

IRA D. GARARD

April 1976

v

Contents

CHAPTER 1

Origins of Food Chemistry

Chemistry is the science of the composition of matter and the changes in composition that occur under changed conditions. Food chemistry, then, concerns the foods for man and animal and includes air and water insofar as they affect food.

The word "food" has many meanings, such as food for thought, plant food, dog food and so on. Our interest here, however, is in animal food, and especially human food, which may be defined as any natural product, fresh or processed, that can nourish the animal body. Definitions differ; some define food as anything that contributes to the growth or physiological functions of the body, but this definition includes drugs. It is practically impossible to make a concise statement that includes all foods and excludes everything else.

In the various industries, food intended for farm animals is called "feed"—horse feed, hog feed, chicken feed; but household pets eat "food"— cat food, dog food, fish food. However, all animal life requires about the same substances for nourishment and so the chemist's work is much the same whether the products with which he works are intended for food or feed.

For the purposes of this book, food is anything that can nourish an animal regardless of the amount of processing required to separate the nutrient from its natural source and prepare it for consumption. The processing itself may be the concern of the chemist. Of course, a chemist is not concerned with the harvesting and threshing of wheat, but he is very much concerned with the grinding of flour and the making of bread.

Space does not permit a history of food chemistry, but the highlights are essential to a realization of the scope of the field.

SOURCES OF FOOD CHEMISTRY

The various discoveries in nutrition and in food processing have all served as matters of new concern to the food chemist.

Preservation of Food

The preservation of food is of ancient origin, and the early processes, such as drying and smoking, were undoubtedly discovered by accident. Edibility and spoilage were discovered by trial and error, but after several thousand years, chemistry came along and enlarged the field of food preservation.

Energy

There was no science of chemistry before the 1770s, but as soon as Lavoisier had established the principles of the science, he turned his attention to the production of human energy, which then became an active topic of research for the next hundred years before the production of energy in man and animals became well understood.

Composition of Foods

As soon as it became evident that animal energy was produced by the oxidation of the carbon and hydrogen in food, the composition of the natural foods came under consideration.

The starches and sugars of foods were the first to be studied. The chemical nature of these substances was not clarified until 1890; they were named *carbohydrates,* however, in 1844.

The chemical nature of fats and their general properties were well established by Chevreul in his publication of 1823.

Proteins were more bothersome. They were soon found to contain nitrogen and the first amino acid was isolated in 1820, but they were called albuminoid substances until 1838 when they were named *proteins.* No clear concept of their chemical nature was established until 1900.

Dietary Requirement

As the composition of the common foods and the chemistry of their components became better known, the problem became: which of these components is essential in the diet and how much of them does a person require? This, of course, was a problem for the physiologist, but the chemist had to provide the pure food components and analyze the secretions and excretions.

Since all the experiments, both biological and chemical, were generally carried out by the same person, the proper name for these scientists was *physiological chemists,* now called *biochemists.*

Since carbohydrates, fats and proteins occur in foods in various proportions, it was more practical to analyze the foods so that the composition of an experimental diet could be calculated. This work was begun about 1840 and still continues when a new food is discovered or manufactured.

It was soon discovered that calcium, phosphorus and iron are also dietary essentials, which added to the chemists' duties.

But the worst was yet to come, for the physiologists found that some proteins are better than others and so, around the turn of the century, chemists were busy isolating pure proteins from foods and trying to find the cause of the difference. It turned out that proteins are composed of about 20 amino acids and by 1940 it was found that eight of them were the real protein essentials. This discovery added the amino acid composition of proteins to the task of the chemists.

Then, about 1906 the physiologists decided that a proper diet also included very small quantities of some organic substances that were present in some foods and not in others. About 1920 these were named *vitamins* and they provided a real challenge to the chemist because of their small concentration in foods and, for years, their properties and sources were entirely a matter of biological experiments. All that the chemists knew about them was that some were soluble in water, and others, which were insoluble in water, were soluble in fats.

By 1931, vitamin C was isolated from lemon juice and a year or so later, a Hungarian chemist obtained 450 gm from peppers. From these beginnings, various vitamins have been isolated, identified and synthesized. As each one was characterized, chemists developed methods for its detection and quantitative determination and used these methods to determine the vitamin content of foods.

Miscellaneous Components

Many chemists were not content with the determination of the nutrient composition of foods but proceeded to find what the other components were, such as acids, pigments, colors and toxins.

LEGAL PROBLEMS

In addition to the numerous problems provided by the nutritionists, another source appeared about 1900. State chemists had discovered that a number of foods sold to the public were adulterated. Consequently, as a result of publicity and activity on the part of these chemists and others, the first federal food law of a general nature was passed. This opened a vast field for food analysts. Not only the purity of foods had to be determined, but also the identity of many of them.

Analysis for purity and identity are not limited to official chemists. A food processor must make sure that his products meet the legal requirements in the geographical area where they are to be sold, otherwise, he may have a large shipment seized by some government agency and may even be subjected to a fine and imprisonment, so it behooves his chemist to learn what the requirements are in his market and see that the products meet them, which frequently requires repeated analysis. Food processes are not too reliable even with the best of intentions on the part of the operator. For example, a food processor made baking powder for use in his product. It was made in 400 lb batches. The man assigned to the work weighed the soda, phosphate and starch and mixed them for a specified time. Every batch had to be analyzed, for the weights of the three components were different and it was a banner day when the operator did not fail to change the weights on his scale for at least one batch.

The establishment of the identity of food products is the concern of both government and processors. Tariff rates on imported oils are different for different oils, consequently, the customs officer must know the identity of an oil to know what the tariff should be. Many food processors purchase foods to be used in making some composite product. If a manufacturer of an olive-oil salad dressing or a corn-oil margarine were informed by some government agency that his product was made from peanut oil, it would be embarrassing at least, and so the chemist must make sure that the oil he is buying is the one he claims to be using. The government chemist is also concerned that the consumer gets the product he pays for.

OFFICIAL ANALYSIS

It is practically impossible to get the same result by different methods of analysis for the same sample. Unlike inorganic material, all foods are unstable and most of them are of complicated composition. The determination of water in a sample of cheese would seem like a simple procedure: weigh a sample, dry it and weigh it again. The loss is the weight of water in the sample. But how does one tell when it is dry? After it is dried and weighed, put it back in the oven for a half hour, then cool it and weigh it again. If it has lost more weight, it was obviously not dry the first time and may not be dry even now, and so, repeat the process until the last two weights are the same. Alas! They probably never will be, for both the carbohydrates and proteins of the cheese decompose slowly at the 100° temperature of the oven. This is not the only difficulty. The cheese will likely melt and then form a crust on top, so the water in the under portion cannot escape.

Many other analytical results differ with the procedure, and when a food product is bought on specification, such as the case of a baker who buys flour from a miller, the seller and the buyer must agree on the method of analysis.

Because of the many complications in analytical work, on September 9, 1884, two chemists from the U. S. Department of Agriculture and state chemists from Connecticut, Georgia, Mississippi, North Carolina, South Carolina and Virginia met in Philadelphia to discuss their problems in analytical chemistry. At the meeting, they organized a society that they called the Association of Official Agricultural Chemists, the AOAC for short. The preamble of its constitution read:

"Membership in the Association is institutional and includes the State Departments of Agriculture, the State Agricultural Colleges and Experiment Stations, the Federal Department of Agriculture and the federal, state and city officers charged with the enforcement of food, feed, drug, fertilizer, insecticide and fungicide control laws."

The Association adopted an elaborate plan for testing each analytical method before it became official. The field of analysis broadened in 1938 with the legislative control of drugs, cosmetics and hazardous substances. Consequently, in 1965 the name was changed to the Association of Official Analytical Chemists, but it is still the AOAC. It has no list of members and no dues. At present, it has over 600 chemists studying analytical methods in 50 different categories. These chemists specialize. There is a Committee on Dairy Products with several subcommittees to deal with fresh milk, butter, cheese and other products of the dairy industry. Sometimes a subcommittee is concerned with a single determination.

Before a method is finally adopted it is subjected to a collaborative study by several chemists who analyze the same sample by the method, and perhaps by other methods to determine the convenience, accuracy and reproducibility of the method.

The Association also has cooperative arrangements with about a dozen other societies that are concerned with analysis in special fields of common interest such as the International Dairy Federation, American Society of Brewery Chemists and the American Public Health Association.

There was nothing "official" about the methods originally—they were simply the methods the official chemists used. However, on June 30, 1906, the first general federal food law was signed by the President and it specified these methods for the determination of adulteration of food, and thereby made the methods actually official.

The AOAC holds annual meetings at which they discuss their problems and they also publish The Journal of the Association of Official Analytical Chemists. In the early years of the Association, the methods were published as a Bulletin of the Department of Agriculture, but since 1915 the Association has published a volume called Official Methods of Analysis of the Association of Official Analytical Chemists. The book is revised every five years; the present edition is the 12th published in 1975. Since some of the methods are dropped or others added at each annual meeting, the Association publishes an annual supplement. It also publishes a Manual of Cosmetic Analysis and books relating to drug analysis.

Any food chemist must have the book of Official Methods, for when he needs to make an analysis these are the methods to use, especially if there is a possibility of litigation. The chemist may find or devise a simpler method than the official one for controlling his processes, but to appear in court with any method other than the official, one is more than likely to lose the case.

SOCIETIES

The food chemist is dependent on the literature in his field, and should belong to the professional societies in order to keep up to date. The three principal ones in the United States are: The Institute of Food Technologists

(IFT), The American Oil Chemists Society (AOCS) and The American Association of Cereal Chemists (AACC). Chemists in a special field may need to belong to one of the societies that are more specialized. There are several of these, here and abroad, that deal with a single industry—dairy industry, brewing and so on. Each society publishes a journal. The IFT publishes two, *Food Technology* and *The Journal of Food Science*. The former contains articles on food processing and the latter on the scientific aspects of food. The AOCS and the AACC publish articles in the chemistry of their respective fields.

Attendance at a meeting of one of these societies and getting acquainted with the members is often a quick and easy way to learn what is new in your field of interest. Any member probably knows something that you do not.

LIBRARY

The student should survey the shelves in the food section of his university library to become acquainted with the books on special topics in the field of food chemistry, for there are many books that deal with a small area of the food field. There are books on margarine, cheese, wheat grain and many other special topics as well as the more general ones, such as dairy chemistry, fats and oils, and baking. The alert chemist must know the literature of his field.

When the student finishes school and goes to work in a large laboratory, either official or industrial, he will probably find an adequate library. If he goes into a small laboratory where he is alone, he will need to acquire a small library of books and journals that treat the subjects he has to deal with.

The journals contain reviews and advertisements of new books, and the chemist should at least scan them for anything new in his field of interest.

The references at the end of the subsequent chapters will serve as a starting point in the literature and may be adequate for most work.

MISCELLANEOUS INFORMATION

One source of information not to be overlooked is advertising. Even the weekly ad of a supermarket may announce an item you didn't know existed, and it may be a competitor of one of your own products.

Salesmen are another source of information not to be ignored. A good one has a surprising knowledge of technical information and of commercial practice. If he is selling equipment, he knows what it will accomplish and who is using it; if he is selling a basic food product, he knows who is using it and for what purpose.

There are usually visits to food plants during the annual meetings of the societies. These are worthwhile, for the manager of the plant assigns a guide who is familiar with the process and, although he will not disclose any secrets, he will give you a much better idea of his portion of the food industry than you can obtain from a book.

If you travel on your vacation, visit any food industry you come upon. It will take no more than an hour or so. Even unfamiliar agriculture is worthwhile. I once saw some men and women working in a vegetable field. They didn't seem to be harvesting the crop and so I stopped and inquired what they were doing. They were pulling weeds out of the rows of spinach, which disclosed a problem of vegetable processing that I had not thought of—in case you are involved with the canning or freezing of spinach or other greens, look out for weeds.

RESEARCH

The chemist in the laboratory may have problems that do not come from either the nutritionists or the law.

Originally, food was grown on the farm and processed in the kitchen, but as towns and cities grew, more and more processing was done in industrial plants and the product sold in the stores. Not only was wheat ground, cattle butchered, sugar refined and such, but new products were developed, such as salad dressings, table syrups, frozen desserts and even frozen full dinners.

A food processor may wish to market a new product to add to his list, and the chemist must formulate it and be sure that it has an acceptable color, flavor and texture and a suitable shelf life. This often presents a real problem and the chemist must know the chemistry of everything that he includes in the product.

Very often the chemist must undertake research because his training, experience and library do not solve his problem. He should then go to the nearest library and consult *Chemical Abstracts* to see if there is any literature on the subject that he does not know about, for library work is much faster and cheaper than laboratory work unless the nearest library is many miles away. There may be an important article in a foreign journal, which can be rented from *Chemical Abstracts* if the abstract is not sufficient. The abstractor will probably be willing to make a translation of the article, or, in many localities one can find an immigrant who can read the language. I once encountered an article in Polish that I needed to read. The college personnel director sent me a student major in my own department who, to my amazement, sat down by my desk and read the article in English while I made notes. Personnel officers know a good deal about your colleagues and if they fail, there is always the want ad section of your newspaper.

A textbook cannot hope to cover more than the general principles of food chemistry and to indicate their general application to food problems. Once out of school, the chemist is on his own, and I know of no field of chemistry that offers more diversity or a greater challenge.

In this chapter, I have tried to indicate that food chemistry does not yield to a single textbook or laboratory manual but that it is connected with nutrition, the law and with a huge and varied industry. I have also tried to indicate to the new chemist how to go about his duties in case he finds himself without a director.

In the chapters that follow, I shall present the chemistry of those substances that make up our food. Once he has mastered the chemistry presented here and the basic laboratory operations, the chemist should have no difficulty reading journal articles or detailed books on special subjects. His big problem will not be how to carry out laboratory processes, but what ones to carry out and when they are not necessary. Not a single problem mentioned in this chapter has been laid to rest; they often turn up in the most unexpected places. The solution of one problem invariably creates one or more others.

BIBLIOGRAPHY

ASSOC. OFFIC. ANAL. CHEMISTS. 1975. Official Methods of Analysis, 12th Edition. Assoc. Offic. Anal. Chemists, Washington, D.C.

BEATON, G. H. 1964A. Nutrition, Vol. 1, Macro-Nutrients and Nutrient Elements. Academic Press, New York.

BEATON, G. H. 1964B. Nutrition, Vol. 2, Vitamins, Nutrient Requirements and Food Selection. Academic Press, New York.

BEATON, G. H. 1966. Nutrition, Vol. 3, Nutritional Status, Assessment and Applications. Academic Press, New York.

GOLDBLITH, S. A., and JOSLYN, M. A. 1964. Milestones in Nutrition. Avi Publishing Co., Westport, Conn.

HARRIS, R. S., and KARMAS, E. 1975. Nutritional Evaluation of Food Processing. Avi Publishing Co., Westport, Conn.

HEID, J. L., and JOSLYN, M. A. 1967. Fundamentals of Food Processing Operations. Avi Publishing Co., Westport, Conn.

KRAMER, A. 1973. Food and the Consumer. Avi Publishing Co., Westport, Conn.

KRAMER, A., and TWIGG, B. A. 1970. Quality Control for the Food Industry, 3rd Edition. Vol. 1, Fundamentals. Avi Publishing Co., Westport, Conn.

KRAMER, A., and TWIGG, B. A. 1973. Quality Control for the Food Industry, 3rd Edition. Vol. 2, Applications. Avi Publishing Co., Westport, Conn.

MARGEN, S. 1971. Progress in Human Nutrition, Vol. 1. Avi Publishing Co., Westport, Conn.

MCCOLLUM, E. V. 1957. A History of Nutrition. Houghton Mifflin Co., Boston.

POTTER, N. N. 1973. Food Science, 2nd Edition. Avi Publishing Co., Westport, Conn.

PRESCOTT, S. C., and PROCTOR, B. E. 1937. Food Technology. McGraw-Hill Book Co., New York.

SHERMAN, H. C. 1948. Food Products. Macmillan Co., New York.

VON LOESECKE, H. W. 1949. Outlines of Food Technology. Reinhold Publishing Corp., New York.

WATT, B. K., and MERRILL, A. L. 1963. Agriculture Handbook No. 8, Composition of foods. ARS-USDA, Washington, D.C.

WHITE, A., HANDLER, P., and SMITH, E. C. 1973. Principles of Biochemistry, 5th Edition. McGraw Hill Book Co., New York.

CHAPTER 2

Composition and Properties of the Fats

The American food supply contains 11.5 billion lb of fats annually; therefore, the chemical and physical properties of these substances are of the utmost importance to the nutritionist and the food chemist.

FATTY ACIDS

The fats are mixtures of *glycerides*, which are esters of the triatomic alcohol glycerol, $C_3H_5(OH)_3$, and some of the fatty acids. The fatty acids of the paraffin series, the saturated series, each contains an even number of carbon atoms and conforms to the general formula, $C_nH_{2n+1}COOH$. The other acids that occur in fats also have an even number of carbon atoms, but they are unsaturated and belong to the ethylene, diethylene or other unsaturated series of compounds.

Saturated Acids

The structure of the simplest acid obtained from fats is

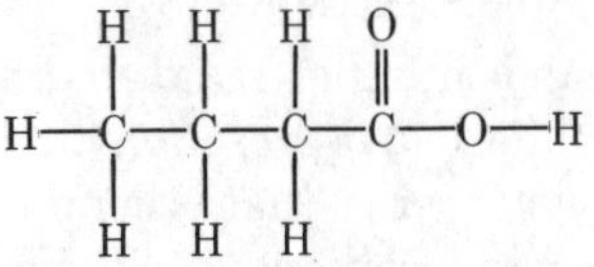

$$\begin{array}{ccccccc} & H & H & H & O & & \\ & | & | & | & \| & & \\ H- & C- & C- & C- & C & -O- & H \\ & | & | & | & & & \\ & H & H & H & & & \end{array}$$

Butyric Acid

Add two CH_2 groups to the formula of butyric acid and you have the formula of the next one, which is caproic

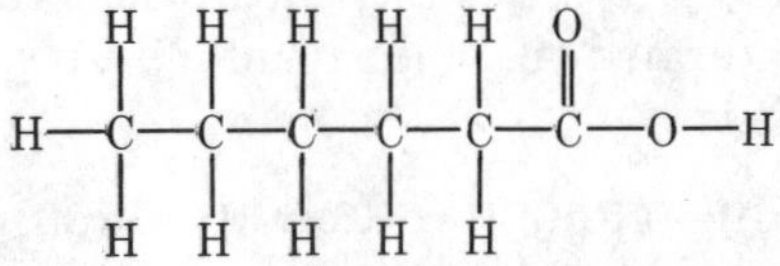

$$\begin{array}{ccccccccc} & H & H & H & H & H & O & & \\ & | & | & | & | & | & \| & & \\ H- & C- & C- & C- & C- & C- & C & -O- & H \\ & | & | & | & | & | & & & \\ & H & H & H & H & H & & & \end{array}$$

Caproic Acid

Lengthening the chain in this manner produces the structural formulas of all the other acids in the series that occur in fats. To get the formula of a given acid, however, it is easier to use the above general formula. For example, for the acid with 10 carbon atoms, N = 9; therefore, C = 9, H = 19, and so the formula is $C_9H_{19}COOH$.

TABLE 2.1

THE FATTY ACIDS

Common Name	Systematic Name	Formula	Melting Point (°C)	Neutralization Value
Butyric	butanoic	C_3H_7COOH	−8	636
Caproic	hexanoic	$C_5H_{11}COOH$	−1	483
Caprylic	octanoic	$C_7H_{15}COOH$	16	389
Capric	decanoic	$C_9H_{19}COOH$	31.3	326
Lauric	dodecanoic	$C_{11}H_{23}COOH$	43.5	280
Myristic	tetradecanoic	$C_{13}H_{27}COOH$	54.4	246
Palmitic	hexadecanoic	$C_{15}H_{31}COOH$	62.9	219
Stearic	octadecanoic	$C_{17}H_{35}COOH$	69.6	197
Arachidic	eicosanoic	$C_{19}H_{39}COOH$	75.4	180
Oleic	9-octadecenoic	$C_{17}H_{33}COOH$	14	198
Linoleic	9,12-octadecadienoic	$C_{17}H_{31}COOH$	—	200
Linolenic	9,12,15-octadecatrienoic	$C_{17}H_{29}COOH$	—	202
Arachidonic	5,8,11,14-eicosatetraenoic	$C_{19}H_{31}COOH$	—	184

Unsaturated Acids

The first three of the unsaturated acids are related to stearic—they all contain 18 carbon atoms. *Oleic,* the first one, has a double bond in the center and is 9-octadecenoic acid

$$CH_3(CH_2)_7CH{=}CH(CH_2)_7COOH$$
Oleic Acid

The C of the COOH group is number 1 and so the double bond occurs between carbons 9 and 10. The acid is present in all the edible fats and is the only member of the ethylene series that is common in fats.

Linoleic acid, or 9,12-octadecadienoic acid, has the formula

$$CH_3(CH_2)_4CH{=}CHCH_2CH{=}CH(CH_2)_7COOH$$
Linoleic Acid

This acid is common to many of the food fats, particularly the oils.

Linolenic acid, or 9,12,15-octadecatrienoic acid occurs in many food oils and in the more unsaturated oils of the paint industry, such as linseed and tung oils. Its formula is

$$CH_3CH_2CH{=}CHCH_2CH{=}CHCH_2CH{=}CH(CH_2)_7COOH$$
Linolenic Acid

Arachidonic acid has recently become important in nutrition and the amount of it in the common fats has not yet been extensively determined, but it is known to be present in several of them. Its empirical formula is $C_{20}H_{32}O_2$. From the C_{20} and the low H content, it is obvious that it is related to arachidic and is highly unsaturated; it contains four double bonds, which are at positions 5, 8, 11 and 14.

The unsaturated acids with two or more double bonds are the *polyunsaturated* acids. Linoleic and arachidonic are essential nutrients. Linoleic cannot be synthesized in the animal body but occurs abundantly in food oils with the exception of olive and coconut. The body can synthesize arachidonic from linoleic.

The unsaturated acids in fats are very difficult to isolate and purify. Furthermore, there are several possible geometric isomers of each of them. Of oleic with only one double bond there are two, a *cis* and a *trans* form; linoleic has two *cis* and two *trans* forms and linolenic has three of each. Arachidonic may have eight geometrical isomers.

Chemical constants of the acids may be calculated from their formulas. For example

$$C_3H_7COOH + KOH \rightarrow C_3H_7COOK + H_2O$$

$$\begin{array}{cc} 88 \text{ gm} & 56\,000 \text{ mg} \\ 1 \text{ gm} & x \end{array}$$

$x = 636$ mg per gm of acid, which is the neutralization value of butyric acid.

Physical constants cannot be calculated and few have been determined because of the difficulty in getting the pure acid.

Nomenclature

The common names of the acids were derived from the name of the source from which they were first obtained. Butyric was prepared from butter; caproic, caprylic and capric were isolated from the fat of goat's milk (*L. capra*, a goat). Lauric was first obtained from the seed fat of the laurel tree and myristic from nutmeg oil; the nutmeg tree is *Myristica fragrance*. Palmitic came from palm oil, stearic from stearin, which is the Greek word for tallow and arachidic from peanuts, *Arachis hypogaea*. Linoleic and linolenic acids were isolated from linseed oil.

More recently, the acids have been given chemical names, which are shown in Table 2.1, that indicate the number of carbon atoms and the number of double bonds. For the latter, *an* = 0, *ene* = 1 and they continue with *diene*, *triene*, etc.

GLYCERIDES

There are three kinds of glycerides; mono-, di-, and triglycerides. Monobutyrin is an example of the first type

$$
\begin{array}{c}
\text{H} \quad\quad\quad \text{O} \\
| \quad\quad\quad\quad || \\
\text{H}-\text{C}-\text{O}-\text{C}-\text{C}_3\text{H}_7 \\
| \\
\text{H}-\text{C}-\text{OH} \\
| \\
\text{H}-\text{C}-\text{OH} \\
| \\
\text{H}
\end{array}
$$

Monobutyrin

TABLE 2.2

ACID CONTENT OF FATS

Fat	Percent of Total Acids				
	C_{14}	C_{16}	C_{18}	Oleic	Linoleic
Butter[1]	10.5	27.6	8.4	28.9	3.8
Chicken[2]	0.2	25.6	7.0	38.4	21.3
Lard	0.9	26.7	13.3	49.0	5.3
Tallow	4.2	30.0	14.3	49.6	2.5
Cocoa butter	—	24.4	35.4	38.1	2.1
Coconut[3]	18.0	9.2	2.0	7.6	1.6
Corn	—	7.8	3.5	46.3	41.8
Cottonseed	—	20.2	2.0	30.1	44.5
Olive	0.5	10.4	2.1	83.7	4.3
Peanut[4]	—	8.0	4.0	60.0	22.0
Safflower[5]	—	4.2	1.6	26.3	67.4
Soybean[5]	1.0	7.0	6.0	26.0	54.0

[1] Butter fat also contains butyric 3.4, caproic 1.6, caprylic 1.2, capric 2.7 and lauric 3.8.
[2] Chicken fat contains hexadecanoic 7.
[3] Coconut fat contains caproic 0.5, caprylic 7.9, capric 7.5 and lauric 45.0.
[4] Peanut oil contains arachidic 4.0.
[5] Safflower oil contains 0.1 and soybean 6.0 linolenic.

If two of the OH groups are esterified, the product is a diglyceride

$$CH_2O\!-\!CO\!-\!C_3H_7$$
$$CHO\!-\!CO\!-\!C_3H_7$$
$$CH_2OH$$

Dibutyrin

If all three are esterified the product is a triglyceride

$$CH_2O\!-\!CO\!-\!C_3H_7$$
$$CHO\!-\!CO\!-\!C_3H_7$$
$$CH_2O\!-\!CO\!-\!C_5H_{11}$$

Caprodibutyrin

The di- and triglycerides may be either simple or mixed—simple if all the acid radicals are alike, mixed if they are not. The triglycerides may have three different radicals, for example, oleopalmitostearin.

COMPOSITION OF FATS

The natural fats are mixtures of mixed glycerides. Simple glycerides are scarce in natural fats; milk fat contains all the low acids in Table 2.1, but there are no simple glycerides of any of the first three acids.

TABLE 2.3

UNSAPONIFIABLE CONTENT OF FATS

Fat	%	Fat	%
Butter	0.3–0.6	Lard	0.6–0.8
Cocoa butter	0.3–0.8	Olive oil	0.6–1.3
Coconut oil	0.2–0.4	Peanut oil	0.2–0.8
Corn oil	1.5–2.8	Safflower oil	0.5–1.3
Cottonseed oil	0.7–1.5	Sesame oil	0.7–1.3

It is very difficult to determine the percentage of the various glycerides present in a natural fat, nor is it important to the food chemist to do so. The individual acids present are of more importance and the acid content of a few food fats is shown in Table 2.2.

The acid composition of fats reported in the literature varies considerably. The methods of determination are not very precise and different samples of the same fat have somewhat different compositions. Animal fats, both milk and body, differ with the diet of the animal. Fats of vegetable origin differ with the soil and climate, with the variety and the stage of maturity. The values in Table 2.2 are average values compiled from several sources.

Unsaponifiables

The unsaponifiable content of a fat is defined as those compounds that are soluble in ether but not saponified by strong alkali.

The Official Method for their determination saponifies a sample of the fat with a solution of KOH in alcohol. Water is then added and the unsaponified matter is extracted from the soap solution with ether. The ether solution is then washed well with water, the ether evaporated and the residue weighed. Considerable attention to detail is necessary.

Table 2.3 shows the range of unsaponifiable matter to be expected in a food fat.

The small percentage of unsaponifiable matter led to its neglect for many years; it was not toxic and did not affect the use of the fats for any purpose. More recently, research in the chemistry of food and nutrition has indicated the importance of several components of this fraction of the fats.

Sterols.—Cholesterol is present in animal fats and, in recent years, it has been shown to be associated with diseases of the heart and arteries. However, the human body makes cholesterol from the saturated fatty acids, and whether the amount in foods is of real importance is doubtful at present. To be on the safe side, doctors are recommending avoidance by heart patients of foods that contain this sterol. The subject of the role of cholesterol in nutrition is an active topic of research. It is a well-known constituent of brain, nerve and other body tissues.

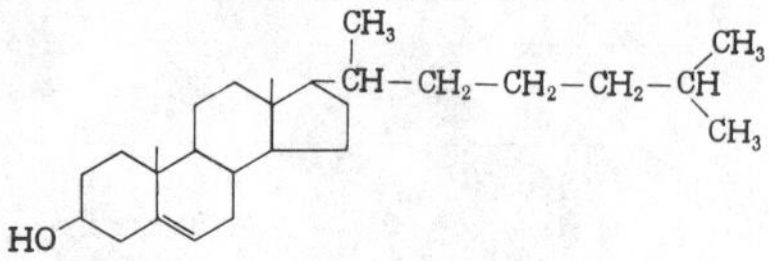

Cholesterol

Cholesterol occurs in animal fats only and serves as a basis of an analytical method to distinguish between animal fats and those of vegetable origin. The Official Methods include procedures for its determination in baked products, eggs, macaroni and mixed animal and vegetable fats.

A qualitative test converts the sterols to the acetate by heating with acetic anhydride. The purified cholesterol acetate melts at 114°C; the acetates of the various phytosterols melt at temperatures from 126°C to 144°C.

The quantitative method depends on the addition of digitonin, $C_{56}H_{92}O_{29}$, which forms a precipitate with the sterol in a 95% alcoholic solution. The method is used in medical practice to determine cholesterol in blood and also in food chemistry to identify animal fats.

There are several sterols present in fats of vegetable origin that are collectively known as *phytosterols*. All the sterols found in fats have the basic phenanthrene structure with a 5-carbon ring fused on

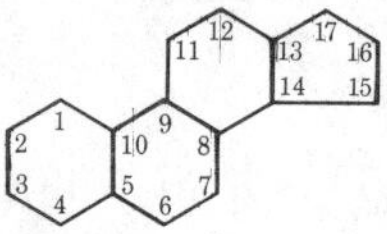

Basic Structure of Sterols

The OH group at 3 and the methyl groups at 10 and 13 are common to all of them; they differ in the aliphatic side-chain attached to carbon 17.

Sitosterol has the side-chain

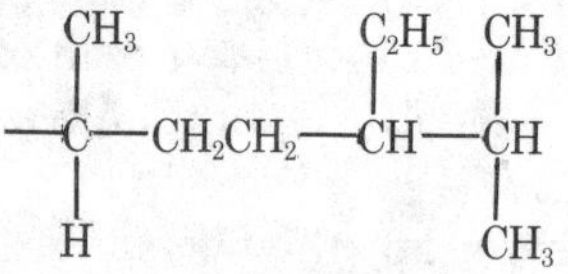

Ergosterol is present in yeast and is important in nutrition because it becomes vitamin D when it is irradiated. Its basic structure differs from that of the other sterols by having a double bond between carbons 5 and 6 and between 7 and 8.

The irradiation of ergosterol produces some profound changes in the structure of the molecule, which will be apparent from a comparison of the two formulas.

Hydrocarbons.—In addition to the sterols, there are several hydrocarbons that may be present in the unsaponifiable portion of a fat.

Carotene is present in several fats of vegetable origin and the cow concentrates it, from the grass she eats, in the milk fat. Since the sheep and the goat do not concentrate this hydrocarbon in the milk fat, a carotene determination will distinguish between the milk of these animals.

Carotene has the empirical formula $C_{40}H_{56}$. Its structure is

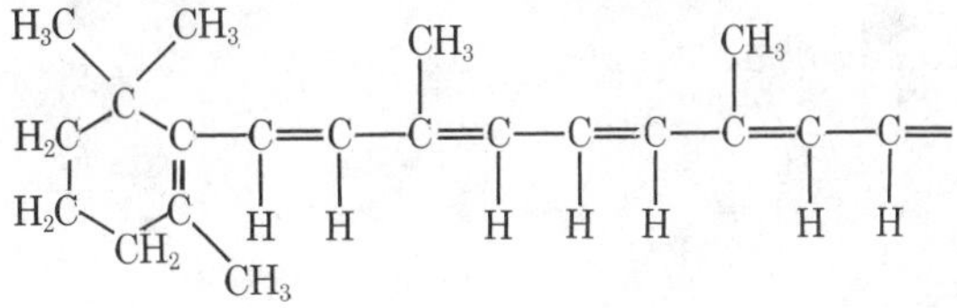

Carotene
½ molecule

Natural carotene is a mixture of three isomers, alpha, beta and gamma. The natural mixture usually contains 85% beta, 15% alpha and a trace of gamma. The above formula is that of the beta form. The molecule is symmetrical and the animal mechanism splits it in the middle and oxidizes the free end to a primary alcohol, which is vitamin A

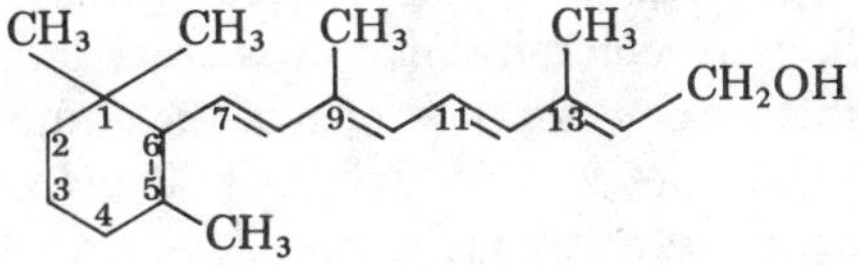

Retinol (vitamin A_1)

Carotene is therefore a precursor of vitamin A, which also occurs already formed in several edible fats and in green leaves that also contain carotene.

Squalene, $C_{30}H_{50}$ is another hydrocarbon that occurs in some fats. It has the structure

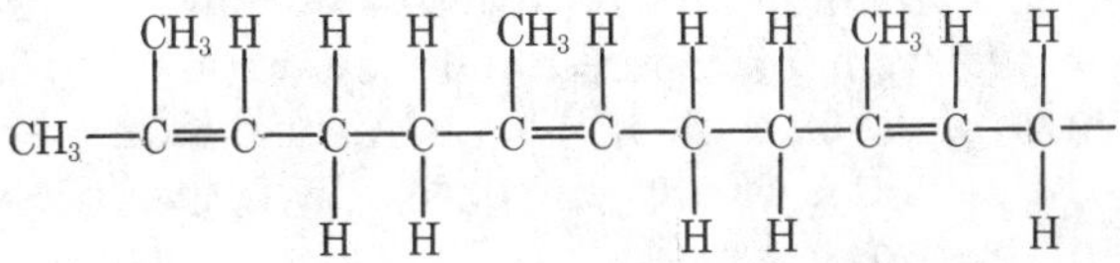

Half of squalene formula

This hydrocarbon is also symmetrical; the formula shows only half of the molecule, the other half continues with the first half in reverse.

Squalene occurs in fish liver oils and in olive oil. Its determination is useful for detecting olive oil in other vegetable oils. The Official Methods contain procedures for its determination.

Chlorophyll

Chlorophyll, the green pigment of vegetation, is found in olive oil, avocado oil, soybean oil from green beans and in a few other sources.

TABLE 2.4

FAT SOLVENTS

Solvent	Boiling Point (°C)
Acetone	57
Benzene	80
Carbon disulfide	46
Carbon tetrachloride	77
Chloroform	61
Ether	35
Dichloroethylene	52
N-Hexane	69
N-Pentane	36
Petroleum ether	60–90

PHYSICAL PROPERTIES OF THE FATS

Solubility

The fats are all insoluble in water but, above their melting points, they are soluble without limit in many organic solvents; at low temperatures they are less soluble. The most common fat solvents are listed in Table 2.4.

The fats are soluble without limit in most of the liquids in Table 2.4. The table contains only the more common solvents. There is a large number of hydrocarbons, chlorinated hydrocarbons, ethers, and esters that dissolve fats.

The fats are only slightly soluble in the alcohols of low molecular weight, but increase in solubility as the temperature rises. Figure 2.1 shows the solubility of three fats in 95% ethyl alcohol.

In general, the fats increase in solubility with the molecular weight of the alcohol, the temperature and the degree of unsaturation of the fat.

The fat solvents are important for two reasons: they are used commercially to extract fats from their sources and they are used by the analyst for the determination of the fat content of foods. An analytical method specifies the solvent to be used. It is usually anhydrous ethyl ether, petroleum ether, or a mixture of the two.

Ether.—Ethyl ether is a hazard in the laboratory. In the determination of fat in food, the fat is extracted and then the ether is evaporated in an open flask. This cannot be done over a burner or a hot plate as one evaporates water, for the ether vapor is much heavier than air (74 versus 29) and will descend to the burner or the hot plate and ignite. Nor can the flask be put into an electric drying oven while there is still ether in it, for many such ovens have the make and break inside the drying chamber and the spark will ignite an ether-air mixture.

Furthermore, ether forms unstable peroxides. They are less volatile than

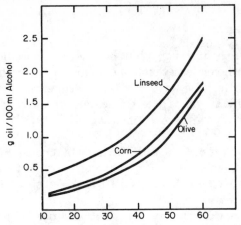

FIG. 2.1. SOLUBILITY OF FATS IN ALCOHOL

the ether, concentrate as the ether evaporates and will explode when the ether is removed. Ether should be tested for peroxides by adding a few milliliters of an acidified 2% solution of potassium iodide to an equal volume of ether in a test tube. The peroxides oxidize the iodide to free iodine, which can be seen or made more prominent by adding a few drops of a starch solution. Copper and iron prevent or retard the formation of these peroxides and manufacturers often add one or the other of these metals to cans of ether. If ether is allowed to stand in a glass bottle, it should contain a small coil of copper wire, but even then the ether should be tested before it is used for any purpose that involves removing it by evaporation. One method of removing peroxides is to shake the ether in a separatory funnel with a solution of ferrous sulfate. The Official Methods include a procedure for preparing anhydrous ether.

Commercial Solvents.—Several factors enter into the selection of a fat solvent for the commercial extraction of a food fat from seeds or other vegetable matter. Cost is important. Petroleum fractions have been the cheapest, but this may not continue to be so. The boiling point and flammability are also important; the solvent must be easily removed from the fat—some solvents are toxic and some, carbon disulfide for example, have a vile odor.

The solvent is usually removed by a steam distillation; the steam and the vapor of the solvent are condensed and form two layers since most of the commercial solvents are insoluble in water. Acetone, however, is completely soluble in water and would have to be recovered by a fractional distillation, which is an expensive operation.

ANALYTICAL CONSTANTS

Specific Gravity

All the fats are lighter than water, but the exact specific gravity is useful to the chemist who has occasion to distinguish between them. Fats expand considerably with rising temperature, consequently, the temperature at which the specific gravity is determined must be specified. Coconut oil, for example, has the specific gravity 0.926 at 15°C, 0.9188 at 25°C and 0.9150 at 30°C.

Specific gravity is one of the items called "fat constants." They are not very constant, however, and "fat characteristics" is now often used; the specific gravity of olive oil is 0.914–0.918 and corn oil is 0.921–0.928; all the other oils show similar variations.

The fats do not differ greatly from each other in specific gravity. In general, the values increase with increasing unsaturation and with diminishing molecular weight of the glycerides; the specific gravity of peanut oil is 0.914–0.926 at 15° and its iodine value is 92–100, while linseed oil has the specific gravity 0.930–0.938 at 15° and an iodine value of 175–202. The iodine value is a measure of the unsaturation of the oil. Likewise, tallow has the specific gravity 0.896 and a saponification value of 196–200 while coconut oil has the specific gravity 0.920 and a saponification value of 253–262. The saponification number is a measure of the molecular weights of the glycerides; the larger the number, the lower the weight.

The determination of the specific gravity of a fat is very simple in principle. The method calls for a special pyknometer of thin glass that holds about 25 ml. The pyknometer is cleaned, dried and weighed. It is then filled with recently boiled, distilled water, adjusted to temperature and weighed again. This gives the weight of the water it holds. Then, the pyknometer is dried, filled with the fat, again adjusted to temperature and weighed, which supplies the weight of the fat. The exact volume of the pyknometer does not enter into the determination since the two volumes are the same; the specific gravity is the ratio of the weight of the fat to that of the water at the same, or some specified temperature.

A specific gravity determination is tedious because the pyknometer and water and also the pyknometer and oil must be adjusted accurately to temperature. The Official Method makes the determination at 25°C for both the water and the fat. In Europe, the laboratories are much colder and so the chemists use a lower temperature, and many chemists in this country have followed their example, consequently, many specific gravity tables report the values at 15.5°C/15.5°C. The Official Methods make the determination at 25°C/25°C.

In order to compare results, the Official Methods offer the following formula:

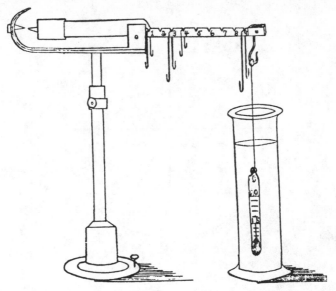

FIG. 2.2.   THE WESTPHAL BALANCE

$G = G' + 0.00064 \, (T - 25°\text{C})$
G = sp gr at 25°/25°C
G' = sp gr at T/25°C
T = temperature at which sp gr is measured.

Because of the time required to determine the specific gravity of a fat by the above method, a Westphal balance is sometimes used in routine work. The method takes much less time and the precision is within ±0.0005 if the temperature is carefully adjusted. The Official Methods do not recognize the method.

Index of Refraction

When light passes at an angle from one medium to another, its direction is changed at the interface of the two media. Here we are concerned with the change in direction when light passes from air into a fat. The actual change depends on the angle at which the light strikes the fat and, in order to get a significant value, the chemists have resorted to trigonometry. In Fig. 2.3, A,B represents the surface of the fat and C,D is a line perpendicular to the surface. E,O is the incident ray of light and F,O the refracted ray; consequently, I is the *angle of incidence* and R the *angle of refraction*. The index of refraction is the ratio of the sines of these angles and is indicated by the letter N,

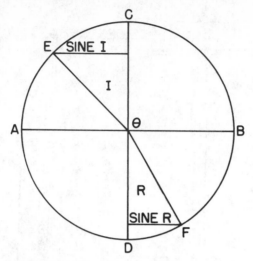

FIG. 2.3.   REFRACTIVE INDEX

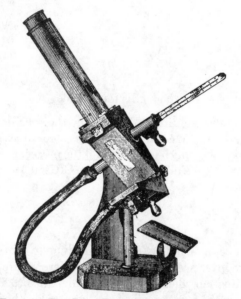

FIG. 2.4.   THE ZEISS BUTYRO-REFRACTOMETER

$$N_D = \frac{\text{Sine I}}{\text{Sine R}}$$

For precise work, the wavelength of the light must be specified; sodium light, the D line of the spectrum, is generally used and so the tables report

TABLE 2.5

SPECIFIC GRAVITY AND INDEX OF REFRACTION OF FATS

Fat	Specific Gravity 15°/15° C	Index of Refraction 25° C
Corn	0.921–0.926	1.4733–1.4742
Cottonseed	0.917–0.925	1.4660–1.4680
Linseed	0.930–0.938	1.4786–1.4810
Olive	0.914–0.919	1.4657–1.4690
Peanut	0.917–0.926	1.4620–1.4706
Safflower	0.925–0.928	1.4769–1.4775
Sesame	0.919–0.924	1.4704–1.4730
Soybean	0.922–0.925	1.4722–1.4750
Sunflower	0.922–0.927	1.4658–1.4720

N_D. The temperature makes a big difference in the value and should be recorded, $N_D.^{40°}$ The Official Methods use 25°C for oils and 40°C for plastic fats.

The determination may be made with a standard Abbe refractometer or a butyrorefractometer. Both instruments have jackets around the prisms for the circulation of water to adjust the temperature. If results are to be compared with those in a table, the temperature must be controlled.

The index of refraction varies with the specific gravity and in the same direction (Table 2.5). The determination is made very quickly and requires only a few drops of the fat.

Melting Point

The term "solid fat" is a misnomer and is used only as a matter of convenience. Lard and tallow are plastic masses consisting of solid crystals of glycerides and some that are liquids. The solid glycerides have different melting points and, as the temperature rises, the fat softens as one glyceride after another melts. Pure solids, such as ice, have a sharp melting point, but fats, like all other mixtures, melt over a range of temperature.

The melting range of the common salad and cooking oils is hard to determine and of little use; it is, however, applicable for the plastic fats. There are two Official Methods for its determination. In one, the melted fat is drawn into a capillary tube and the bottom of the tube sealed. The tubes are then refrigerated at 4°C–10°C for about 16 hr. A tube is attached to a thermometer with the fat opposite the bulb and heated slowly in a beaker of water.

The Wiley method is unique. The melted fat is dropped onto a cake of ice floating in water. The disk formed should be about a centimeter in diameter and weigh about 200 mg. The determination depends on the insolubility of fats in dilute alcohol and also on the fact that fats are lighter than water but heavier than alcohol. After the disk has remained in contact with

TABLE 2.6

MELTING POINTS AND TITER NUMBERS

Fat	Melting Range	Titer Value
Beef tallow	44—46	38—46
Butter fat	28—35	38—42
Cocoa butter	30—34	48—50
Coconut oil	23—26	22—25
Lard	36—46	38—46
Palm oil	27—50	42—45
Palm kernel oil	24—28	20—26
Corn oil	—10——20	18—20
Cottonseed oil	—12——13	32—36
Peanut oil	—20	30—39
Safflower oil	—13——18	11—17
Sesame oil	—4——6	25—35
Sunflower seed oil	—17	22—24
Olive oil	—	26—30
Teaseed oil	—	13—15

the ice for about 2 hr, a large test tube is filled half full of recently boiled, distilled water and then 95% alcohol is added as carefully as possible until the tube is nearly full. A disk of fat is added and sinks to the level where the diluted alcohol has the same specific gravity as the fat. A thermometer is inserted with the bulb near the disk and the tube is heated in a beaker of water. At the melting point the opaque disk collapses to a transparent sphere.

Crystals of glycerides are polyphoric, that is, crystals of different structure form under different conditions and each form has a different melting point. If tristearin is cooled rapidly from the liquid state, the crystals melt at 54.5°C, but if the liquid is cooled slowly, they melt at 65°C and if the cooling is very slow, they melt at 71.5°C.

The expression "fats and oils" is a loose one that appears in the literature and depends on the melting point. If a fat is a liquid at ambient temperature it is called an oil; the solid or plastic fat has no special name—it is just a fat. Coconut oil is an oil in its native tropics but a fat in an American laboratory.

Titer Test

The titer value is the melting point of the acids from a fat. The fat is saponified with KOH in glycerol and the acids freed with H_2SO_4. The melting point procedure and the preparation of the acids are detailed in the Official Methods. The values are sometimes more useful than the other constants for the identification of a fat (Table 2.6).

CHEMICAL PROPERTIES

There are five chemical properties of fats that are of interest to the food chemist: hydrolysis, unsaturation, combustion, slow oxidation and decomposition.

Hydrolysis

Since the fats are esters, their outstanding property is hydrolysis. Heated with water and a catalyst, they split into glycerol and the component acids. The reaction is reversible and the acids must be removed to bring the reaction to completion. This is mainly an industrial reaction, and the food chemist seldom has occasion to use it.

Enzymes likewise hydrolyze fats to glycerol and the acids, which is the reaction of the digestion of fats by man and other animals. Fats in storage may also deteriorate by enzymic hydrolysis.

If a fat is heated with a strong alkali, the acids are neutralized as rapidly as they are released and thus removed from the reverse reaction. This is the reaction by which fats are converted into soap and is, therefore, called *saponification*.

$$C_3H_5(OCOC_{17}H_{33})_3 + 3NaOH \rightarrow C_3H_5(OH)_3 + 3NaOCOC_{17}H_{33}$$

Triolein   Lye   Glycerol   Sodium oleate

The food chemist is not interested in saponification on a large scale unless his company is also in the soap business, but he does find the reaction very useful as a laboratory process.

Saponification Value.—The number of milligrams of KOH required to saponify 1 gm of fat is called the *saponification number*, or *value*. To determine the number, 2.5–3 gm of the fat are saponified with an alcoholic solution of KOH. Alcohol is used as the solvent because fats are more soluble in it than in water and so the reaction is faster than it is in an aqueous medium. The solution is boiled to saponify the fat quickly, cooled and titrated with 0.5 N HCl. The KOH solution is not standardized and so an equal portion of it (25 ml) is titrated. The difference between the two titrations is the volume of 0.5 N KOH required to saponify that sample of fat. From the weight of the fat and the volume of 0.5 N KOH, the value can be calculated, since 1 ml of 0.5 N KOH contains 28 mg KOH.

The saponification number is a measure of the weighted average of the molecular weights of the acid radicals of the fat.

Reichert-Meissl and Polenske Values.—The Reichert-Meissl value is the number of milliliters of 0.1 N alkali required to neutralize the volatile, soluble acids from 5 gm of fat; the Polenske value is the number required to neutralize the volatile insoluble acids. The volatile acids are butyric, caproic, caprylic, capric and lauric. There is no method of separating these

TABLE 2.7

HYDROLYTIC VALUES OF FATS

Fat	Saponification	Reichert-Meissl	Polenske
Butter	210–235	20–34	1.5–3.0
Coconut	245–260	6–8	14–18
Cod liver	171–189	6.6–7.5	—
Corn	188–193	4.3	—
Cottonseed	191–198	0.22	—
Olive	190–194	0.6–1.5	—
Palm	196–205	0.1–0.2	0.2–0.3
Palm kernel	220–250	4.8–7.0	9.4–11
Peanut	186–194	0.4	—
Safflower	188–194	0.2	0.1
Sesame	188–193	1.1–1.2	—
Soybean	190–194	0.5–2.8	—
Teaseed	190–195	0.3–1.1	—

acids quantitatively from the higher acids present in a fat. Consequently, the analytical methods are empirical, but if the directions in the analytical manuals are followed exactly, the results on a given sample of fat will be consistent.

The two values are obtained in the same operation. The fat is saponified with KOH dissolved in glycerol, a nonvolatile solvent in which fats are slightly soluble. After the saponification is complete, the acids are set free by the addition of water and sulfuric acid. The aqueous solution is then distilled under specified conditions of apparatus, rate of distillation, and amount of distillate collected. The insoluble acids collect in a layer on top of the aqueous distillate and, since these acids are volatile, the temperature and rate of flow of the condenser water are important.

The strictest attention must be paid to details. The Official Methods specify 15 pieces of SiC to prevent bumping during the distillation. Older methods used pumice of unspecified amount or particle size. The author found powdered pumice worthless and a variation in value with the number and size of the larger pieces.

The time of the distillation is specified as 30 ± 2 min. If the process exceeds a variation of more than 2 min, the result is questionable.

The soluble and insoluble acids are separated by filtration. The kind of filter paper used is important; the insoluble acids pass through a very porous paper and, furthermore, if the insoluble acids strike the paper before it is wet, they will pass through the specified paper.

The importance of detail is indicated by the following note at the end of the procedure in the Official Methods: "Unless these directions are followed in every detail, satisfactory results cannot be obtained."

The usefulness of the Reichert-Meissl and Polenske values is very limited, but valuable within those limits. Milk fats, coconut oil and a few other oils from palm sources are the only fats that contain the low molecular weight acids, but the chemist who has occasion to work with one or more such fats will find these methods essential. Table 2.7 contains hydrolytic values for some of the more common fats.

In addition to the usefulness with the fats mentioned above, the Reichert-Meissl and Polenske values are sometimes helpful in the solution of special problems. For example, the milk fat of ewes and goats has a higher Polenske value than that of cow's milk and so the value is of possible use in distinguishing between those milks, since cow's milk is much cheaper than the other two. It is also useful in distinguishing between cheeses made from these milk fats, such as between Roquefort and blue cheese.

In cases involving unusual palm or marine oils, these methods may be useful; the body fat of the porpoise has the Reichert-Meissl value 46.

Unsaturation

All fats contain oleic acid and most of them contain other unsaturated acids. The presence of these acids results in low melting point for the fats and the variation in the unsaturated acid content makes the unsaturation a useful property for the identification of a fat.

Iodine Number.—Since fats contain differing amounts of unsaturated acids, they differ considerably in the amount of reagents they will absorb. The selection of an additive reagent for use in fat analysis is not easy; it must be of low volatility, react quantitatively and be easily measured.

The three halogens, Cl_2, Br_2, and I_2 in various combinations have been selected by the analyst as the best for the purpose. Chlorine and bromine cannot be used alone; they add readily but they also undergo substitution reactions with the formation of HCl or HBr and thus halogenate the saturated acids. Iodine adds too slowly and not quantitatively.

There are two Official Methods for the Iodine Value. The Wijs Method uses iodine chloride, ICl, and the Hanus Method, iodine bromide, IBr. In both methods the reagent is dissolved in glacial acetic acid.

A weighed sample of the fat is put into a special flask, the reagent added and allowed to stand for a half hour. Then a solution of potassium iodide is added and reacts at once with the reagent that has not been added to the fat.

$$ICl + 2KI \rightarrow 2KCl + I_2$$
$$IBr + 2KI \rightarrow 2KBr + I_2$$

The free iodine is titrated with standard sodium thiosulfate solution. A blank that contains the amount of reagent used on the fat is then titrated; the difference between the two titrations represents the amount of the reagent absorbed by the fat. The above equations show that one molecule of the ICl or IBr equals 1 molecule of I_2. Therefore, the number of grams of

TABLE 2.8

IODINE VALUES OF FATS

Fat	Iodine Value	Fat	Iodine Value
Nondrying		Semidrying	
Beef tallow	35–42	Corn oil	116–130
Butter fat	26–38	Cottonseed oil	103–111
Chicken fat	66–71	Sesame oil	103–117
Cocoa butter	34–40	Sunflower seed oil	120–130
Coconut oil	8–10	Drying	
Lard	47–66	English walnut oil	132–153
Olive oil	77–94	Linseed oil	175–209
Peanut oil	83–95	Safflower seed oil	140–150
Teaseed oil	78–93	Soybean oil	124–136

iodine required by the sample is easily calculated from the standard of the thiosulfate solution, which is expressed in grams of iodine per ml of solution. The Iodine Value is the number of grams of iodine absorbed by 100 gm of the fat.

The two methods do not always give the same value for the same sample of fat, especially if the fat is highly unsaturated. However, they agree closely for the food fats, but the name of the method should always be recorded with the value, for example, Iodine Value 112 (Hanus).

Volatility of iodine and of the reagent is a problem, and the determination is always carried out in a flask of special design and closed by a glass stopper to avoid loss of these vapors. The determination requires special attention to details and should always be made at least in duplicate; triplicate is better if the analyst has had little experience with the procedure.

It is convenient to classify oils on the basis of their unsaturation. The usual classes are: *nondrying* with iodine values under 100, *semidrying* with iodine values between 100 and 130 and *drying* for the oils with values above 130.

Hydrogenation.—Oleic acid, $C_{17}H_{33}COOH$, has one double bond in its formula and if 2 hydrogens are added to it we have the formula of stearic acid, $C_{17}H_{35}COOH$. Oleic acid melts at 13°C and stearic at 69.6°C. The glycerides of these two acids have a similar difference in properties; triolein melts at 49°C, trilinolein at 12°C, trilinolenin at −23°C and tristearin at 71.5°C. From these data it is clear that the consequence of hydrogenation is to increase the melting point of the fat, perhaps convert an oil to a plastic fat.

In 1897, Professor Paul Sabatier (1854–1941) of the University of Toulouse discovered that when the vapor of an unsaturated organic compound is mixed with hydrogen and the mixture passed over a nickel catalyst, the hydrogen adds and the compound becomes saturated. The process was first applied to oils in 1902.

Hydrogenation is primarily an industrial process and will be further described in Chapter 12.

Rancidity

The word *rancidity* is not a chemical term, but a general one that means the development of an unpleasant odor and taste in a fat regardless of its cause. I have heard acrimonious discussions of rancidity caused by failure to define the term.

Absorptive.—Rancidity by absorption is common. Many substances are very soluble in fats and, if a fat is exposed to the vapor of such a substance, the odor is absorbed. This is common in industry and commerce. Butter or other fat exposed to onions, paint, gasoline or even fruits will soon absorb enough of the odor to become inedible. During World War II, when steel drums became scarce, some lard was shipped in second-hand wooden barrels. Some of the barrels had evidently been used to ship vanillin or a similar compound. Several of these barrels were received by a customer with the lard so odorous that it could not be used.

In another instance, cheese was packed in wooden pails that had been treated with creosote without the cheese maker knowing it; the cheese was a total loss.

Eggs have been ruined by storage with citrus fruit; egg yolk is about 30% fat and the eggs absorbed the essential oils from the skins of the fruit.

Butter absorbs terpenes whether they come from turpentine or pine wood. Absorption of odors in storage is not the only source of absorptive rancidity in butter. If the cow eats wild onions, garlic or other odorous weeds, she concentrates the odor in the milk fat.

Spraying a warehouse with an insecticide can ruin all the fat stored there unless it is in airtight containers. Beware of any noticeable odor in a room where fats are stored.

Absorptive rancidity is presumably purely physical, although in some cases the absorbed substance may react with the fat. Since this type of rancidity seldom involves chemical action, some chemists do not consider it rancidity at all.

Hydrolytic.—This type of rancidity is limited to those fats that contain acids of low molecular weight, C_4 to C_{10}. These acids have a strong disagreeable odor and taste and anything that frees them causes rancidity, but, practically speaking, it is limited largely to butter, coconut oil and some other fats of palm origin, any fat with an appreciable Reichert-Meissl or Polenske value.

Tissues that produce fats generally produce lipases to hydrolyze them and some organs produce lipases even though they do not produce fats; the lipases of the digestive system are familiar examples. Lipases are also produced by bacteria and other microorganisms. Enzymes are inactivated by heat, but the manufacture of butter does not employ heat; therefore, butter fat is particularly susceptible to hydrolysis. Refrigeration retards the hydrolysis but does not stop it. Nair (1930) proved that milk contains a lipase that hydrolyzes fat rapidly at 37°C.

More than a dozen species of bacteria have been found to produce lipases. The common molds and some yeasts also produce them. Moisture and nitrogenous material are essential for the growth of these organisms; consequently, pure, dry fats do not support them, but 0.3% water is sufficient and only a trace of protein is required. Butter contains about 15.5% water and 0.6% protein.

Coconut oil, lard and other fats contain traces of protein, and moisture may accumulate on the surface because of condensation of the moisture in the head space of the drum or tank during refrigeration, and organisms will soon attack the fat.

Any food that contains fat, such as cakes and processed milk products contain both water and protein. All such products may develop hydrolytic rancidity if they contain milk fat, coconut oil or palm kernel oil.

There is no chemical test for hydrolytic rancidity. Odor and taste are the usual criteria and they vary from person to person. People have different threshold values for odors and tastes, and so a fat that smells or tastes rancid to one person may not do so to another. However, if a fat smells rancid to one person, for all practical purposes, it is a rancid fat.

Oxidative.—All fats oxidize but the extent differs with the degree of unsaturation, exposure to air and the presence of accelerators or inhibitors.

The point of unsaturation is the point of attack. The simplest case is the oxidation of oleic acid. The first step in the oxidation is the addition of a molecule of oxygen to the double bond

$$\underset{\overset{|}{C}}{\overset{H}{}}=\underset{\overset{|}{C}}{\overset{H}{}} + O_2 \rightarrow \underset{\underset{O}{\overset{|}{}}}{\overset{H}{\underset{|}{C}}}-\underset{\underset{O}{\overset{|}{}}}{\overset{H}{\underset{|}{C}}}$$

By some mechanism which is not well understood, the peroxide splits at the double bond with the formation of aldehydes that can be further oxidized to acids. The structure of the peroxide as well as the mechanism are uncertain, but the overall results are clear

$$CH_3(CH_2)_7\underset{\underset{O-}{\overset{|}{}}}{\overset{H}{\underset{|}{C}}}-\underset{\underset{-O}{\overset{|}{}}}{\overset{H}{\underset{|}{C}}}(CH_2)_7COOH \rightarrow CH_3(CH_2)_7CHO + HOC(CH_2)_7COOH$$

Pelargonic   Azelaic
aldehyde   aldehyde

These low molecular weight aldehydes and their corresponding acids constitute the odorous substances of oxidative rancidity. Furthermore, most food fats contain linoleic acid and some contain linolenic acid. The more highly unsaturated fats are attacked by the oxygen and split into several low-carbon aldehydes and acids.

The chemistry of oxidative rancidity is very complex. Much research has been done on the subject, but the mechanism of the reaction has not been entirely clarified. The course of the reaction differs with the temperature. Among the compounds that have been identified in rancid fats are azelaic aldehyde, azelaic acid, pelargonic and heptylic aldehydes, the aliphatic acids from acetic to nonylic and the corresponding aldehydes.

The analytical constants of the fats are changed slightly by rancidity; the iodine number decreases and the Reichert-Meissl and Polenske values increase. These changes are to be expected, but they are of little value in determining the extent of the rancidity. There is no Official Method for the detection of rancidity, but several tests have been suggested. One that shows the beginning of the process is the *iodine test*. An acidified solution of potassium iodide is shaken with the fat and the free iodine identified with starch. This test has been used for the quantitative measure of the rancidity by titrating the free iodine with standard thiosulfate solution or by measuring it colorometrically.

The *Schiff test* for aldehydes has been used as a test for rancidity. The Schiff reagent is made by decolorizing an acid solution of fuchsin with sulfite. When this solution is shaken with a rancid fat, the red color reappears.

The *Kreis test* has been most widely used since 1902. The fat is shaken with concentrated HCl, then a solution of phloroglucinol is added and the mixture shaken again; the aqueous layer becomes pink or red. According to Lea (1938), the Kreis test depends on the presence of both peroxides and acrolein, CH_2=CHCHO. The latter is oxidized to epinephrine aldehyde,

$$\overset{\displaystyle O}{\overset{\displaystyle /\backslash}{CH_2-CHCHO}}$$

which then condenses with the phloroglucinol to give the red product.

Other tests have been devised, but none has been entirely satisfactory because of the nature of the rancidity process. A fat can contain peroxides and not be rancid because the peroxides do not have a disagreeable odor and taste. Or the peroxides may all be decomposed and the fat rancid. The aldehydes are the rancid substances and so the aldehyde test is probably the most satisfactory of all the chemical tests.

Since rancidity is the presence of a disagreeable odor and taste, an organoleptic test is the most satisfactory, and it is the test that the consumer uses. If a housewife buys butter, lard or a cake that smells rancid, it *is* rancid as far as she is concerned and she returns it to the grocer who must refund her money or lose her trade. A buyer of fat in large quantity, such as a baker, has a different problem. The seller may not agree with him that a shipment of fat is rancid for there is no standard for odor and taste, and these qualities vary from person to person.

A baker can use a fat that is slightly rancid, because the rancidity is caused by volatile substances that are driven off by the heat, but the manufacturer of pie crust or muffin mixes must have a fat that is not rancid and will not become so for several months, and a test has been devised that indicates the resistance of a fat to oxidative rancidity.

Several substances and conditions accelerate rancidity and several retard it.

Accelerators.—All chemical reactions are accelerated by a rise in temperature, and so all fats, either alone or in mixtures, should be refrigerated. Light also accelerates the rancidity reactions, therefore fats should not be exposed to any more light than is necessary. Air is essential for oxidative rancidity and is very effective when fat is mixed with flour and the other ingredients of a composite food. Oxidizing agents, such as hydrogen peroxide, benzoyl peroxide, nitrous acid, and many organic aldehydes and nitro compounds are accelerators. Dyes and flavors may also be accelerators.

Metals are accelerators of rancidity. Tests made on several metals, by immersing them in corn oil, cottonseed oil and lard, showed that copper is the most active accelerator, tin and aluminum the least active, with iron, lead and zinc intermediate. Tests with solutions of the metal in the fat also indicated copper to be the most active followed by iron, chromium, and nickel with tin and aluminum the least active.

The effect of metals is very important to the food chemist in the selection of packaging material and machinery that comes into contact with the fat. Copper and its alloys are obviously to be avoided and any unfamiliar metal or alloy must be tested before it is used.

Antioxidants.—When a fat oxidizes, there is a long period of slow peroxide formation before the rancid odors appear; this is called the *induction period.* Then when the odors appear, the rate of oxidation increases enormously. Theories have been advanced to explain this phenomenon, but none of them is generally accepted. One explanation is that the fat contains a natural antioxidant which is first oxidized, and when it is all destroyed, the oxidation of the fat is free to occur. Another explanation is the formation of some substance during the reaction that acts as a catalyst of the oxidation.

Whatever the correct explanation may be, the induction period is different for different fats and for different samples of the same fat. Many vegetable fats contain tocopherols, which are antioxidants, and the amount of these present in an oil may be a factor in the length of the induction period. There are several tocopherols. One of them has the formula:

α-tocopherol

Alphatocopherol is vitamin E. The tocopherols differ in the groups at positions 1, 2 and 4 in the benzene ring: in alpha they are CH_3, CH_3 and CH_3, in beta CH_3, H, CH_3, in gamma CH_3, CH_3, H and in delta CH_3, H and H.

Wheat germ oil contains from 150 to 200 mg of tocopherols in 100 gm of the oil. The other common vegetable oils contain 2 mg to 100 mg in 100 gm. Since the induction period is not proportional to the tocopherol content, it is thought that there may be other antioxidants in the vegetable oils.

Most antioxidants are phenolic compounds. Butylated hydroxyanisole (BHA), butylated hydroxytoluene (BHT), gum guaiac, propyl gallate and the tocopherols are at present on the approved list of food additives. The Official Methods include procedures for the determination of antioxidants.

Butylhydroxytoluene (BHT)   Butylhydroxyanisole (BHA mixture)

The rancidity problem differs with the condition of the fat; pure hydrogenated oil keeps longer than lard, but mix the fat with flour and the rancidity delay is reversed.

The actual percentage of the fat involved in rancidity is small, probably less than 0.1%.

Complete Oxidation

Fats will burn, and differ somewhat in heat of combustion. A fair average is 9.45 kcal per gm. This property is important in nutrition, for the fats are oxidized in the body and produce greater energy than either carbohydrates or proteins. They do not supply 9.45 kcal, however, because their oxidation is preceded by incomplete digestion.

Digestion occurs in aqueous solution, but fats are not soluble in water, and, unless they are highly emulsified in the food, they must be emulsified in the intestine, which is done by the bile acids. This enables them to pass through the walls of the intestines and also presents a larger surface for the lipases.

The fat is first hydrolyzed and then some of the acids go into the synthesis of body fat, the remaining acids and the glycerol being oxidized. Nutritionists estimate that 95% of the fat eaten is oxidized and therefore report the energy value as 9 kcal per gm. Individual fats may supply more or less energy than the average. Watt and Merrill (1963) reported: lard, 9 kcal; salad and cooking oils, 8.8; beef fat, 7.3; and butter fat, 8.8.

Thermal Stability

When fats are heated, three stages in the effects are recognized: smoke point, flash point, and fire point. The products of decomposition that constitute the smoke vary·with the temperature, but few of them have been identified. Apparently the first effect of a high temperature is to split the fats into glycerol and the constituent acids.

The temperature of the smoke point differs enormously with the percentage of free fatty acids in the fat. Bailey (1945) reported that 20% free fatty acids reduced the smoke point of cottonseed oil and peanut oil from 450° to 300°C.

One of the well-known components of the thermal decomposition of fats is acrolein, which is formed by the dehydration of the glycerol

$$C_3H_5(OH)_3 \rightarrow CH_2{=}CHCHO + 2H_2O$$

Acrolein is a colorless liquid that boils at 52.5°C. It has a strong penetrating odor, which is very prominent in kitchens or in plants where food is fried and in restaurants that serve "sizzling steaks."

The smoke contains some fat, some fatty acids in various stages of decomposition, and a little free carbon.

The acids are more volatile than the glycerides and so both the flash point and the fire point of a fat are reduced by the presence of free acids. These points are determined by passing a lighted taper over the surface of the oil in a special cup. As the temperature of the oil is being slowly raised, at a certain temperature the vapor will ignite, but the flame will immediately go out. At a higher temperature, the flame will continue to burn.

Since all three thermal points vary so much with the condition of the fat, they are not among the fat constants. However, they should be determined for a fat that is purchased for use in frying because of the fire hazard.

Heated fats also oxidize and polymerize. Several studies on fats heated with and without air have been made and the nutritional value of the residual fat investigated. Growth of experimental animals was retarded, possibly because of the destruction of vitamins by the heat (Perkins 1960). Few studies have been published on the deterioration and nutritional effect of heating fat under commercial conditions of deep fat frying. The chemical and nutritional effect of heating fats under conditions of use needs further investigation.

OTHER LIPIDS

The fats are members of a large class of compounds called *lipids*. Other than the fats, the phosphatides are the chief members of the class that concern the food chemist. Of these, the *lecithins* and the *cephalins* are the most important. They are diglycerides with a phosphoric acid link to a nitrogenous compound in the third glycerol position

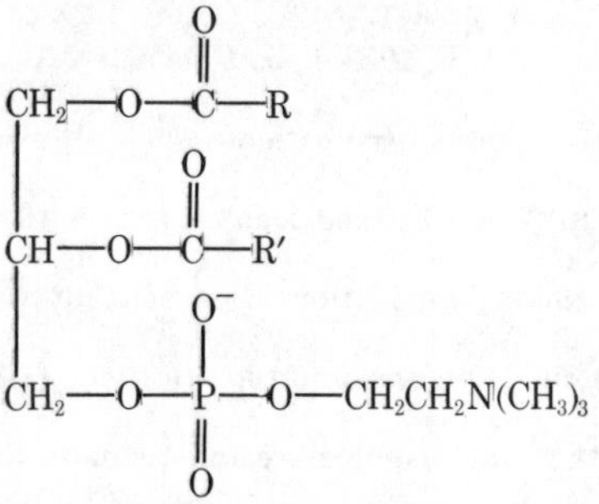

$$
\begin{array}{l}
\overset{\displaystyle O}{\underset{\displaystyle \|}{}} \\
CH_2-O-C-R \\
\overset{\displaystyle O}{\underset{\displaystyle \|}{}} \\
CH-O-C-R' \\
\overset{\displaystyle O^-}{\underset{\displaystyle |}{}} \\
CH_2-O-P-O-CH_2CH_2N(CH_3)_3 \\
\underset{\displaystyle O}{\underset{\displaystyle \|}{}}
\end{array}
$$

Alpha-lecithin

The nitrogenous group in lecithin is choline. If the phosphate link is on the middle carbon of the glycerol, the compound is beta-lecithin. The acid groups R and R′ may be palmitic, stearic, oleic or arachidonic.

Lecithins

These liquids are yellow, waxy substances that turn brown on exposure to light. They are soluble in alcohol, benzene and chloroform, but not in acetone or ethyl acetate. Heating with acids hydrolyzes them to the constituent units. They are of biological importance as constituents of nerve tissue. They also occur in liver and in egg yolk. Soybeans are a commercial source. They are used in the food industry as emulsifiers and as antioxidants.

Cephalins

The *cephalins* differ from the lecithins by having ethanolamine as the nitrogenous group.

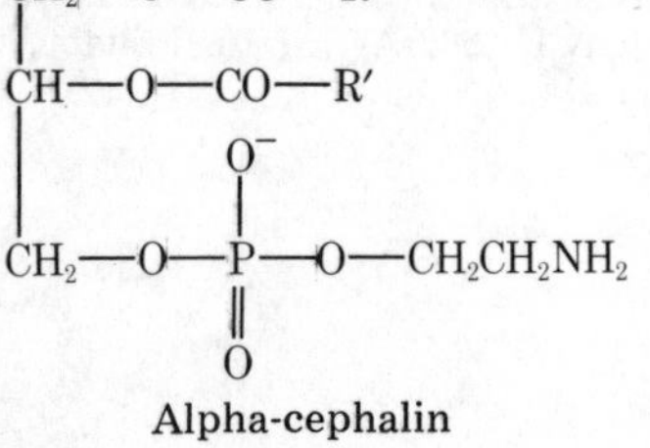

$$
\begin{array}{l}
CH_2-O-CO-R \\
CH-O-CO-R' \\
\overset{\displaystyle O^-}{\underset{\displaystyle |}{}} \\
CH_2-O-P-O-CH_2CH_2NH_2 \\
\underset{\displaystyle O}{\underset{\displaystyle \|}{}}
\end{array}
$$

Alpha-cephalin

The acid radicals R and R′ are stearic, oleic, linoleic and arachidonic.

The "lecithin" from soybeans is 35% lecithin and 65% cephalin. Soybeans contain about 1.8% crude lecithin and butter fat contains about 1.4%. The cephalins are also of biological importance.

Several other complex lipids are known, but none of them is of general importance to the food chemist.

BIBLIOGRAPHY

ASSOC. OFFIC. ANAL. CHEMISTS. 1975. Official Methods of Analysis, 12th Edition. Assoc. Offic. Anal. Chemists, Washington, D.C.

AURAND, L. W., and WOODS, A. E. 1973. Food Chemistry. Avi Publishing Co., Westport, Conn.

BITTENBENDER, C. D. 1970. Fat determination—a new physical method. J. Food Sci. 35, 460–463.

BRINK, N. F., and KRITCHEVSKY, D. 1968. Dairy Lipids and Lipid Metabolism. Avi Publishing Co., Westport, Conn.

CORT, W. M. 1974. Hemoglobin peroxidation test screens antioxidants. Food. Tech. 28, No. 10, 60–66.

ECKEY, E. W. 1956. Esterification and interesterification. J. Am. Oil Chemists' Soc. 33, 575–579.

EICHBERG, J. 1939. Lecithin—its manufacture and use in the fat and oil industry. Oil and Soap, 16, 51–54.

GARARD, I. D., MINSKY, A., BAKER, J. H., and PASCALE, V. 1937. Identification of Roquefort cheese. Ind. Eng. Chem. 29, 1169–1171.

HILDITCH, T. P., and WILLIAMS, P. N. 1964. The Chemical Constitution of Natural Fats, 4th Edition. Chapman and Hall, London.

JAMIESON, G. S. 1943. Vegetable Fats and Oils, 2nd Edition. Reinhold Publishing Corp., New York.

KUMEROW, F. A. 1964. The role of polyunsaturated fats in nutrition. Food Tech. 18, No. 6, 49–53.

LEA, C. H. 1938. Rancidity in Edible Fats. Her Magesty's Stationery Office, London.

LEE, F. A. Basic Food Chemistry. Avi Publishing Co., Westport, Conn.

MARKLEY, K. S. 1960–1969. Fatty Acids, 2nd Edition. Parts 1–5. John Wiley & Sons, New York.

MOREAU, J. R, and LAVOIE, M. T. 1971. An emulsion method for rapid determination of fat in raw meats. J. Food Sci. 36, 760–763.

NAIR, J. H. 1930. Lipase in raw, heated and desiccated milk. Ind. Eng. Chem. 22, 42–45.

PERKINS, E. G. 1960. Nutritional and chemical changes occurring in heated fats. Food Tech. 14, 508–514.

SCHULTZ, H. W. 1962. Lipids and Their Oxidation. Avi Publishing Co., Westport, Conn.

STAMLER, J. 1959. Experimental studies on dietary fats, cholesterol and atherosclerosis. Food Tech. 13, 50–57.

STUCKEY, B. N. 1962. Antioxidants. In Symposium on Foods; Lipids and Their Oxidation. H. W. Schultz, E. A. Day, and R. O. Sinnhuber (Editors). Avi Publishing Co., Westport, Conn.

SWERN, D. 1964. Bailey's Industrial Oil and Fat Products. John Wiley & Sons, New York.

WATT, B. K., and MERRILL, A. L. 1963. Agriculture Handbook No. 8, Composition of foods. ARS-USDA, Washington, D.C.

CHAPTER 3

Carbohydrates

After scientific chemistry got under way, about 1800, chemists soon discovered that the starches and sugars were compounds of carbon, hydrogen and oxygen and that the last two were in the ratio of 2 to 1, which made a formula that could be written $C_x(H_2O)_y$. With this in mind, Professor Carl Schmidt of the University of Dormat suggested the name *carbohydrate* for the group, but it was another eighty years before Emil Fischer (1857–1919) put together his own research and that of his predecessors to establish the chemical structure of the sugars.

CLASSIFICATION

The carbohydrates may be classified as:
A. Monosaccharides
    I. Pentoses—arabinose, rhamnose, ribose, xylose
    II. Hexoses
        1. Aldohexoses—galactose, glucose, mannose
        2. Ketohexoses—fructose, sorbose
B. Oligosaccharides
    I. Disaccharides
        1. Reducing—lactose, maltose, cellobiose, melibiose
        2. Nonreducing—sucrose
    II. Trisaccharides—melezitose, raffinose
    III. Tetrasaccharides—stachyose
C. Polysaccharides
    I. Homo
        1. Pentosans—araban, xylan
        2. Hexosans
            a. Glucosans—cellulose, dextrin, glycogen, starch
            b. Fructosans—inulin
    II. Hetero—gums, mucilages, pectins

MONOSACCHARIDES

The simplest of all sugars is glycolic aldehyde

$$CH_2OH$$
$$|$$
$$CHO$$

It is of no consequence as a sugar, except to show the two essential groups of a monosaccharide, an alcohol group, —OH and a carbonyl group, =CO.

35

The generic names of the monosaccharides are formed with a prefix indicating the number of carbon atoms attached to oxygen and a suffix, *ose*.

Structure

Glycolic aldehyde is the only diose. There are two trioses, glyceric aldehyde and dihydroxyacetone

$$
\begin{array}{c}
\text{CHO} \\
| \\
\text{H}\!-\!\text{C}\!-\!\text{OH} \\
| \\
\text{CH}_2\text{OH}
\end{array}
\qquad\qquad
\begin{array}{c}
\text{CH}_2\text{OH} \\
| \\
\text{C}\!=\!\text{O} \\
| \\
\text{CH}_2\text{OH}
\end{array}
$$

Glyceric aldehyde   Dihydroxyacetone
Melting point 142°C   Melting point 75°C

The formulas of these two trioses suggest an addition to the classification of the monosaccharides: glyceric aldehyde is an *aldose* and dihydroxyacetone is a *ketose*.

Another feature of the sugars is indicated by the formula of glyceric aldehyde; the molecule is asymmetric, which means that there are optical isomers, a D- and an L- form. The formula of one is the mirror image of that of the other.

$$
\begin{array}{c}
\text{CHO} \\
| \\
\text{H}\!-\!\text{C}\!-\!\text{OH} \\
| \\
\text{CH}_2\text{OH}
\end{array}
\qquad\qquad
\begin{array}{c}
\text{CHO} \\
| \\
\text{HO}\!-\!\text{C}\!-\!\text{H} \\
| \\
\text{CH}_2\text{OH}
\end{array}
$$

D-Glyceric Aldehyde   L-Glyceric Aldehyde

Monosaccharides exist in these two forms and, in each case, the secondary alcohol groups of the one are written as the mirror images of those of the other. Fischer arbitrarily selected the formula for the D- form as that with the —OH group on the carbon, next the —CH$_2$OH group on the right when the formula is written with the —CHO group at the top and the —CH$_2$OH group at the bottom. This is carried through all the monosaccharides regardless of whether they are dextrorotatory or levorotatory. D-Fructose, for example, is levorotatory. The direction of the rotation is indicated by + for dextro and − for levo.

There are four aldopentoses

$$
\begin{array}{c}
\text{CHO} \\
| \\
\text{H}\!-\!\text{C}\!-\!\text{OH} \\
| \\
\text{HO}\!-\!\text{C}\!-\!\text{H} \\
| \\
\text{HO}\!-\!\text{C}\!-\!\text{H} \\
| \\
\text{CH}_2\text{OH}
\end{array}
\quad
\begin{array}{c}
\text{CHO} \\
| \\
\text{H}\!-\!\text{C}\!-\!\text{OH} \\
| \\
\text{HO}\!-\!\text{C}\!-\!\text{H} \\
| \\
\text{H}\!-\!\text{C}\!-\!\text{OH} \\
| \\
\text{CH}_2\text{OH}
\end{array}
\quad
\begin{array}{c}
\text{CHO} \\
| \\
\text{H}\!-\!\text{C}\!-\!\text{OH} \\
| \\
\text{H}\!-\!\text{C}\!-\!\text{OH} \\
| \\
\text{H}\!-\!\text{C}\!-\!\text{OH} \\
| \\
\text{CH}_2\text{OH}
\end{array}
\quad
\begin{array}{c}
\text{CHO} \\
| \\
\text{H}\!-\!\text{C}\!-\!\text{OH} \\
| \\
\text{H}\!-\!\text{C}\!-\!\text{OH} \\
| \\
\text{HO}\!-\!\text{C}\!-\!\text{H} \\
| \\
\text{HO}\!-\!\text{C}\!-\!\text{H} \\
| \\
\text{CH}_3
\end{array}
$$

L-Arabinose(+)   D-Xylose(+)   D-Ribose(−)

L-Rhamnose(+)

These sugars are seldom found free in nature, but are very plentiful as units in polysaccharides.

Rhamnose has only five carbon atoms attached to oxygen and is, therefore, classified as a pentose. It is called a *methylpentose*. Taking into account the D- and L- forms, there are 2 aldotrioses, 4 aldotetroses, 8 aldopentoses, without the alkyl derivatives, and 16 aldohexoses.

INDIVIDUAL SUGARS

Fortunately for the student of food chemistry, the only monosaccharides of individual importance are D-glucose, D-fructose, D-galactose and occasionally D-mannose and L-sorbose

CHO	CHO	CHO	CH₂OH	CH₂OH

(structural formulas)

D-Glucose(+)   D-Galactose(+)   D-Mannose(+)   D-Fructose(−)   L-Sorbose(−)

In each case, the formula of the L sugar is the mirror image of that of the D form.

Glucose.—Glucose is also called *dextrose* and *grape sugar*. It is the sugar oxidized in the body to produce energy. The blood always contains glucose, which varies in amount from 0.08% to 0.15%. Just after carbohydrates are eaten and in cases of diabetes, the sugar content of the blood is high above normal; concentrations as high as 12% have been reported.

Glucose occurs widely in fruits and vegetables and is manufactured in quantity from starch. It crystallizes with one molecule of water of hydration, which it loses when it is heated to 110°C. The hydrate melts at 86°C and the anhydrous form at 146°C. The solubility of glucose increases rapidly with rise in temperature; a saturated solution contains 27% glucose at 20°C and 80% at 80°C. It is only slightly soluble in 95% alcohol and in ether.

Galactose.—This sugar is seldom found free in nature, but is obtained by the hydrolysis of lactose, and it also occurs as a unit in several polysaccharides. It melts at 168° and is only slightly soluble in water. Concentrated nitric acid oxidizes it to mucic acid which is insoluble. This constitutes a test for galactose, either free or as a constituent of an oligosaccharide or polysaccharide. It has also been used as a semiquantitative method. Mucic acid is almost insoluble in weak acid solution, but the yield from the oxidation of galactose is only 75% of the weight of the sugar. However, the results are reproducible if specified conditions are used (Bates 1942). The other saccharic acids are soluble.

$$
\begin{array}{c}
\text{COOH} \\
| \\
\text{H—C—OH} \\
| \\
\text{HO—C—H} \\
| \\
\text{HO—C—H} \\
| \\
\text{H—C—OH} \\
| \\
\text{COOH}
\end{array}
$$

Mucic Acid

Mannose.—Mannose has been identified in oranges, olives and molasses, but it does not occur in quantity in nature in the free state. It is found as a unit in several polysaccharides. The sugar may be obtained by the hydrolysis of ivory nuts. It melts at 133°C.

Fructose.—The most important ketose sugar is fructose, also called *levulose* and *fruit sugar*. It occurs in several fruit juices and in honey and is prepared commercially from other sources. It is the sweetest of all the sugars. It is very soluble: 100 ml of water dissolves 375 gm at 20°C—the saturated solution is 79% sugar; at 50°C, 100 ml of water dissolves 666 gm of the sugar—the saturated solution is 87% sugar.

Fructose is very difficult to crystallize, but crystals may be obtained from alcohol. Absolute alcohol must be used for the final dehydration. The crystals melt at 104°C.

Sorbose.—This ketose is not common in foods, but is an intermediate in the synthesis of ascorbic acid. The sugar is prepared by bacterial fermentation of sorbitol, which in turn may be made by the reduction of the aldehyde group of glucose to the primary alcohol. The crystals melt at 160°C.

PROPERTIES OF THE SUGARS

The monosaccharides and the oligosaccharides constitute the sugars. They are all crystalline solids, soluble in water and have a sweet taste. There are several other properties which are of value to the analyst.

Optical Rotation

Aside from the melting points, there is only one physical property of the sugars of much value to the analyst. They are all optically active and differ considerably in the extent to which they rotate light. In order to establish constants for the sugars, the conditions must be specified. The measurement is always made in aqueous solution, and so the concentration of the solution, the wavelength of the light, the length of the path of light through the solution, and the temperature all affect the degree of rotation. The result is reported as the *specific rotation* and is abbreviated by α (alpha D). The formula is

$$(\alpha)_D^t = \frac{100a}{l \times c}$$

a = degrees of rotation measured
l = length of the tube in decimeters
c = concentration of the solution in grams of sugar in 100 ml of solution
D = wavelength of the light
t = Celsius temperature

The measurement of the specific rotation of a sugar is simpler than the explanation of it. Ten grams of sugar are weighed, dissolved in water and made up to a volume of 100 ml. Any dust particles floating in the solution will interfere with the polarimeter reading and so the solution is filtered directly into the polarimeter tube. These tubes come in 1, 2 and 3 decimeter lengths; the 2 dm tube is the most commonly used.

The filled tube is placed in the polarimeter, faced with a sodium light or a white light if the instrument has a light filter, and the scale read. Any analytical manual gives detailed instructions for obtaining precise results. If the reading were 10° at a room temperature of 23°C, then

$$(\alpha)_D{}^{23°} = \frac{100 \times 10}{2 \times 10} = 50°$$

Many polarimeters are equipped with a light filter consisting of a short tube filled with a solution of potassium dichromate, so that an ordinary white light bulb can be used as the source of light. Also, some tubes have a jacket for the circulation of water around the solution to control the temperature, which is essential when fructose is involved. The rotation of the other sugars changes very little with changes in temperature. Fructose changes considerably and the change is used in one method for the analytical determination of the sugar (Bates 1942) although the exact degree of change is uncertain; the literature contains values from 0.0341 to 0.0362 for the change of one degree of temperature of 1 gm of fructose.

Some laboratories, such as those in sugar refineries, where the principal task is the determination of the amount of sucrose in samples of raw sugar, use a special instrument called a *saccharimeter*, which has a scale that reads directly in the percentage of sucrose when the sample is 26 gm in 100 ml of solution, the reading taken at 20°C with a 2 dm tube and a dichromate light filter.

Mutarotation.—Some of the properties of the sugars cannot be explained by either the aldehydic or the ketonic formula; one of them is mutarotation, which means changing rotation and may be illustrated by the behavior of glucose.

If glucose is dissolved in water and polarized immediately, the specific rotation is 112.2°. If measurements are made at intervals over several hours, the value slowly falls to 52.7° and then there is no further change. If the solution is heated, or if a few drops of an alkali are added, the change occurs at once, but two or more readings should always be taken to be sure the change is complete when making an analysis.

Mutarotation was first noticed in 1846 but remained unexplained for 50 years, when Tanret discovered beta-D-glucose.

Glucose is ordinarily crystallized from cold water or water and alcohol mixed, but Tanret crystallized it from water at 100°C, and the product had a specific rotation of 19°; it also showed mutarotation, rising slowly to 52.7° where it stopped. This supplied the explanation of mutarotation. There are two forms of glucose, which have been named alpha-D-glucose and beta-D-glucose. If either of them is dissolved in water, it changes gradually to the other until an equilibrium mixture is reached, which has the rotation 52.7°.

Glucose shows the chemical properties of an aldehyde as indicated by the Fischer formula, but the mutarotation cannot be explained by that formula and so another had to be devised—a ring formula that does not show any aldehyde group, but does have an extra asymmetric carbon atom. In order to explain all the properties of a glucose solution, it is necessary to indicate an equilibrium among three forms of it

α-D-Glucose   Aldehydic Glucose   β-D-Glucose

It is now customary to write the ring structure of the sugars as hexagons in a horizontal position with the H and —OH groups above and below the ring

α-D-Glucose

TABLE 3.1

SPECIFIC ROTATION OF SUGARS

Sugar	$(\alpha)_D^{20°}$	Sugar	$(\alpha)_D^{20°}$
L-Arabinose	104.5	Mannose	14.2
Fructose	−92.4	Raffinose	105.2
Galactose	80.2	Rhamnose	8.2
Glucose	52.7	Ribose	−23.7
Invert sugar	−21.0	L-Sorbose	−43.4
Lactose	52.6	Sucrose	66.5
Maltose	130.4	Xylose	18.8

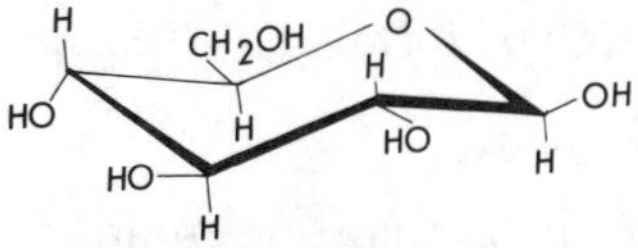

β-D-Glucose

Glycosides

In writing formulas or equations, the chemist uses whichever kind serves his purpose best; for example, if gaseous HCl is passed into a solution of glucose in methanol, two methyl derivatives are produced that are called glucosides and have the formulas

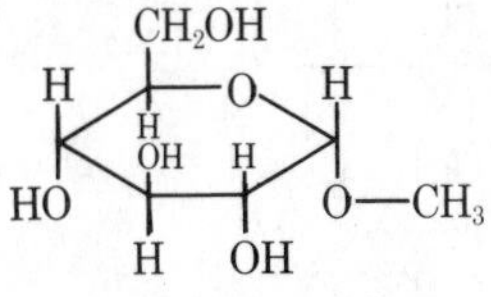

α-Methylglucoside 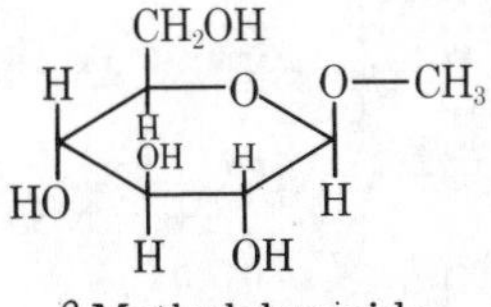   β-Methylglucoside

All the sugars that have aldehydic properties form compounds of this type. The generic name of such compounds is *glycosides*—the disaccharides are glycosides.

CHEMICAL PROPERTIES

All the monosaccharides and maltose and lactose, all the sugars that show mutarotation, also show aldehydic properties; many of them are useful for the identification or determination of individual sugars. All the sugars have some of the properties of alcohols.

Acetylation

Like alcohols in general, the sugars react with acetyl chloride or acetic anhydride to form esters. The aldopentoses form tetraacetates and the aldohexoses, pentacetates.

Aldehyde Reactions

The monosaccharides and some of the oligosaccharides undergo the reactions of aldehydes in general. The —CHO group is reduced to a primary alcohol group —CH_2OH by sodium amalgam and alcohol.

The aldehyde group adds HCN to lengthen the carbon chain, which is part of a method of synthesizing one sugar from another, a hexose from a pentose, for example.

A mild oxidizing agent, such as bromine water, oxidizes the aldehyde group to —COOH forming a *glyconic* acid. A stronger agent, such as concentrated nitric acid, oxidizes both end groups to —COOH, forming a *saccharic* acid. Indirect methods can oxidize the —CH_2OH group but not the —CHO group to —COOH forming a *glycuronic* acid.

Molecular Reactions

Some reactions involve more than one group of the sugar molecule and some involve the entire molecule.

Osazones.—Those sugars with an aldehydic or ketonic group react with phenylhydrazine to form a yellow crystalline solid called an *osazone*. The mechanism of the reaction is uncertain, but it apparently takes place in at least three stages, as indicated by Emil Fischer, who discovered the reaction. His explanation was indicated by the following formulas

$$
\begin{array}{ll}
\begin{array}{c}
\text{H—C}{=}\text{O} \\
| \\
\text{H—C—OH} \\
| \\
\text{HO—C—H} \\
| \\
\text{H—C—OH} \\
| \\
\text{H—C—OH} \\
| \\
\text{CH}_2\text{OH}
\end{array}
& + \text{H}_2\text{NHNC}_6\text{H}_5 \longrightarrow
\begin{array}{c}
\text{H—C}{=}\text{N—NH—C}_6\text{H}_5 \\
| \\
\text{H—C—OH} \\
| \\
\text{HO—C—H} \\
| \\
\text{H—C—OH} \\
| \\
\text{H—C—OH} \\
| \\
\text{CH}_2\text{OH}
\end{array}
\end{array}
$$

D-Glucose   D-Glucose Phenylhydrazone

This hydrozone formation has been well established; the following steps by which it is converted into the osazone were proposed by Fischer, but are now questioned.

The osazone reaction is accomplished by dissolving the sugar in water and adding phenylhydrazine hydrochloride or acetate, and sodium acetate, and then heating the mixture in a test tube in a beaker of water. In a few minutes the yellow crystals precipitate. The osazones of lactose and maltose, however, are soluble in hot water and so the mixture must be cooled before the crystals appear.

The osazone reaction occurs only in a slightly acid solution. Phenylhydrazine is strongly alkaline and the hydrochloride and acetate are acid in

$$
\begin{array}{l}
\text{H—C=N—NH—C}_6\text{H}_5 \\
\text{H—C—OH} \\
\text{HO—C—H} \\
\quad \text{etc.}
\end{array}
\;+\; \text{C}_6\text{H}_5\text{NHNH}_2 \;\longrightarrow\;
\begin{array}{l}
\text{H—C=NNHC}_6\text{H}_5 \\
\text{C=O} \\
\text{H—C—OH} \\
\quad \text{etc.}
\end{array}
\;+\; \text{C}_6\text{H}_5\text{NH}_2 \;+\; \text{NH}_3
$$

$$
\begin{array}{l}
\text{H—C=NNHC}_6\text{H}_5 \\
\text{C=O} \\
\text{HO—C—H} \\
\text{H—C—OH} \\
\text{H—C—OH} \\
\text{CH}_2\text{OH}
\end{array}
\;+\; \text{C}_2\text{NNHC}_6\text{H}_5 \;\longrightarrow\;
\begin{array}{l}
\text{H—C=NNHC}_6\text{H}_5 \\
\text{C=NNHC}_6\text{H}_5 \\
\text{HO—C—H} \\
\text{H—C—OH} \\
\text{H—C—OH} \\
\text{CH}_2\text{OH}
\end{array}
$$

Phenylglucosazone

aqueous solution; therefore, sodium acetate is added to the solution of one of these salts to repress the ionization and bring the pH into the necessary range.[1]

The osazone reaction is not very useful for the identification of a sugar because the osazones all melt within ten degrees of 200°C, and even when well purified, they melt with some decomposition. However, the length of time required for the osazone to form differs with the sugar (Garard and Sherman 1918) and a method based on this difference has a limited usefulness. The osazone crystals are different and a microscope will help in their identification. Mannose does not form an osazone and sucrose does not, but sucrose is hydrolyzed by the acid and the invert sugar slowly gives a precipitate. Glucose and fructose, of course, give the same osazone. The reaction is not in the Official Methods.

Copper Reduction.—The sugars that contain the carbonyl group, which includes all the monosaccharides and lactose and maltose are called *reducing sugars*. They reduce alkaline solutions of silver and copper salts. Silver solutions such as $Ag(NH_3)_2OH$, are reduced to metallic silver, which may deposit on the wall of the vessel in the form of a mirror.

Both silver and copper hydroxides are insoluble; therefore, something must be added to the alkaline solution to form a soluble complex ion with the metal. Ammonium hydroxide, which contains free NH_3, forms such a complex with both metals, $Ag(NH_3)_2{}^+$ and $Cu(NH_3)_4{}^{++}$, but these ammonia complexes are unstable because the ammonia escapes from the so-

---

[1] Phenylhydrazine is very poisonous. Fischer suffered from its effects from 1891 until his death in 1919.

lution and the base precipitates when the solution is allowed to stand for a short time.

In 1846, Hermann Fehling (1812–1885) developed a quantitative method for the determination of reducing sugars. His solution contained $CuSO_4$, NaOH and sodium potassium tartrate, $NaKC_4H_4O_6$. Since the invention of the original method, several chemists have modified it. In the 12th edition of the Official Methods there are several methods that use the copper reduction principle. The copper is reduced to the cuprous state and precipitates as the red cuprous oxide, Cu_2O.

After standing for a few weeks, Fehling's solution and its modifications yield a precipitate when heated alone. The tartrate ion is changed by the alkali and acts as the reducing agent, consequently, the solutions are stored unmixed; the copper sulfate in one bottle and the alkaline tartrate in another. The solutions are mixed at the moment of use. Keeping the solutions separate is extremely important. Mixed solutions have been sold to doctors to use as a test for glucose in urine and they often found it when there was none there. Many doctors changed to Benedict's solution, which contains $CuSO_4$, Na_2CO_3 and sodium citrate. The citrate does not reduce the copper ion and the solution can be stored in one bottle.

Reducing sugars do not reduce the copper ion unless the solution is alkaline although the alkali reduces the concentration of the copper ion. The alkali causes the sugar to enolize and split into fragments that contain from 1 to 5 carbon atoms, for example

In this illustration, an aldopentose and formaldehyde are formed, but enolization also occurs between carbons 2 and 3 and 3 and 4 so that the effect of alkali on sugars is very complex and a number of compounds are formed; 116 have been predicted (Nef 1914). The actual compounds differ with the alkalinity and the temperature, but most of them are strong reducing agents.

Fehling's solution does not oxidize the sugars to CO_2 and water; the extent of the oxidation depends on the time and the temperature, the concentration of copper and alkali and possibly other factors, consequently, all the quantitative methods specify the exact conditions to be used. In the Munson–Walker procedure, which is one of the Official Methods, the copper sulfate solution contains 34.6 gm $CuSO_45H_2O$ in 500 ml of solution and the copper is determined by electrolysis and adjusted to a concentration of 440.9 mg Cu in 25 ml of solution. The alkaline solution contains 173 gm of $NaKC_4H_4O_6 \cdot 4H_2O$ and 50 gm NaOH in 500 ml of solution.

For the sugar determination, 25 ml of each of the two solutions are transferred to a 400 ml beaker of alkali-resistant glass and 50 ml of the sugar solution are added. The solution is heated on an asbestos gauze over a bunsen burner so that boiling begins in 4 min and is continued for exactly 2 min, then filtered at once through an asbestos pad in a Gooch crucible. The Cu_2O is washed with water at 60°C.

The Cu_2O may be dried and weighed or dissolved in acid and titrated with thiosulfate or potassium permanganate, or the copper may be determined electrolytically.

After the weight of the copper or cuprous oxide has been obtained, the sugar equivalent is found in a table that contains the equivalent weight of glucose, fructose, invert sugar, lactose, and maltose. For example, 50 mg Cu or 56.3 Cu_2O are equivalent to 24.1 mg glucose, 26.5 mg fructose, 25.5 mg invert sugar, 38.3 mg lactose, or 44.6 mg maltose. Obviously the chemist must know the identity of the sugar in the solution. In case it is unknown, or a mixture, it is calculated as glucose and reported as *reducing sugars.*

The Official Methods specify the procedure to be used for the analysis of several foods that contain sugar; molasses, confectionery, honey, commercial glucose, maple sugar and syrup, and hydrolytic starch products all have prescribed methods for sugar determination. There are also methods for special problems in the determination of fructose and invert sugar.

The chemist must choose a method from the Official Methods, from a manual of food analysis or from a journal for any special problem he may encounter. In fact, he may need to try several methods before he can make a selection for routine use. Time, convenience, and the degree of precision required are the main considerations.

Ferricyanide Reduction.—In 1861, it was discovered that reducing sugars reduce an alkaline solution of ferricyanide to ferrocyanide, $Fe(CN)_6^{---}$ to $Fe(CN)_6^{----}$, but the reaction was not adapted to the determination of such sugars until 1923. It has been used to determine glucose in blood, fruit juices and cereal products. In general, the method is less precise than the copper reduction methods, but it is much more convenient. In the Official Methods it is applied to cereal foods.

DISACCHARIDES

The disaccharides of most interest to the food chemist are lactose, maltose and sucrose.

Lactose

The sugar of milk is the least soluble of the three disaccharides. At 20°C a saturated solution of lactose contains 16% of the hydrate; at 50°C it contains 30%; and at 100°C the solution contains 61% of anhydrous lactose. It is insoluble in alcohol and ether.

Lactose crystallizes from water as the monohydrate $C_{12}H_{22}O_{11} \cdot H_2O$, below 92°C, and in anhydrous form above that temperature. If the crystals of the hydrate are heated above 130°C, they lose the hydrate water and melt at 202°C.

The sweetness of lactose is estimated to be about one fourth that of sucrose.

Lactose is hydrolyzed by both acids and enzymes to two simple sugars

$$C_{12}H_{22}O_{11} + H_2O \rightarrow C_6H_{12}O_6 + C_6H_{12}O_6$$

Lactose   D-Glucose D-Galactose

Its structure is that of a glucose-4-β-galactoside

Aside from its importance in nutrition, lactose is a component of whey, which is somewhat of a problem in the cheese industry.

Maltose

This sugar rarely occurs in nature, but it is obtained by enzyme action on starch.

Maltose

The empirical formula of maltose is also $C_{12}H_{22}O_{11}$ and it is likewise hydrolyzed to two monosaccharide molecules by acids and enzymes; both monosaccharide molecules are glucose.

Maltose crystallizes in fine white needles with one molecule of water that it loses at 100°C. It is extremely soluble in water, somewhat soluble in alcohol, and insoluble in ether.

The structural formula of maltose is that of a glucose-α-glucoside.

Sucrose

The sugar of the sugar cane, sugar beet, and maple tree is sucrose. It is also one of the sugars in many fruits and vegetables.

Sucrose crystallizes readily from water without hydration. The crystals are very fine or very large (rock candy) according to the conditions of crystallization; the slower the formation of the crystals, the larger they are.

The solubility of sucrose in water has been determined by several chemists and the results do not agree very closely, consequently, the values recorded in Table 3.2 are approximations within the limits of the published values.

TABLE 3.2

SOLUBILITY OF SUCROSE

°C	Sucrose %	Sucrose in 100 Gm Water	°C	Sucrose %	Sucrose in 100 Gm Water
10	65	190	60	74	287
20	67	204	70	76	320
30	69	219	80	78	362
40	70	238	90	80	415
50	72	260	100	82	487

Like lactose and maltose, sucrose is hydrolyzed by enzymes and acids to two monosaccharide molecules. The empirical equation is the same but the products are D-glucose and D-fructose.

Sucrose differs from lactose and maltose in that it does not contain a carbonyl group and, therefore, its solutions do not show mutarotation, do not form osazones, and do not reduce alkaline silver and copper solutions. The structural formula shows why sucrose differs in chemical properties from maltose and lactose. The carbonyl groups of the constituent monosaccharides are tied up by the position of the oxygen bridge between the glucose and fructose. Its structure may be considered either glucose-1-β-fructoside or fructose-2-α-glucoside

Sucrose

Sucrose hydrolyzes more easily than either lactose or maltose. Invertase, or slight acidity, catalyzes the reaction effectively. Another unusual feature of the hydrolysis is the fact that the sign of the optical rotation changes. The specific rotation of sucrose is $+66.5°$ and that of the hydrolytic product is $-21.6°$. This change in sign has resulted in the term *inversion* for the hydrolysis of sucrose, and *invert sugar* for the equimolar mixture of glucose and fructose produced by the inversion.

In 1849, the French chemist, Clerget, devised a method for the determination of sucrose in mixtures by using the inversion principle. A solution of unknown composition is polarized and then inverted with acid or invertase and polarized again. The percentage of sucrose is calculated from the extent of the change in rotation.

Sucrose may also be determined by measuring reducing sugars before and after inversion, since sucrose does not reduce copper solutions and both fructose and glucose do. The increase in copper reduced by the inversion is a measure of the invert sugar formed, and the weight of sucrose is 95% of that of the invert sugar.

$$C_{12}H_{22}O_{11} + H_2O \rightarrow C_6H_{12}O_6 + C_6H_{12}O_6$$

Sucrose	Glucose	Fructose
342	180	18

Whether the amount of sucrose is determined by polarimetric or copper reducing methods, the hydrolysis must be carried out with great care in order to avoid the hydrolysis of other substances that may be present. The invertase method is the more reliable of the two because it is specific for the hydrolysis of sucrose, whereas acid will catalyze the hydrolysis of all the oligosaccharides and polysaccharides if it is strong enough and in contact long enough. The Official Methods contain procedures for the use of both acid and invertase and for both the polarimetric and copper reduction methods for the determination of sucrose.

In addition to the importance of sucrose inversion in analytical chemistry, it is also important in technology. Sucrose crystallizes readily and invert sugar crystallizes with the greatest difficulty. If a solution of invert sugar is evaporated, it becomes a solid mass with the consistency of asphalt. This feature is beneficial or a nuisance in technology depending on the problem. The general public has known for at least a century that if a little vinegar is added to cooking taffy, it will prevent the taffy from "going to sugar." Taffy, like glass, is a supercooled solution and, in time, the sugar crystallizes and changes the taffy to a powder. From this ancient observation, it is obvious that inversion is of use in the confectionery industry, for invert sugar not only fails to crystallize, but it also retards the crystallization of the sucrose and, furthermore, the invert sugar is sweeter than the sucrose.

In the manufacture of granulated sugar, inversion must be avoided in order to get the highest yield possible of crystalline raw sugar.

TRISACCHARIDES

Several trisaccharides are known, but raffinose and melezitose are the only ones that are likely to come to the attention of the food chemist.

Raffinose

This trisaccharide occurs in sugar beets, molasses, some fruit syrups and in cottonseeds. It crystallizes with 5 molecules of water of hydration. The crystals resemble those of sucrose. They melt at 80°C. The formula of the anhydrous sugar is: $C_{18}H_{32}O_{16}$.

If raffinose is hydrolyzed completely with acid, it yields 1 molecule of glucose, 1 of fructose and 1 of galactose. If the hydrolysis is catalyzed with invertase, the products are fructose and a disaccharide, *melibiose:* if the enzyme emulsin is used, the products are sucrose and galactose. A partial structural formula, therefore, must be

$$C_6H_{11}O_5—O—C_6H_{10}O_4—O—C_6H_{11}O_5$$

| Fructose | Glucose | Galactose |

| Sucrose | Melibiose |

The food chemist's chief concern with raffinose is the special methods that must be used to determine sucrose in its presence.

Melezitose

This trisaccharide also has the formula $C_{18}H_{32}O_{16}$. It crystallizes with 2 molecules of water in its composition. It occurs in the exudation from the leaves of certain trees (honey dew) and is sometimes found in honey. It melts at 154°C and hydrolyzes to 1 molecule of fructose and 2 of glucose.

TETRASACCHARIDES

There are several tetrasaccharides, especially among the hydrolytic products of starch, but few of them have been identified and named.

Stachyose

The formula for stachyose is $C_{24}H_{42}O_{21}$. Its complete hydrolysis yields 1 molecule of fructose, 1 of glucose and 2 of galactose.

POLYSACCHARIDES

As the name indicates, polysaccharides are carbohydrates that hydrolyze to a large number of monosaccharide units. The exact number is seldom if ever known. The *simple* polysaccharides hydrolyze to monosaccharides only; the conjugated contain other units.

The polysaccharides are very different from the sugars; the natural ones are insoluble in all common solvents, are not crystalline, do not form osazones and have little or no reducing action on alkaline silver and copper solutions. Cellulose and starch are the two natural, simple polysaccharides of most importance to the food chemist.

Cellulose

All plants contain cellulose in their cell walls. It is the presence of cellulose that distinguishes plants from animals among the lower forms of life.

Absorbent cotton, linen fiber and tissue or filter paper are the most familiar forms of cellulose. The bast fibers of celery and the stems and veins of leaves are mostly cellulose. Less obvious is the cellulose in the walls of the other plant cells.

When heated for a long time with a strong acid, cellulose is hydrolyzed stepwise through cellotetroses, cellotrioses and finally to α-D-glucose. *Cellobiose* is a reducing sugar similar to maltose and its structure gives a clue to the structure of cellulose. It is a glucose-4-glucoside and so the structure of cellulose is apparently

SEGMENT OF A CELLULOSE MOLECULE

The formula shows that the cellulose molecule consists of β-D-glucose units joined together by an oxygen bridge from the 4 carbon of one unit to the 1 carbon of the next. This ties up the H, OH structure on the 1 carbon and thus prevents the display of copper reduction and the aldehydic properties. However, the purest cellulose has a copper number of 0.05, that is, 100 gm of cellulose reduces 0.05 gm of copper from Fehling's solution, but whether this reduction is due to end groups or to slight hydrolysis to cellobiose is not known.

The number of glucose units in the cellulose molecule is not known because, cellulose being insoluble, it is impossible to measure its molecular weight by the usual cryoscopic, ebullioscopic or osmotic methods. Indirect methods give results from 20,000 to 2,000,000. This variation is too great to be due to error and so it is likely that there are celluloses of different chain length. The cellulose of a young plant is somewhat different from that of a mature one of the same species. It may be that the chain length of the molecule increases with the age of the plant.

Cellulose reacts to form ethers and esters. The acetyl compound is a well-known textile (Celanese) and the nitrate is a textile, a plastic and a high explosive.

Although cellulose is insoluble, it reacts with ammoniacal cupric hydroxide (Schweitzer's reagent) and with NaOH and CS_2 to form soluble complexes that are intermediates in the manufacture of rayon and cellophane.

The human digestive tract contains no enzymes that hydrolyze cellulose and therefore it is not a food. Nutritionists refer to it as *roughage,* and analytical chemists call it *crude fiber.*

Ruminants, however, can digest it because it is hydrolyzed by bacteria or other organisms during storage in the stomach.

There is a recent development in cellulose technology of interest to the food chemist. The U.S. Army's research laboratory has developed an enzymatic method for the hydrolysis of cellulose, and has it to the pilot plant stage (1974). The cellulase is produced by the fungus *Trichoderma viride.* The conversion of the cellulose to glucose is carried out at 50°C and produces a glucose syrup of a concentration from 2% to 10%. Newspapers have been used as the source of cellulose and the process promises to be useful both as a means of disposing of cellulose wastes and as a contribution to the food supply. Engineering details are now under study.

Starch

The most common polysaccharide in food chemistry is starch. It occurs almost as widely in the plant world as cellulose. Unlike cellulose, it is not a structural unit, but serves the plant as a means of food storage to furnish energy and building material for the young plant, consequently, it occurs mostly in the seeds, tubers, or roots of the higher plants. When a seed is moistened, enzymes develop and hydrolyze the starch to soluble sugars that supply energy and carbon compounds for the construction of the sprout until it develops roots and leaves and becomes able to produce its own food and use the energy of sunlight. This statement is oversimplified because the plant also uses protein and fat stored in the seed.

The potato is a tuber, or modified stem, and its starch content and tendency to sprout are well known.

Sweet potatoes, parsnips, turnips, carrots and cassava are all roots that contain an abundance of starch.

One important characteristic of starch is that it occurs in the plant in granules that differ in size and shape with the source of the starch. This provides the chemist with a means of identifying raw starches from the various sources. Reichert (1913) has published photomicrographs of an immense number of starches, but the number of commercial starches is not great, and the chemist who has occasion to identify starches can provide himself with small vials of authentic samples for microscopic comparison with an unknown starch, for the starch grains themselves are far more useful for comparison than any picture.

Starch is insoluble in cold water, which is apparently due to the physical structure of the starch grain, for the natural starches are all composed of two substances, one of them soluble in water, the other insoluble. That starch consists of two fractions with different properties was first observed in 1829, and since then, the separation and properties of the two fractions have been the subject of study.

Each fraction of starch has received two names. The insoluble fraction is called *alpha amylose* and *amylopectin;* the soluble fraction, *beta amylose* and simply *amylose.*

The amylose fraction is the simpler of the two in structure. It consists of a straight chain of D-glucose units connected by an oxygen bridge from the 1 carbon of one unit to the 4 carbon of the next.

NON-REDUCING END   AMYLOSE CHAIN   REDUCING END

SYMBOLIC REPRESENTATION OF AN AMYLOSE CHAIN

Amylose is soluble in water and gives a deep blue color with iodine. Beta amylase hydrolyzes it completely to maltose; acids to glucose. It does not reduce alkaline copper solutions and does not react with phenylhydrazine to form an osazone. It is not fermented by yeast.

The molecular weight of amylose is 10,000 to 50,000. It has a specific rotation of $+200°$.

Amylopectin is insoluble in water and gives a violet color with iodine. It also fails to show the reactions of the sugars. It is partly hydrolyzed to maltose by beta amylase. Its molecular weight is 50,000 to 1,000,000.

The structure of amylopectin consists of chains like those of amylose, but it also contains branches from the primary alcohol groups, carbon 6 of the chain, to carbon 1 of the branch. This linkage is not attacked by beta amylase and so only 65% of the molecule is hydrolyzed.

Several methods have been devised for the separation of the two fractions of starch, but unfortunately they do not give consistent results. That the proportions of the two fractions vary considerably with the source of the starch is obvious from the properties of the various starches. Table 3.3 contains some experimental values for the composition of some common starches.

TABLE 3.3

PERCENTAGE OF AMYLOSE IN STARCHES

Source	%	Source	%
Arrowroot	21	Sago	27
Barley	27	Sweet potato	20
Corn	26	Tapioca	18
Potato	23	Waxy corn	0.06
Rice	17	Wheat	25

Starch, like cellulose, can be methylated, acetylated and nitrated with three groups to the glucose unit; these reactions were useful in the establishment of the chemical structure of starch.

Although the structure of starch as presented here is that of a simple polysaccharide, there is some evidence that the molecule contains a very small amount of phosphate and of fatty acids. The amount is so small, however, that it does not enter into the problems of the food chemist.

When starch is hydrolyzed, it passes through several stages, some of which are indicated by the following equation:

$$(C_6H_{10}O_5)_x + H_2O \rightarrow (C_6H_{10}O_5)_y + H_2O \rightarrow$$
$$\text{Starch} \qquad\qquad \text{Amylodextrin}$$
$$(C_6H_{10}O_5)_z + H_2O \rightarrow (C_6H_{10}O_5)_u + H_2O \rightarrow$$
$$\text{Erythrodextrin} \qquad\qquad \text{Achroodextrin}$$
$$C_{12}H_{22}O_{11} + H_2O \rightarrow C_6H_{12}O_6$$
$$\text{Maltose} \qquad\qquad \text{Glucose}$$

Dextrins

Partial hydrolytic products of starch may be prepared by heating starch or by hydrolyzing it with acids or enzymes. Three dextrins are recognized. The first is *amylodextrin*. It is soluble in water and gives a blue color with iodine. The second is *erythrodextrin* and takes its name from the red color it gives with iodine. The third is *achroodextrin*, which is named from the fact that it gives no color with iodine.

All the dextrins contain some carbonyl groups and, therefore, reduce Fehling's solution. They are dextrorotatory and precipitated by alcohol.

The dextrins are not only intermediate products in the hydrolysis of starch, but also intermediates in its natural synthesis. They occur in the leaves and fruits of plants and also in honey.

Dextrins are used in paper adhesives, such as that on envelopes, labels and postage stamps. They are a component of corn syrups and are used in confectionery and other food products.

Inulin

Inulin is a polysaccharide of potential value in the food industry. It occurs in the bulb of the Jerusalem artichoke, the dahlia, dandelion roots, and in other bulbs. It is an amorphous, white powder that gives no color with iodine. Its molecule consists of fructose units in a beta (2 to 1) linkage. Acids hydrolyze it to D-fructose, but the human digestive tract does not contain any enzyme that will hydrolyze it and, therefore, it has no nutritional value unless it is first hydrolyzed to the sugar.

Glycogen

This polysaccharide is sometimes called animal starch. It occurs in the liver and muscles of the higher animals and in scallops and other mollusks. It is not exclusively an animal product, however, because it also occurs in yeasts and mushrooms.

Glycogen dissolves in water to form an opalescent dispersion that is dextrorotatory. It gives a red-brown color with iodine and does not reduce Fehling's solution. Its structure is similar to that of amylopectin; it consists of branched chains of α-D-Glucose units joined by a 1–4 linkage with branches attached by a 1–6 linkage. The individual chains are shorter than those of amylopectin.

Hemicelluloses

The name of this group of substances is misleading; they are not half cellulose. In fact, there is no good agreement as to what they are. They have been described as substances that are associated with cellulose and can be extracted with alkali, apparently meaning xylan; it can be extracted from wood and hydrolyzed to the pentose, xylose.

Another hemicellulose occurs in straw, the hulls of grain and corncobs. It also hydrolyzes to xylose.

This group is difficult to define. Their structure is not completely known and among them, they hydrolyze to give xylose, arabinose, galactose, galacturonic acid and sometimes glucose. They are seldom, if ever, human food, for the human digestive tract contains no enzymes that attack them.

Gums and Mucilages

The Official Methods make no distinction between gums and mucilages. However, the latter are usually considered to be those heterosaccharides that form gels, such as agar and Irish moss.

Agar.—Agar is extracted from seaweed of the genus *Getidium* with boiling water. When pure, it is a white powder insoluble in cold water and alcohol. Its chemical structure is not definitely known, but in part it consists of a polysaccharide chain of both D- and L-galactose units joined by a 1–3 link and some units joined by a 1–4 link. It also contains considerable sulfate and is strongly acid when freed from calcium or other positive ions.

Although it is insoluble in cold water, hot water dissolves it to a viscous solution that sets to a stiff gel; as little as 1.5% forms a firm gel.

Agar has no food value for lack of a digestive enzyme. In fact, it is often used as a laxative because of its ability to swell and add bulk to the intestinal contents.

The chemist is interested in agar because of its physical properties rather than its chemical composition. It is used in confectionery and other products to produce texture.

Carageenan, or Irish Moss.—Another seaweed polysaccharide is obtained from the genus *Chondrud*. It is a galactan similar to agar in its properties. It is used in chocolate products, soups and other products as a protective colloid or to increase the "body" of the product.

Gums

The gums are products of land flora. Many species produce them; peach, plum and cherry trees exude them when the bark is injured. Only a few of the many gums are commercialized.

Quince Seed Gum.—The seed of the quince is surrounded by a gum that is gathered on a commercial scale as a by-product of the quince fruit industry.

Gum Arabic.—One of the most common gums in industry is gum arabic. It is obtained from a tree of the Acacia family. It is soluble in water.

Gum arabic is the calcium, magnesium or potassium salt of arabic acid. A 1% solution of the acid has the pH 2.70, but a solution of the gum itself has the pH 4.5–5.5. Arabic acid hydrolyzes to L-arabinose, D-galactose, L-rhamnose and D-glucuronic acid.

Gum Tragacanth.—This gum is obtained from the genus *Astrolagus*. On hydrolysis it yields arabinose and glucuronic acid.

Other gums that are commercialized are karaya, carob bean and locust bean.

Pectins

Both constitution and nomenclature are more definite in the pectic substances than in the groups just mentioned. The American Chemical Society has decided that the word *pectin* designates those water-soluble pectinic acids of varying methyl ester content and degree of neutralization which are capable of forming gels with sugar and acid under suitable conditions. This general definition recognizes the fact that there are several pectins.

Pectinic acids are colloidal polysaccharides containing more than a negligible proportion of methyl ester groups.

Pectic substances is the name of a group of those complex, colloidal carbohydrates that occur in or are prepared from plants and contain a large proportion of anhydrogalacturonic acids which are thought to exist in chain-like formation. The carboxyl group of these galacturonic acids may be partly esterified by methyl groups and partly or completely neutralized by one or more bases. These definitions will be clear from the following basic formula, which is that of pectic acid:

Pectic acid

Portion of the pectin molecule

From the formula, it is clear that the various pectic substances consist of this basic structure, which has an acid group on each unit, and some or all these groups may be esterified to the methyl ester —$COOCH_3$ or neutralized to form salts. Since the molecular weight of the pectins ranges from 10,000 to 400,000 it is obvious that the pectins also vary in chain length.

The pectins form viscous solutions in water. The viscosity increases with the length of the chain and the degree of esterification. With sugar and acid they form gels, which is their important characteristic in the jam and jelly industry.

Dilute acids hydrolyze both the methyl esters and the glycosidic chain, so that over cooking may destroy the gelling property entirely.

Pectin occurs in the plant in the form of protopectin, which is insoluble in water and is probably a compound with cellulose or other polysaccharide. In addition to the pectin in fruit and vegetable juices, the albedo of citrus fruits contains a protopectin from which pectin is prepared commercially. One method used to release the pectin is heating the albedo short of boiling with dilute sulfurous acid.

BIBLIOGRAPHY

ANON. 1974. Waste cellulose possible glucose source. Chem. Eng. News. *52*, 20.

ASSOC. OFFIC. ANAL. CHEMISTS. 1975. Official Methods of Analysis, 12th Edition. Assoc. Offic. Anal. Chemists, Washington, D.C.

AURAND, L. W., and WOODS, A. E. 1973. Food Chemistry. Avi Publishing Co., Westport, Conn.

BANKS, W., and GREENWOOD, C. T. 1975. Starch and its Components. John Wiley & Sons, New York.

BATES, F. J. 1942. Polarimetry, Saccharimetry and the Sugars. Govt. Printing Off., Washington, D.C.

CONN, E. E., and STUMPF, F. K. 1966. Outlines of Biochemistry. John Wiley & Sons, New York.

GARARD, I. D. 1974. The Story of Food. Avi Publishing Co., Westport, Conn.

GARARD, I. D., and SHERMAN, H. C. 1918. A study of the glucosazone reaction. J. Am. Chem. Soc. *40*, 955–969.

HULLINGER, C. H., PATTEN, E. V., and FRECK, J. A. 1973. Food applications of high-amylose starches. Food Tech. *27*, No. 3, 22–24.

IHDE, A. J. 1964. Development of Modern Chemistry. Harper and Row, New York.

IFT COMMITTEE. 1959. Pectin standardization. Food Tech. *13*, 496–500.

JOSLYN, M. A. 1960. The chemistry of protopectin. *In* Advances in Food Research, Vol. 2. Academic Press, New York.

JUNK, W. R., and PANCOST, H. M. 1973. Handbook of Sugars. Avi Publishing Co., Westport, Conn.

KERTEZ, Z. I. 1963. The Pectic Substances. John Wiley & Sons, New York.

LAWRENCE, A. A. 1973. Edible Gums and Related Substances. Noyes Data Corp., Park Ridge, N.J.

LEE, F. A. 1975. Basic Food Chemistry. Avi Publishing Co., Westport, Conn.

MOUSSERI, J., STEINBERG, M. P., NELSON, A. I., and WEI, L. S. 1974. Bound water capacity of corn starch and its derivatives by NMR. J. Food Sci. *39*, 114–116.

NATL. RES. COUNCIL. 1972. Food Chemicals Codex, 2nd Edition. Natl. Acad. Sci., Washington, D.C.

NEF, J. U. 1914. Dissociation in the Sugar group. Ann. 403, 204–383.

PETERSON, N. B. 1975. Edible Starches and Starch-derived Syrups. Noyes Data Corp., Park Ridge, N.J.

REICHERT, E. T. 1913. The differentiation, and specificity of starches, genera, species etc. Pub. 173, Parts 1 and 2. Carnegie Institution of Washington. Washington, D.C.

SCHOCH, T. J. 1941. Physical aspects of starch behavior. Cereal Chem. *18*, 121–128.

SCHOCH, T. J. 1942. Fractionation of starch by selective precipitation with butanol. J. Am. Chem. Soc. *64*, 2957–2961.

SCHULTZ, H. W. 1969. Carbohydrates and Their Roles. Avi Publishing Co., Westport, Conn.

SCHWARTZ, M. E. 1974. Confections and Candy Technology. Noyes Data Corp., Park Ridge, N.J.

SHALLENBERGER, R. S., and BIRCH, G. G. 1975. Sugar Chemistry. Avi Publishing Co., Westport, Conn.

SHERMAN, H. C. 1952. Chemistry of Food and Nutrition. Macmillan Co., New York.

SMILEY, E. L. 1966. Microbial polysaccharides. Food Tech. 20, No. 9, 112–115.

SMITH, F., and MONTGOMERY, R. 1959. The Chemistry of Plant Gums and Mucilages, and some Related Polysaccharides. ACS Monograph 141. Reinhold Publishing Corp., New York.

STANICK, J., CERNY, M., KOCOUREK, J., and PACAK, J. 1963. Monosaccharides. Academic Press, New York.

WALDT, L. M. and KEHOE, D. 1959. Starch chemistry for the food technologist. Food Tech. *13*, 1–4.

CHAPTER 4

Proteins

Proteins have been recognized as animal substances since ancient times; Pliny (23–79) called egg white *albumen.* The animal substances gave off a strong, characteristic odor when they were burned. Since egg white when dried and burned gave the characteristic odor, all the animal substances came to be called albumens, and the name still appears occasionally in the literature.

After 1775, chemists began to find that all living matter appeared to contain albuminous substances and in 1838, G. J. Mulder (1802–1888), professor of Chemistry at the University of Rotterdam, coined the word *protein.*

The proteins are very complex substances, and it is not surprising that it took chemists so long to unravel their chemical nature. In 1811, a French chemist found that proteins contain nitrogen and that not all proteins are alike.

Gelatin was the first protein to be isolated. It was obtained in the 18th century by boiling bones in water. It is today the only pure protein in the food market, and it has come in for a lot of study.

In 1820, Henri Braconnet (1781–1855) hydrolyzed gelatin by boiling it for a long time with dilute sulfuric acid. He obtained a white crystalline substance that he called *glycine,* which turned out to be the simplest unit in the structure of the proteins. He then got *leucine* from wool by the same process and, in 1846, a German chemist obtained tyrosine from cheese in the same manner. All these results were empirical, for the chemistry of the first half of the 19th century was rather primitive and mostly inorganic.

ANALYTICAL METHODS

One difficulty was the lack of good analytical methods. The Dumas method for the determination of nitrogen was a combustion method; it was very tedious and required considerable skill and experience.

In 1883, a Danish chemist, J. G. T. C. Kjeldahl, devised the method that bears his name, and variations of the method are still in use and included in the Official Methods.

By the early years of the 20th century, several proteins had been isolated from their natural sources in a reasonably pure state and analyzed for their nitrogen content. The results varied from 15% to 18% with an average value of 16%. This work became the basis for the determination of protein in foods, feeds and other biological products.

The Kjeldahl method and its modifications are invariably used in food analysis for the determination of protein. A weighed sample is digested with concentrated sulfuric acid. The products of the digestion are CO_2, H_2O and NH_3. The acid is reduced to SO_2, and the ammonia remains in the flask as ammonium sulfate, $(NH_4)_2SO_4$. Water is added, the solution made alkaline with NaOH and then distilled. The NH_3 distils with the water into an accurately measured standard solution of HCl and the excess acid is titrated with standard NaOH. The difference between the original HCl and the amount titrated was neutralized by the NH_3. If this value were 10 ml of 0.1 N HCl, then the sample contained enough nitrogen to make 10 ml of 0.1 N nitrogen solution or 0.14 gm. Since proteins average 16% nitrogen, the protein is 100/16, or 6.25 × 0.14 gm N or 0.875 gm of protein.

Many modifications of the Kjeldahl method have been made to improve its precision or convenience but, aside from any inaccuracies in the procedure, there are some fundamental defects in the method. The two main ones are: (1) there may be nitrogen compounds in the sample other than protein, and (2) the protein in the sample under analysis may not contain 16% nitrogen. Efforts have been made to correct the second defect. In the analysis of wheat flour, for example, the Official Methods use the factor 5.7 instead of 6.25 because of the known higher nitrogen content of the wheat proteins.

Protein is so essential in nutrition that the amount of it in a food product is extremely important and the inaccuracies in the determination are usually much less than the variation from one sample to another of the same food, whether the sample is an apple, a loaf of bread or a barrel of flour.

AMINO ACIDS

The chemical nature of the proteins began to be unraveled with the preparation of glycine, leucine and tyrosine from protein molecules, and by 1901, 16 of these products had been obtained from various proteins. Meantime, the chemical structure of these compounds had been determined. They were all alpha amino acids; glycine is amino acetic.

Glycine

By 1935, the number of these acids obtained from proteins had exceeded twenty. The amino acids are white crystalline solids soluble in water but not in many organic solvents.

The acids all have high melting points and most of them decompose

TABLE 4.1

AMINO ACIDS FROM PROTEINS

Acid	Formula	Melting Point °C
Glycine	H_2NCH_2COOH	232°
Alanine	$CH_3CH(NH_2)COOH$	298
Valine	$(CH_3)_2CHCH(NH_2)COOH$	315
Leucine	$(CH_3)_2CHCH_2CH(NH_2)COOH$	
Isoleucine	$CH_3CH_2CH(CH_3)CH(NH_2)COOH$	280
Serine	$HOCH_2CH(NH_2)COOH$	228
Threonine	$CH_3CH(OH)CH(NH_2)COOH$	—
Cysteine	$HSCH_2CH(NH_2)COOH$	—
Cystine	$\begin{array}{l} S\text{-}CH_2CH(NH_2)COOH \\ \mid \\ S\text{-}CH_2CH(NH_2)COOH \end{array}$	260
Methionine	$CH_3SCH_2CH_2CH(NH_2)COOH$	—
Aspartic acid	$HOOCCH_2CH(NH_2)COOH$	251
Glutamic acid	$HOOCCH_2CH_2CH(NH_2)COOH$	206
Arginine	$\begin{array}{l} HN{=}C\text{---}NH\text{---}(CH_2)_3CH(NH_2)COOH \\ \qquad\mid \\ \qquad NH_2 \end{array}$	238
Lysine	$H_2N(CH_2)_4CH(NH_2)COOH$	224
Phenylalanine	$C_6H_5CH_2CH(NH_2)COOH$	280
Tyrosine	$HOC_6H_4CH_2CH(NH_2)COOH$	318
Tryptophane	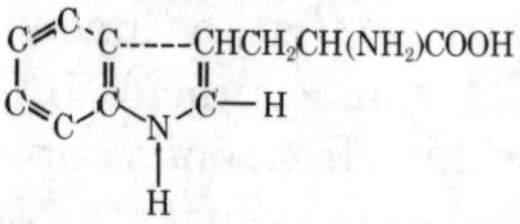	290
Histidine	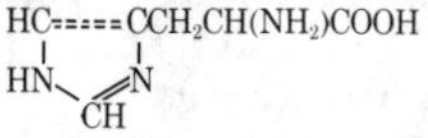	235
Proline	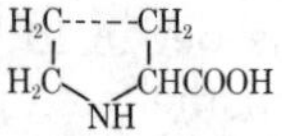	—
Hydroxyproline	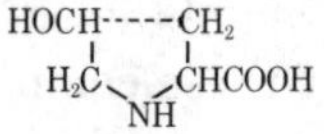	270

during the melting. This may seem unusual for organic compounds of low molecular weight. For example, acetic acid CH_3COOH (mol wt 60) melts at 17°C, monochloroacetic $ClCH_2COOH$ (mol wt 94.5) melts at 63°C and glycine (mol wt 75) melts at 232°C.

The explanation of the high melting points is that each acid contains the basic —NH_2 group and the acidic —COOH group; consequently, they are amphoteric and of the nature of internal salts. They react with HCl to form amino hydrochlorides and with NaOH to form the sodium salt.

An inspection of the formulas in Table 4.1 discloses at least one asymmetric carbon in all of them except glycine, and so they are optically active. Those obtained from proteins are the levo isomers.

STRUCTURE OF THE PROTEINS

Since the hydrolysis of a protein yields amino acids almost exclusively, the natural question is how these acids are joined together in the protein.

The first step in providing the answer to the question was the preparation and proof of structure of glycylglycine by Emil Fischer in 1901.

$$H_2N—\overset{\overset{\displaystyle H}{|}}{\underset{\underset{\displaystyle H}{|}}{C}}—\overset{\overset{\displaystyle}{\|}}{\underset{\underset{\displaystyle O}{}}{C}}—\overset{}{\underset{\underset{\displaystyle H}{|}}{N}}—CH_2COOH$$

Glycylglycine

The compound was synthesized by several methods, but the final result was the elimination of a molecule of water by removing the —OH group from one molecule of the acid and an H from the —NH$_2$ of another, and thus joining the two acid molecules. The

$$--\overset{\overset{\displaystyle}{\|}}{\underset{\underset{\displaystyle O}{}}{C}}---\overset{}{\underset{\underset{\displaystyle H}{|}}{N}}---$$

group formed is called a *peptide* linkage.

Glycylglycine has an amino group (—NH$_2$) at one end and the —COOH group at the other so that the dipeptide can react further in either or both directions to form a chain of indefinite length. By 1907, Fischer had synthesized an octadecapeptide from glycine and leucine consisting of 3 leucyl and 15 glycyl units. It was leucyl-triglycyl-leucyl-triglycil-leucyl-octoglycylglycine.

In 1916, Emil Abderhalden, who had been a student of Fischer's, synthesized a polypeptide of 19 units. He also used glycine and leucine but connected them in a different order; the molecular weight was 1326.

These polypeptides had some of the properties of proteins, but they contained only 2 acids and the molecular weight was far less than that of a protein.

There is considerable evidence for Fischer's theory of protein structure and it is still generally accepted. For example, (1) there are very few free —NH$_2$ or —COOH groups in a protein, indicating that they are tied up in some way, which the peptide linkage does; (2) hydrolysis of proteins results in an equal number of free —NH$_2$ and —COOH groups, and (3) several dipeptides have been found in partly hydrolyzed proteins.

The structure of the protein, however, is not as simple as that of the octadecyl peptide. Diamino acids, dicarboxyclic acids, hydroxyl groups and sulfhydril groups all offer opportunities for branching

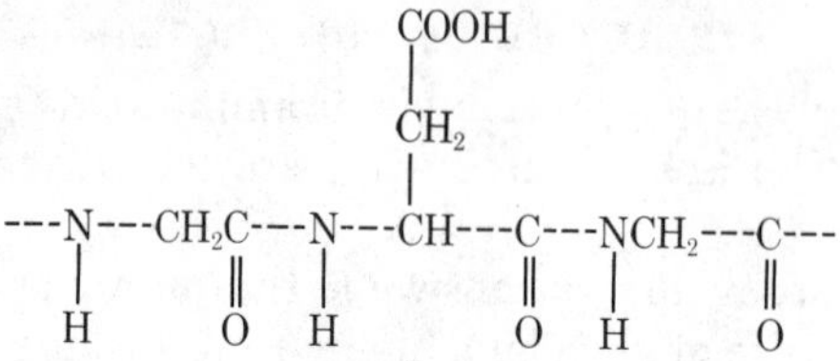

$$
\begin{array}{c}
\text{COOH} \\
| \\
\text{CH}_2 \\
|
\end{array}
$$

---N---CH₂C---N---CH---C---NCH₂---C---
　　|　　‖　|　　‖　|　　‖
　　H　　O　H　　O　H　　O

Glycine　Aspartic acid　Glycine

In the above equation, the —COOH group of the aspartic acid offers opportunity for a branch by reaction of the —NH₂ group of any amino acid. Threonine, cysteine, glutamic acid, arginine, lysine, tyrosine, histidine, proline and hydroxyproline all invite branches. These second —NH₂, —COOH, —OH groups and hydrogen bonds may also join two chains together.

With 20 acids that may be combined in any order of succession, with branches and with a different number of units in the chain, the number of proteins is infinite. That such a variety exists is supported by the fact that chemists have not found two proteins alike. Fibroin, the protein of silk fiber, is the simplest protein that has been analyzed; it contains glycine, alanine, tyrosine and about 1% of three other acids.

The *fibrous* proteins, such as the elastin of muscles, the collagen of connective tissue and the keratin of wool, appear to consist of long chains of amino acid units, but the *globular* proteins, such as egg albumen and the gliadin of wheat, are more complicated in spatial arrangement and may be spherical or in some other compact form.

Molecular Weights

The molecular weights of the proteins are difficult to measure because of the size of the molecule. The results of measurements are recorded in Table 4.2, but they do not claim an accuracy within less than 1,000.

TABLE 4.2

MOLECULAR WEIGHTS OF PROTEINS

Protein	Molecular Weight	Protein	Molecular Weight
Casein	75,000	Lactalbumen	41,500
Egg albumen	310,000	Pepsin	36,000
Gliadin	27,500	Tobacco Mosaic	15,000,000
Insulin	46,000	Virus	
Hemoglobin	67,000	Zein	40,000

Source: Thomas (1934).

Isoelectric Points

The proteins are amphoteric because of the free alkaline —NH_2 and acid —COOH groups at the ends of the chains; some are on the basic side and some are acidic. If an acidic protein is put into an electric cell, the protein migrates to the positive electrode. Likewise, a basic protein migrates to the negative electrode.

If an alkali is added slowly to an acid protein, an acidity is reached at which the protein does not migrate in either direction. The pH of this solution is called the *isoelectric point*. Table 4.3 contains these values for some common proteins.

TABLE 4.3

ISOELECTRIC POINTS OF PROTEINS

Protein	Isoelectric Point	Protein	Isoelectric Point
Albumen (cow's milk)	4.6	Glutenin	5.4
Albumen (egg)	4.8	Hemoglobin	6.8
Casein (cow's milk)	4.6–4.9	Histone	8.5
Gelatin	4.4–5.6	Pepsin	2.5–3.3
Gliadin	5.8–6.6	Trypsin	7.0–8.0

Source: Thomas (1934).

Composition

There are ten essential amino acids, and it is important for the food chemist to know the best sources of them. The necessity for these acids in the diet presents a challenge for the food chemist. The soybean meal used for poultry feed, for example, is deficient in methionine, and that acid is now isolated from fish meal and also synthesized primarily for supplementing the poultry diet. It is synthesized commercially from hydrogen cyanide, acrolein and methyl mercaptan. It is estimated that 60 million lb of the acid will be required in North America in 1975.

Other common food and feed proteins are deficient in one or more amino acids and, like the vitamins, need to be added to deficient diets. Table 4.4 shows the content of some common proteins in these acids.

Denaturation

Almost any form of energy will denature some of the proteins and all of them, with the possible exception of gelatin, can be denatured. Even shaking will denature some of them, also the addition of an acid or a base or ethyl alcohol. The effect of heat is of most concern to the chemist. If milk (lactalbumin) protein is heated to 80°C, it is denatured and the milk has a cooked taste. This is important in the pasteurization and condensation of

TABLE 4.4

ESSENTIAL AMINO ACIDS IN PROTEINS

Protein	Gm per 100 Gm Protein									
	Arginine	Histidine	Isoleucine	Leucine	Lysine	Methionine	Phenylalanine	Threonine	Tryptophane	Valine
Arachin	13.50	2.6	4.55	7.61	2.72	0.24	6.96	2.89	0.90	4.85
Casein	3.81	2.65	5.63	9.22	8.20	2.58	4.89	4.80	1.20	6.90
Gelatin	9.11	0.77	1.25	3.30	4.44	0.78	2.33	2.10	0.04	2.90
Lactalbumen	3.42	1.50	6.12	12.90	10.80	1.62	3.59	5.37	1.72	5.82
Ovalbumen	6.03	2.06	7.41	9.43	6.32	4.43	7.17	4.48	1.30	7.54
Ox muscle	7.87	1.98	6.50	9.37	10.00	2.75	4.58	5.80	1.30	5.85
Phaseolin from beans	5.97	2.24	6.69	10.50	7.20	1.12	8.04	4.16	0.50	6.00
protein										
Zein	1.95	0.76	5.03	21.10	0.21	1.41	7.30	2.62	0.03	3.98
Corn, whole yellow	4.68	2.23	4.39	14.33	2.30	1.44	5.26	3.89	0.48	5.30
Egg, whole dried	8.50	2.30	8.73	12.55	7.88	2.74	7.31	5.91	1.71	6.99
Milk, dry skim	2.81	2.40	6.12	10.35	7.10	2.17	4.54	5.25	1.12	6.93
Oatmeal	7.31	2.20	4.40	7.76	2.96	1.04	4.58	3.60	1.04	5.22
Peanut flour	12.27	2.35	3.42	6.72	2.93	0.82	5.26	3.05	0.88	4.14
Soybean flour	7.79	2.38	4.98	7.49	7.16	0.99	4.76	4.55	1.12	4.90
Whole wheat	4.48	2.03	3.60	5.89	2.50	0.99	4.06	2.97	0.83	4.10
Yeast, dried	4.70	1.62	4.93	0.76	7.28	1.10	3.72	5.28	0.94	5.28

milk. The effect of heat on egg albumin is well known, and it improves the digestibility of the protein.

Denaturation has been described as any change in the protein except hydrolysis. Just what the chemical changes are is not known and they probably differ from one protein to another.

Heat is said to destroy certain amino acids, but they are probably not destroyed in the usual sense of the word, but rendered inactive towards enzymes, that is, the enzymes do not release them by hydrolysis. Lysine, threonine and methionine are especially sensitive to heat and the loss during cooking may reach 40% of the lysine, 20% of the threonine and 10% of the methionine.

Amino acids react with other components of the food, particularly, the aldehyde sugars. In the Maillard reaction, or browning, that occurs when foods are subjected to dry heat (roasting, baking, frying) there is a reaction between amino acids and sugars; histidine, threonine, phenylalanine, tryptophane and lysine appear to be the most likely aicds in the combination.

In one experiment, Harris and von Loesecke (1960) report that a mixture of casein and glucose held at 37°C lost $\frac{2}{3}$ of its lysine in 5 days. In 30 days, it lost 90% of its lysine, 70% arginine, 50% methionine and 10% of its tyrosine to enzymatic release.

QUALITATIVE TESTS FOR PROTEINS

The Official Methods (12th Edition, Chapter 7) include 5 methods for the detection of proteins. These are all color reactions, and there are several others reported in the literature that may be useful at times.

Biuret Reaction

This reaction is quite general as it is a test for the peptide bond, —CONH—, although histidine alone shows the test.

A dilute solution of copper sulfate is added to a solution of the protein and then a dilute solution of NaOH is added. A violet color is produced by the formation of biuret, $H_2NCONHCONH_2$, which reacts with the cupric hydroxide to form a complex, colored compound.

Ninhydrin Reaction

This reaction is characteristic of the alpha amino group, and, therefore, indicates proteins or any of their hydrolytic products.

Ninhydrin (triketohydrindene hydrate) is added to a solution of the protein and a reaction with an amino group proceeds as shown in the equation,

Ninhydrin Reaction

The color of the compound formed varies somewhat with the source of the —NH_2 group from deep blue to a reddish violet. The method has been used as a colorimetric quantitative method.

Folin's Reagent

This is a test for amino acids and also for ammonia. The latter is easily eliminated from a protein test for the test is made in a strongly alkaline solution. Alkaloids and aniline also give the test but are seldom present to interfere in a test for the amino acids.

Beta naphthoquinone and sulfuric acid are added to an alkaline solution of a protein and produce a bright red color.

Millon Reagent

This is a test that depends on the presence of a phenolic group and is therefore a test for tyrosine, which is the only natural amino acid that contains the group.

The reagent must be prepared daily because it does not keep well. Mercury is dissolved in concentrated nitric acid and the acidity adjusted. When a protein is heated with the reagent, a brick-red color develops.

Xanthoproteic Reaction

This simple test consists in adding concentrated nitric acid to protein, which produces a bright yellow color. The addition of ammonium hydroxide turns the color to a deep orange. Many an amateur chemist has proved by this reaction that his skin contains protein.

The test depends on the nitration of the benzene ring and is therefore limited to those proteins that contain tryptophane, tyrosine or phenylalanine, but there are few proteins that lack all three of these acids.

Adamkiewicz Reaction

Acetic acid is added to a protein in a test tube, then concentrated sulfuric acid is added carefully to the tilted tube so that the acid runs down the side and forms a separate layer at the bottom. A reddish violet ring forms at the juncture of the two layers.

This is a test for tryptophane. It is caused by the traces of glyoxylic acid formed from the acetic acid by the sulfuric acid. It is then decomposed to formaldehyde, which reacts with the tryptophane.

Acree-Rosenheim Reaction

This is another test for tryptophane, and is similar to the Adamkiewicz reaction. A drop of formalin is added to a solution of the amino acid or protein and then concentrated HCl; a violet color is produced.

Other Tests

There are other color reactions, but the above tests will probably be adequate for the purposes of the food chemist.

CLASSIFICATION

Biologists and chemists have struggled for years with a classification of the proteins. The main classes are easy: (1) simple proteins, (2) conjugated proteins, and (3) derived proteins. Within classes 1 and 3, the only basis of classification is their physical properties, and all the proteins in class 1 are very much alike. However, the following subdivisions have been decided on.

I. Simple Proteins

These are the natural proteins that yield only amino acids when hydrolyzed, although traces of carbohydrate have been found in some of them.

A. *Albumins* are simple proteins soluble in water and coagulable by heat. Examples are egg albumin, serum albumin (blood), legumelin (peas), and lactalbumin.

B. *Globulins* are insoluble in water but soluble in neutral solutions of salts of strong acids and strong bases. Examples are myosin (muscles), serum globulin (blood), edestin (hemp seed), legumin (peas), tuberin (potato), amandin (almonds), and arachin (peanuts).

C. *Glutelins* are insoluble in all neutral solvents, but soluble in dilute acids and alkalis. Examples are glutenin (wheat), and oryzenin (rice).

D. *Prolamins* are soluble in 50% to 80% alcohol, but insoluble in water, absolute alcohol and other neutral solvents. These are all from plant sources. Examples are gliadin (wheat), zein (corn), hordein (barley), and kafirin (kaffir corn).

E. *Scleroproteins,* or *Albuminoids* are insoluble in neutral solvents and occur mainly in the structural or protective tissues of animals. Examples are collagen (connective tissue), keratin (hair, nails, hoofs, horns), and the fibroin of silk.

F. *Histones* are soluble in water and insoluble in very dilute ammonia. A coagulum is formed by heating, which is soluble in dilute acids. On hydrolysis, they yield a high percentage of basic amino acids (arginine, lysine, histidine).

G. *Protamines* are lower in molecular weight than the histones. They are soluble in water and dilute ammonia, are strongly basic and not coagulated by heat. They yield basic amino acids when hydrolyzed, mostly arginine. They are of little importance to the food chemist. Examples include protein of the sperm of salmon and other fish.

II. Conjugated Proteins

These proteins are essentially simple proteins attached to a nonprotein compound. Their composition is obvious from the description of the four classes.

A. *Nucleoproteins* are compounds of simple proteins with nucleic acids, which are complex compounds that hydrolyze to phosphoric acid, a sugar, 2 pyrimidine bases and 2 purine bases. They are mainly of biological interest; one has been separated from the thymus gland and one from yeast.

B. *Glycoproteins* have the simple protein unit attached to a substance containing a carbohydrate group other than a nucleic acid. An example is mucins (saliva).

C. *Phosphoproteins* contain phosphorus in organic union with a simple protein other than a nucleic acid or a lecithin.

D. *Lipoproteins* are conjugates of simple proteins and a lipid such as lecithin or a sterol. Examples are proteins from egg yolk, nerve tissue, and cell nuclei.

E. *Chromoproteins* are conjugates of simple proteins and a colored substance. The heme of the hemoglobin of the blood is the most familiar example. The food chemist who works with meat will be concerned with this protein.

III. Derived Proteins

The first two main classes of proteins contain the proteins as they occur in nature, or essentially unchanged by the process of isolation. Class III contains the proteins that have been altered by some agent.

1. *Primary Derivatives*

These are slightly altered natural proteins.

A. *Proteans* are insoluble products produced by the action of water, dilute acids or anzymes. Examples include casein (milk curd) and fibrin (blood clot).

B. *Metaproteins* are the products of the further action of water, acids or enzymes. They are insoluble in neutral solvents, but soluble in dilute acids and alkalis.

C. *Coagulated Proteins* are the result of the action of alcohol or heat on natural proteins. An example is cooked egg white.

2. *Secondary Derivatives*

These are the products of considerable hydrolysis by acids or enzymes.

A. *Proteoses* are soluble in water, not coagulated by heat and precipitated by a saturated solution of ammonium sulfate or zinc sulfate.

B. *Peptones* are hydrolytic products of proteins of lower molecular weight than the proteoses. They are soluble in water, not coagulated by heat and not precipitated by saturating the solution with a salt.

C. *Peptides* are further hydrolytic products of proteins and are compounds of known composition, containing two or more amino acids in combination. The class also includes the synthetic peptides.

The chief advantage of the protein classification to the food chemist is that it enables him to identify the nature of the compound when he encounters the name of one of the classes in the literature.

ENZYMES

At least as early as 1745, chemists were studying the cause of fermentation, and discovered that the glutinous mass that forms on the surface of fermenting grape juice causes the fermentation. This, of course, is yeast.

In 1833, a chemist separated a diastase from malt that would hydrolyze 2000 times its own weight of starch. Other similar substances were soon discovered.

By 1878, it was clear that chemical changes in foods were caused by substances, produced by living organisms, but operative when isolated from the cells that produced them. To all such substances a German physiologist gave the name *enzyme,* which he derived from two Greek words meaning *in yeast* or *in leaven.*

Enzymes may now be defined as catalysts that are produced by living cells.

Osborne (1895) investigated malt amylase and decided that it is a protein. This was the first work on the chemical nature of enzymes. Subsequent work followed slowly, but all enzymes are now thought to be proteins. This has been controversial until recent years, chiefly because a renowned German chemist failed to get a test for a protein in his enzyme solutions, probably because he didn't have enough of the compound to give the usual tests for a protein.

Crystalline Enzymes

The early experimenters followed the success of their purification of enzymes by measuring the weight of substrate the enzyme preparation would hydrolyze. When no further purification would improve the activity, the enzyme was considered pure and, of course, it was as pure as that method of purification would produce. All such preparations, however, were amorphous powders. But in 1926, Sumner crystallized urease, an enzyme that hydrolyzes urea to CO_2 and NH_3. This was followed rapidly by the crystallization of several enzymes by other workers; pepsin in 1930, pancreatic amylase in 1931, trypsin in 1932, papain in 1937 and others since then.

It may be rash to say that all enzymes are proteins because there are thousands of them produced by plant and animal tissues. Those that have been purified are proteins, most of them are simple proteins, some contain other organic groups or metals. However, if chlorophyll is included among enzymes, that is at least one exception, as it is not a protein.

CATALYTIC ACTION

Catalysts are substances that accelerate a reaction but do not enter into it stoichiometrically. For example, a mixture of hydrogen and oxygen will remain just a mixture indefinitely at room temperature, but react instantly if a palladium wire is inserted into the mixture. The product of the reaction is water, and the palladium is still just palladium metal although the surface will change physically if the wire is used repeatedly. General chemistry presents many cases of catalysis and there are various theories to explain the effect.

Here, we are concerned with the catalytic activity of enzymes, particularly those present in foods or those that affect foods such as the digestive enzymes.

In the older literature on catalysis, we often find the statement that a catalyst cannot start a reaction but merely accelerates a reaction that is already in progress. Where this nonsense came from I am uncertain, but it persisted for years although any chemist can name several reactions that show no evidence whatever of progress. The hydrogen–oxygen mentioned above is one example. For another, sterilize a starch paste and seal it up. It remains a starch paste, but add a little amylase and it will give a test for glucose in a few minutes.

Enzymic Catalysis

Catalysis by enzymes follows the same pattern as the inorganic catalysts. If a reaction is reversible, a catalyst does not change the point of equilibrium.

TABLE 4.5

OPTIMUM pH OF ENZYMES

Enzyme	Substrate	pH
Invertase (yeast)	Sucrose	4.4
Invertase (intestinal)	Sucrose	7.0
Pepsin	Gelatin	2.4
Pepsin	Ovalbumin	1.5
Amylase (barley)	Potato starch	4.5

There are several factors that affect the speed of a reaction catalyzed by an enzyme.

Concentration of Enzyme.—As one might expect, the speed of a reaction is proportional to the concentration of the enzyme. Since the catalytic effect depends on contact of the enzyme molecule with the reacting substance, the chance of contact increases with the number of enzyme molecules present. This proportionality assumes an abundance of substrate molecules; as the reaction approaches completion, the rate falls off because of the lack of substrate.

Temperature.—As the temperature rises, the speed of an enzymatic reaction increases up to the temperature that begins to denature the enzyme. Since enzymes are proteins, they are easily denatured. This effect produces an optimum temperature for each enzyme. It varies with the acidity of the substrate and other factors. The optimum is usually between 40° and 50° although some have been reported as high as 65°C. The enzymes of the human digestive tract operate at 37°C, which may be the optimum temperature under the conditions in which they operate. *In vitro* the optimum is usually somewhat higher. Probably all enzymes are inactivated at 100°C, and much lower temperatures will inactivate most of them if they are held for some time, 5 min at 70°C, for example.

Acidity.—Enzymes are sensitive to the acidity of the substrate and the optimum pH may vary with the nature of the substrate and the temperature (Table 4.5).

That temperature also has an effect on pH is indicated by the fact that malt amylase has an optimum pH of 4.3 at 25°C, 5.0 at 45°C and 5.7 at 60°C.

The optimum pH appears to be related to the isoelectric point of the protein substrate and varies with temperature and the source of the enzyme.

General Conditions.—Certain substances increase the activity of enzymes. Among them are K^+, Mg^{++}, Zn^{++}, Fe^{++}, Cu^{++} and Cl^-. In some cases one of these ions must be present to get any action at all.

Several substances inhibit enzyme action and in a few cases a substance has been found that accelerates one enzyme action and inhibits another.

It is impossible to state general conditions for optimum enzyme activity. Each enzyme is specific in the type of reaction it will catalyze. The hydrolytic enzymes are specific to the type of bond they will hydrolyze. An enzyme that will hydrolyze starch will not affect cellulose; a sucrase will not hydrolyze lactose.

Furthermore, an enzyme rarely has occasion to catalyze a reaction of a pure substance, and the rate of the reaction is affected by the temperature, the acidity and the other substances that may be present in the system. Each system must be considered individually.

Nomenclature and Classification

In the early literature, the physiologists gave conventional names to the enzymes; among them were ptyalin, pepsin, trypsin, papain and bromelin, but there is nothing about these names to indicate the nature of the substances they represent. The present system uses the suffix -*ase* to indicate that the substance is an enzyme, and the rest of the name indicates the nature of the enzyme's activity. An *amylase* catalyzes the hydrolysis of starch (*L Amylum,* starch), a *lipase* catalyzes the hydrolysis of a lipid, a *lactase* catalyzes the hydrolysis of lactose and so on. However, even chemists must have their digressions; *taka diastase* is an amylolytic enzyme, named for Dr. Takamine who developed the enzyme commercially.

An adjective often precedes the name to indicate the source of the enzyme such as salivary amylase, intestinal lipase and a fungal amylase.

Food Enzymes

There are several other types of enzymes, but by far the greatest number encountered by the food chemist are hydrolytic, such as the lipases, amylases and proteases of the digestive system. Also, every seed contains each of these three enzymes. The seed stores fats, polysaccharides and proteins and as it sprouts it produces enzymes to hydrolyze them to lower molecular weight compounds that are soluble and can be transported along the sprout to support its growth until it develops leaves.

There are also oxidases in foods. The brown color that develops soon after an apple is cut is caused by an oxidation, and the enzyme in yeast that converts sugar into alcohol and carbon dioxide is also an oxidase.

NUTRITIONAL REQUIREMENTS

The food chemist must know the basic principles of nutrition and where to learn the details.

Food has three functions: to supply energy, to supply materials for growth and replacement of body tissues and to regulate the numerous physiological processes that constitute life.

Energy

The first of these takes by far the greatest part of the food supply on a tonnage basis. The requirement, however, is expressed in energy units and not by weight, because different foods supply different amounts of energy per unit weight, largely because of the great variation in water content. All natural foods that supply energy contain water; some of them as much as 95%, and 75% is probably a fair average.

The amount of energy required by a person depends on the amount of physical work he does. To avoid misunderstanding, work is the action of a force through a distance, such as lifting the body from the ground floor to the next floor above by means of a stairway. The heavier the body and the higher the stairs, the more work it takes. Raising the body a few inches with every step and propelling it forward when walking is work. Running or jogging is more work because the body is lifted higher. Each movement of the body or any part of it requires work.

Even as complete rest as possible, that of lying in a comfortable bed, is work, because the heart goes on beating, the lungs pumping, the stomach churning and each of the other life processes is doing a certain amount of work. The energy of complete rest is called the *basal metabolism*. It is highest per unit weight in young children and least in old age. For the average man or woman between the ages of 22 and 62, the basal metabolism will likely be between 1200 and 1500 kilocalories (kcal). For the average person leading a sedentary life, this will be about 70% of the daily energy requirement; for those who work hard physically such as digging a ditch, emptying garbage cans into a truck or loading watermelons, the basal may be less than half the daily requirement. Books on nutrition must be consulted for further information on energy requirement, but the chemist should know that the energy requirement depends chiefly on body weight and activity. If the individual is too fat, he consumes too much energy, for the surplus food above energy requirements—whether carbohydrate, protein or fat—is converted into human fat and stored against a possible future deficiency in the food supply.

Energy is supplied to the body in the form of carbohydrates, fats and proteins. Carbohydrates and proteins supply an average of 4 kcal per gram and fats, 9 kcal. No other component of the food supplies any appreciable amount of energy except alcohol, which supplies 7 kcal per gram. Not all carbohydrates, fats and proteins are the same, but the variation is not very great and all numerical values in nutrition are approximations. For example, grapefruit is listed in the tables as 84% water and at 41 kcal per 100 gm, but

TABLE 4.6

PROTEIN ALLOWANCES

	Age Yr	RDA Gm/day
Infants	0.0—0.5	kg × 2.2
	0.5—1.0	kg × 2.0
Children	1—3	23
	4—6	30
	7—10	36
Males	11—14	44
	15—22	54
	23—51+	56
Females[1]	11—14	44
	15—18	48
	19—51+	46

[1] Pregnant women require 30 gm more than the amount shown in the table and nursing mothers, 20 gm extra.

the one you are eating may be 90% water and supply only 38 kcal per 100 gm.

Tissue Building and Replacement

Carbohydrates, fats, proteins, or alcohol will supply energy, but the requirement for the building and replacement of tissues is more specific: chiefly protein, because the muscles, nerves, connective and protective tissues consist largely of protein. Even the bones and teeth contain protein. There are other building materials such as calcium and phosphorus that are required, but they will be discussed later.

Children up to the early twenties require the most protein in their diets, but fully grown adults also require protein because all the tissues metabolize protein and, of course, hair and nails, which are protein, continue to grow. Experiments have shown that nitrogen products of metabolism continue to be excreted during fasting. The amount decreases as the fast proceeds, but has continued to the end of the experiments made for the purpose— about a month. Consequently, protein is a dietary requirement at all ages. For the past century, biochemists and physiologists have been trying to determine how much protein the diet requires. The results of the experiments and conclusions vary from 40 to 200 gm daily.

In 1863, Congress chartered the National Academy of Sciences to advise the government in scientific matters. It now has over 1100 noted scientists as members. Then in 1916, the Academy established a branch called the National Research Council as a research organization. In 1940, the Academy established the Food and Nutrition Board (FNB) as a branch of the Council, to study the experimental research in nutrition and recommend

TABLE 4.7

HIGH PROTEIN FOODS

Food	Water	Kcal	Protein Gm/100 Gm
Almonds, dried	4.7	598	18.6
Bass, sea	79.3	93	19.2
Beans, dried	10.9	340	22.3
Beef, 73% lean	60.1	266	18.0
Cheese, cheddar	37.0	398	25.0
Chicken	73.7	117	23.4
Cottonseed flour	6.1	356	48.1
Duck	54.3	326	16.0
Eggs, fresh	73.7	163	12.9
Gelatin, dry	13.0	335	85.6
Lamb, 72% lean	56.3	310	15.4
Lobster	78.5	91	16.0
Milk, dry whole	2.0	502	26.4
Oatmeal	8.7	382	20.5
Peanuts	5.6	564	26.0
Pork, lean	68.0	185	17.3
Salmon	63.6	217	22.5
Shrimp	78.2	91	18.1
Soybeans, dry	10.0	403	34.1
Turkey	64.2	218	20.1
Veal, 76% lean	62.0	248	18.0
Wheat, hard	13.0	330	14.0
Yeast, torula	6.0	277	38.6

Source: Watt and Merrill (1963).

the proper daily amount (allowance) of the various nutrients. These are called the Recommended Daily Allowances (RDA). The values for proteins are shown in Table 4.6.

The amount of protein recommended by the FNB assumes that the protein is obtained from a variety of foods, because by 1935, Professor William C. Rose of the University of Illinois had found that 10 of the amino acids must be supplied by the proteins of the diet although arginine and histidine do not appear to be required by adults (Table 4.4). Of course, all the amino acids are required, but about half of them can be synthesized from the others by the body. The 10 essential acids cannot be synthesized, at least not as fast as required in the synthesis of body proteins, and so they must be supplied by the food. Most common foods contain all the amino acids required, but some of them contain inadequate amounts and some may be destroyed by food processing.

Table 4.7 contains the principal high protein foods together with their water and caloric content.

All the high protein foods from the vegetable kingdom are seeds. All natural foods contain protein but the amount varies from the high values recorded in Table 4.7 to a mere trace. For different cuts of meat, different kinds of fish, and all foods not listed in Table 4.7, the student is referred to the Agriculture Handbook No. 8 from which the data in Table 4.7 were taken.

BIBLIOGRAPHY

ALLISON, J. B., VANNEMACHER, R. W. JR., MIDDLETON, E., and SPOERLEIN, T. 1959. Dietary protein requirements and problems of supplementation. Food Tech. *13*, 597–602.

ALTSCHUL, A. M. 1974. New Protein Foods, Vol. 1A. Academic Press, New York.

ANFINSEN, C. B., ANSON, M. L., and EDSAL, J. T. 1965. Advances in Protein Chemistry, Vol. 20. Academic Press, New York.

ASSOC. OFFIC. ANAL. CHEMISTS. 1975. Official Methods of Analysis, 12th Edition. Assoc. Offic. Anal. Chemists, Washington, D.C.

AURAND, L. W., and WOODS, A. E. 1967. Food Chemistry. Avi Publishing Co., Westport, Conn.

BEATON, G. H., and McHENRY, E. W. 1964. Nutrition. Vol. 1, Macronutrients and Nutrient Elements. Academic Press, New York.

BERGMEYER, H. U. 1974. Methods of Enzymic Analysis, Vol. 1–4. Academic Press, New York.

COHN, E. J. 1922. Studies in the physical chemistry of the proteins, I. The solubility of certain proteins at their isoelectric points. J. Gen. Physiol. *4*, 697–722.

COLE, S. J. 1967. The Maillard reaction in food products: carbon dioxide production. J. Food Sci. *33*, 245–250.

COX, C. B., BINKARD, E. F., and CRIGLER, T. P. 1971. Pectin: keystone of foods. Food Tech. *25*, No. 8, 50–54.

DIXON, M., and WEBB, E. C. 1963. Enzymes, 2nd Edition. Academic Press, New York.

FAO. 1970. Amino Acid Content of Foods and Biological Data on Proteins. FAO, United Nations, New York.

FNB. 1974. Recommended Dietary Allowances. Natl. Acad. Sci., Washington, D.C.

GARARD, I. D. 1974. The Story of Food. Avi Publishing Co., Westport, Conn.

GREGORY, K. F., *et al.* 1976. Conversion of carbohydrates to protein by high temperature fungi. Food Tech. *30*, No. 3, 30–35.

GULFREUND, H. 1965. An Introduction to the Study of Enzymes. Blackwell Scientific Publishers, Oxford, England.

GUTCHO, S. 1973. Proteins from Hydrocarbons. Noyes Data Corp., Park Ridge, N.J.

HANSON, L. P. 1974. Vegetable Protein Processing. Noyes Data Corp., Park Ridge, N.J.

HARRIS, R. S., and VON LOESECKE, H. 1960. Nutritional Evaluation of Food Processing. John Wiley & Sons, New York.

JONES, D. B. 1931. Factors for converting percentages of nitrogen in foods and feeds into percentages of protein. Circular *183*. U.S. Dept. Agr., Washington, D.C.

LEE, F. A. 1975. Basic Food Chemistry. Avi Publishing Co., Westport, Conn.

MAILLARD, L. C. 1912. Action of amino acids on sugars. Formation of meloidin in a methodical way. Compt. Rend. *154*, 66–68.

McCOLLUM, E. V. 1957. A History of Nutrition. Houghton Mifflin Co., Boston.

McWEENY, D. J. 1969. The Maillard reaction and its inhibition by sulfite. J. Food Sci. *34*, 641–643.

MIDVAN, A. S. 1974. Mechanism of Enzyme Action. Am. Rev. Biochem. *43*, 357–399.

MIRSKY, A. E., and PAULING, L. 1936. On the structure of native, denatured and coagulated proteins. Proc. Natl. Acad. Sci. *22*, 439–447.

NEURATH, H. 1963. 1970. The Proteins, Vols. 1–5. Academic Press, New York.

OSBORNE, T. B. 1895. The chemical nature of diastase. J. Am. Chem. Soc. *17*, 603.

REED, G. 1966. Enzymes in Food Processing. Academic Press, New York.

SCHULTZ, H. W. 1960. Food Enzymes. Avi Publishing Co., Westport, Conn.

SCHULTZ, H. W., and ANGLEMIER, A. F. 1964. Proteins and Their Reactions. Avi Publishing Co., Westport, Conn.

SUMNER, J. B. 1926. Isolation and crystallization of the enzyme urease. J. Biol. Chem. *69*, 435–441.

TANNENBAUM, S. R. 1971. Single cell protein. Food Tech. *25*, No. 9, 98–103.

THOMAS, A. W. 1934. Colloid Chemistry. McGraw-Hill Book Co., New York.

WALKER, D. B., HORAN, F. E., and BURKET, R. E. 1971. Oilseed proteins. Food Tech. *25*, No. 8, 55–60.

WATT, B. K., and MERRILL, A. L. 1963. Agriculture Handbook No. 8, Composition of foods. ARS-USDA, Washington, D.C.

WOLF, W. J. 1972. What is soy protein? Food Tech. *26*, No. 5, 44–54.

CHAPTER 5

Minerals

In addition to the major nutrients—carbohydrates, fats, proteins, and water—there are several minor nutrients; for example, minerals, which are minor in the amount required, but by no means minor in importance.

These minor nutrients are in two classes: the minerals and the vitamins. The minerals are inorganic elements or compounds, but vitamins are members of several organic classes of substances.

The minerals supply no energy but perform the other two functions of a nutrient: they supply structural material and perform physiological functions.

CALCIUM

It is estimated that a newborn infant contains 28 gm of calcium and a fully grown man, 1200 gm. Most of this calcium is in the bones and teeth. The blood and soft tissues contain low concentrations; the normal concentration in the blood is 10 mg in 100 ml of blood. The calcium gives rigidity to the bones and so its importance in pregnant and lactating women and in growing children is obvious, but the bones are active tissue and also serve as a reserve supply of the element. If there is a calcium deficiency in the diet of a pregnant woman, the bones become depleted to supply the requirement of the pregnancy. Furthermore, the bones lose calcium in normal metabolism. During a fast that lasted 31 days, calcium was excreted daily and 0.138 gm was excreted on the 31st day of the fast. It is obvious then, that calcium must be in the diet at all ages. The FNB recommends 560 mg for infants under 0.5 year and 540 mg from 0.5–1 yr; children 1–10, 800 mg; 11–18, 1200 mg and those 19–51+, 800 mg with an allowance of 1200 mg for pregnant and nursing women.

Milk is the best source of calcium, but Table 5.1 shows the calcium content of several other foods.

PHOSPHORUS

Calcium and phosphorus are usually considered together because both bones and teeth are forms of calcium phosphate. It is estimated that the body contains about 18 gm of phosphorus at birth and 650 gm in middle age. Aside from bones and teeth, phosphorus occurs in the soft tissues and in the blood. Like calcium, it is excreted daily and is therefore essential in all diets. The FNB recommendations for calcium are the same as those for phosphorus in each age group.

Phosphorus is abundant in animal products (meat, milk, eggs) and in seeds (Table 5.1).

IRON

The importance of iron in nutrition is much greater than the 3 gm present in the body indicates. Nearly 90% of this iron is in the blood; it is the prime oxygen carrier. The remainder is distributed among the soft tissues. Lack of iron causes anemia and the FNB recommends from 10 to 18 mg a day. Meats and leafy vegetables are the main food sources.

IODINE

Seaweed had been used as a remedy for goiter for 3000 years before iodine was discovered in 1811. Some 5 yr later, the element was in use as a cure for that disease. It was not too successful, however, because physicians gave doses that were too large and often did more harm than good. It was not until 1914 that the nutritional requirement of iodine was fully established.

Iodine is part of the thyroid gland and so the amount in the body is small; it is estimated at 25 mg in an adult man.

Iodine is a difficult problem for the food chemist. The allowance recommended by the FNB runs from 35 to 150 micrograms for persons of both sexes and all ages. A microgram is one-millionth of a gram, or 0.001 mg. Analysis for quantities of this order are difficult, but, nevertheless, the Official Methods contain procedures for the determination of iodine in some products.

To add to the chemist's difficulties, the only reliable food sources are the seafoods; oysters and salmon are particularly rich in iodine. Both animal and vegetable food products produced on land vary in iodine content with the amount of iodine in the soil. Iodine does not seem to be a physiological requirement of plant nutrition and there are large areas in the world where there is no iodine in the soil and so none in any plant grown there. Neither do the animals of the region contain any iodine if all their food is grown there and none is added to their salt.

The iodine free areas were known as goiterous areas or goiter belts long before the discovery of iodine deficiency as the cause of goiter. A portion of our Pacific Northwest, the area around the Great Lakes, and the Alps and Andes Mountains are all well-known goiterous regions.

Iodine deficiency in foods led to the addition of iodine to salt as early as 1924. One part in 5000 was used but the concentration was later lowered to 1 in 10,000. Canada requires table salt to contain 1 part of KI in 10,000 of salt and Switzerland requires 1 part in 100,000. In this country, the addition of iodine is optional, and any supermarket contains salt both with and without iodine. This may be of little consequence near the coasts where seafood is abundant, but in inland areas it is a matter for concern.

The main natural sources of iodine are seaweeds and the sodium nitrate deposits in Chile, consequently, the element is expensive and the only economical method of adding it to the diet that has been devised has been the iodization of table salt. Under some conditions, the iodine disappears from the salt and so analysis is often necessary. The AOAC has a method for the analysis.

FLUORINE

As early as 1807, French chemists discovered fluorine in animal tissues, particularly in the bones and teeth, but it was over a century before chemists had any idea of its importance in nutrition. Early in the 20th century, the United States Public Health Service made a survey of the incidence of mottled teeth. In some areas, the enamel of the people's teeth was mottled with white, chalky spots that were unsightly but did not decay as badly as normal teeth. By 1931, it was discovered that the condition was caused by the presence of 4 to 7 mg of fluorine per liter of drinking water (4 to 7 ppm). This discovery led to the analysis of drinking water in many places and by 1938, it was found that smaller amounts of fluorine in the water reduced the decay of the teeth of the inhabitants considerably. Apparently, 0.5 to 1 ppm of fluorine in the water reduced the incidence of dental caries by about 65%.

Here again was the problem of getting the element to the consumer. Furthermore, high concentrations are toxic; even 4 ppm causes mottled teeth and so the addition of fluorine to the water supply became and remains a controversial subject, but many cities adjust the fluorine content of the water supplies to 1 ppm. Fluorine is much cheaper than iodine so the cost of the addition to a city water supply is not excessive. One difficulty is that this method of getting fluorine to the population is of no value to the farmers and villagers who do not have a public water supply.

Some manufacturers of toothpaste include fluoride in their formulas, but there should be a better way of supplying fluorine. Dentures are a nuisance and having a tooth filled is not one of our greater pleasures. There is some fluorine in foods, but not regularly enough to be of much help.

MAGNESIUM

The average adult man contains about 35 gm of magnesium. Of this amount, some 70% is in the bones. The remainder is distributed among the soft tissues; it is an essential part of several of the body enzymes. Its function in the body seems to be complex and magnesium deficiencies have been observed in man. The amount required, however, varies with the amount of protein in the diet and also with the amount of calcium, phosphorus and vitamin D. The recommendation of the FNB starts with 60 mg and increases gradually to 400 for males 15 to 18. The recommendation for adults beyond 18 is 350 mg.

Magnesium is a constituent of the chlorophyll molecule and is therefore a plant nutrient. It is present in all leafy vegetables and in most other foods (Table 5.1) so that a magnesium deficiency on a mixed diet is unlikely.

WATER

The human body is about 65% water. Moreover, the processes of both digestion and metabolism occur only in an aqueous medium. Water also has two other functions: it cools the body and removes metabolic wastes.

The FNB recommends 1 ml of water for each kcal of food consumed. On a diet of 2500 kcal, this amounts to 2.5 liters a day. The absolute requirement varies with the climate and personal activity. With strenuous exercise on a hot day, the requirement may reach 10 liters or more.

Since oxidation in the body produces heat continuously, a coolant is essential and the evaporation of water into the breath and from the skin prevents a rise in temperature.

Salt and urea are the principal wastes removed in solution in the urine and perspiration, but by no means the only ones; there are several products of protein metabolism, such as uric acid, createnine and small amounts of all the minerals.

A function of the water in the body that is seldom realized is its protection of the body against sudden changes in temperature. Water has the highest specific heat of all substances (1 cal./gm). If we go suddenly from a warm room to an outdoor temperature below 0°F, or are plunged into ice water, the body temperature drops very little and increased oxidation soon restores it to normal.

Fortunately, natural foods contain considerable water: milk is 87% water, eggs 74%, bread 35% and cabbage 92%. The consumption of fresh foods supplies a large part of the water requirement, and when fats are oxidized, they produce more than their weight of water and the carbohydrates and proteins produce a smaller amount. Although the foods supply considerable water, it is still necessary to drink some each day. Less than a pint is dangerous even on a frugal diet in a cool climate.

Aside from any connection with nutrition, the food chemist has two other concerns with water: the amount used for washing fruits and vegetables, and the water that goes into the brines and syrups that are added to various foods during processing.

Perhaps the first concern is sanitation. Although sanitation is the primary concern of the bacteriologist, the chemist must understand the principles and hazards involved. City water supplies will likely be sufficiently reliable, but if a company draws its water from a well or a stream, constant testing

TABLE 5.1

MINERALS IN FOODS

	Water	Ca	P	Fe	Na	K	Mg
				Gm/100 Gm			
Food	%	Mg	Mg	Mg	Mg	Mg	Mg
Almonds, roasted	0.7	235	504	4.7	198	773	270
Baby food, cereal	6.6	736	821	53.2	452	413	—
Baking powder, phos.	1.6	6279	9438	—	8220	170	—
Beans, white	10.9	144	425	7.8	19	1196	170
Buttermilk, fluid	90.5	121	95	—	130	140	14
Cabbage, raw	92.9	49	29	0.4	20	233	13
Carrots, raw	88.2	32	36	0.7	47	341	23
Cheese, cheddar	37.0	750	478	0.4	700	82	45
Cheese, cottage	79.0	90	173	0.4	290	72	—
Cherries, raw	83.7	22	19	0.4	2	191	14
Chicken, dark meat	73.7	13	188	1.5	67	250	23
Eggs, chicken	73.7	54	205	2.3	172	129	11
Grapefruit	88.4	16	16	0.4	1	135	12
Grapes	81.6	16	12	0.4	3	158	13
Ice cream	62.1	123	99	0.1	40	112	14
Kale	82.7	249	93	2.7	75	378	37
Milk, whole	87.4	118	93	tr	50	144	13
Oatmeal	8.3	53	405	4.5	2	352	144
Oranges	86.0	41	20	0.4	1	200	10
Peaches	89.1	9	19	0.5	1	202	10
Peanuts	5.6	60	401	2.1	2	316	206
Peas, green	78.0	26	116	1.9	—	170	35
Potatoes	79.8	7	53	0.6	3	407	—
Rice, brown	12.0	32	221	1.6	9	214	88
Shrimp	78.2	63	166	1.6	140	220	42
Spinach	90.7	93	51	3.1	71	470	88
Tomatoes	93.5	13	27	0.5	3	244	14
Whole wheat flour	12.0	41	372	3.3	3	370	113

is essential. In some states, a public health department will make sanitary tests.

The hardness of water is sometimes a problem when the water is added to the food. The calcium and magnesium ions of hard water react with pectin to produce a very tough texture in the skins of peas, green lima beans and some other vegetables. Calcium salts are often added to canned tomatoes to make them more firm.

I scarcely need mention that the presence of hydrogen sulfide in water that goes into food would ruin most products.

SALT

Sodium chloride is a dietary essential and one of the most troublesome; too little may result in heat stroke; men who work at high temperatures often take salt tablets. On the other hand, too much salt results in high blood pressure and heart damage and so doctors often prescribe a low-salt or a salt-free diet.

Determination of the salt requirement is very difficult because it varies with so many factors. The adult normally contains about 4 oz of sodium and 9 oz of potassium; the nutritive requirements of these two elements are interdependent. The FNB recommends 1 gm of salt for each liter of water consumed, which would amount to 2.5 gm for the normal water consumption, but sweat and urine contain from 0.5 gm to 1.25 gm of salt per liter and so with excessive sweating, the salt loss is large. The average American diet is thought to contain from 6 gm to 18 gm of salt and 1.4 to 6.5 gm of potassium daily.

There is ordinarily no lack of salt in the American diet because, aside from the salt content of the natural foods, salt is widely used as a preservative and is also a favorite flavor.

There are several grades of salt on the market and the chemist must fit the grade to his requirements. Pure sodium chloride is expensive, and the calcium and magnesium present in the cheaper grades may ruin a product.

POTASSIUM

Since potassium is one of the elements required by plants, both fruits and vegetables contain it. The fruits average about 200 mg and the vegetables 475 mg in 100 gm of food. The dried products are much higher; dried apricots contain 1260 mg, dried prunes 940 mg and raisins 750 mg in 100 gm. Eggs and dairy products other than butter contain appreciable quantities of potassium, but meats are low in this nutrient.

OTHER MINERALS

There are several elements that are essential in the diet in extremely small amounts. Nutritionists call them "trace elements." Active research on the trace elements is in progress and so others may be added to the list. At present, they are chromium, cobalt, copper, manganese, molybdenum, selenium and zinc.

In working with trace elements in either plant or animal nutrition, it is difficult to establish whether an element found to be present is essential or accidental. There are the further difficulties of analytical methods for such small amounts, and in feeding experiments it is hard to be sure that the control diet is entirely free from the element under investigation.

Another complication that enters into research on trace elements is the fact that practically all of them are toxic in large quantity and the amount that can be tolerated must be determined first.

ANALYTICAL PROCEDURES

In the determination of the mineral elements in food, the first procedure is to burn the food. The analysis of the ash for the various mineral elements

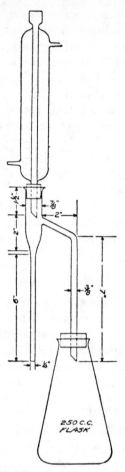

Courtesy of Ind. Eng. Chem. 1925, 17, 147

FIG. 5.1. BIDWELL AND STERLING MOISTURE APPARATUS

is then a problem in inorganic analysis. Sometimes, however, there are special methods because of the small amount of the element present.

The AOAC has Official Methods for the determination of calcium in foods, feeds, fertilizers and pesticides. The usual procedure is to dissolve the ash and precipitate the calcium as the oxalate.

There are Official Methods for phosphorus in foods, feeds and fertilizers, and there are also several methods for iron and magnesium.

There is an Official Method for iodine in color additives, iodized salt and mineral waters, and also for potassium iodate, KIO_3 in wheat flour.

There are methods for the determination of fluorine in baking powders and in other foods. Fluorine is commonly present in phosphate ore and may be carried through the manufacturing process into the food phosphates, which accounts for the method of analysis among the Official Methods.

There are also Official Methods for cobalt, copper, manganese, molybdenum, selenium and zinc. Because of the minute quantities of some of these elements, some of the Official Methods are microchemical and some are spectrographic.

There are methods for the determination of salt in many foods. In most cases, standard silver nitrate solution and nitric acid are added to the food sample and digested until the food is oxidized, so that the AgCl is the only remaining solid. The AgCl is then dried and weighed and the NaCl equivalent calculated.

There are three methods recognized by the AOAC for the determination of water: drying at 100°C or higher in an air oven, drying at 100°C or below in a vacuum oven, so that oxidation of the food is avoided, and a distillation method. The Official Methods indicate which method is to be used for each of several foods. In the first method, fats may oxidize and increase in weight so that the water value will be too low.

The distillation method requires a special glass apparatus, a condenser and a flask. The sample is put into the flask and toluene added. The flask is connected to a reflux condenser by the special apparatus and the toluene boiled. The toluene boils at 110°C and so the water in the sample is evaporated and the mixed vapors of toluene and water pass up into the condenser and drop back into a narrow tube with a milliliter scale on it. The water has a greater density than the toluene and is immiscible with it, consequently, the water collects in the calibrated tube and its volume is read directly; the excess toluene overflows back into the flask (see Fig. 5.1).

The distillation method is somewhat less precise than the drying methods, but it is rapid and adequate for analyses unless the water content is very small.

BIBLIOGRAPHY

ARS-USDA. 1962. Magnesium in human nutrition. Home economics research report No. 19 ARS-USDA, Washington, D. C.

ASSOC. OFFIC. ANAL. CHEMISTS. 1975. Official Methods of Analysis, 12th Edition. Assoc. Offic. Anal. Chemists, Washington, D. C.

AURAND, L. W., and WOODS, A. E. 1973. Food Chemistry, Avi Publishing Co. Westport, Conn.

BEATON, G. H., and McHENRY, E. W. 1964. Nutrition, Vol. I. Academic Press, New York.

DUCKWORTH, R. B. 1975. Water Relations of Foods. Academic Press, New York.

GARARD, I. D. 1974. The Story of Food. Avi Publishing Co., Westport, Conn.

MATZ, S. A. 1965. Water in Foods. Avi Publishing Co., Westport, Conn.

MELOAN, C. E., and POMERANZ, Y. 1973. Food Analysis Laboratory Experiments. Avi Publishing Co., Westport, Conn.

NIELSEN, F. H. 1974. Newer trace elements in human nutrition. Food Tech. 28, No. 1, 38–44.

POMERANZ, Y., and MELOAN, C. E. 1971. Food Analysis. Theory and Practice. Avi Publishing Co., Westport, Conn.

SHERMAN, H. C. 1952. Chemistry of Food and Nutrition. Macmillan Co. New York.

CHAPTER 6

Vitamins

Although they are late arrivals in nutrition, the vitamins are indispensable, even though the requirement is small. As early as 1564, a Dutch physician knew that there was something in oranges that would prevent or cure scurvy. Over two centuries later (1795), the British navy added citrus fruit to the diet to prevent scurvy. After another century, in the 1880s, a Japanese naval officer found that beriberi was the result of a dietary deficiency, but it was not until 1906 that a systematic search for these mysterious nutrients was begun. From that date, information was forthcoming without any long gaps, and research on the subject is still in progress. In 1906, Professor F. Gowland Hopkins of Cambridge University announced that ". . . animals cannot grow upon so-called synthetic diets consisting of mixtures of pure proteins, fats, carbohydrates and salts . . . ."

The literature of the next 14 years contains hosts of papers on the "trace" requirements that Hopkins declared to be in foods. These requirements were given various names, which makes some of the literature of the day confusing. In 1920, Sir Jack Drummond, Professor of Biochemistry in University College, London, proposed that these essential nutrients be called *vitamins* and be distinguished by the letters A, B, C, etc. He said that these names would be adequate until the chemical nature of the substances had been determined and then the chemical names could be used. The simplicity of the system appealed to chemists generally, for all the information at that time was physiological. All the chemical information available was the effect of heat and the solubility; some were soluble in fats and some in water.

Vitamin C was the first to reach chemical identity. In 1931, Professor C. Glenn King of the University of Pittsburgh isolated it from lemon juice and, by the end of 1933, its formula had been established. It was named *ascorbic acid,* and it has been synthesized.

Drummond's prediction came true. Vitamin C is now generally called ascorbic acid by chemists, but most of the other vitamins have such complicated chemical structures that chemists still use the physiological names. Since 1933, the vitamins have been of the greatest concern to the food chemist.

FAT SOLUBLE VITAMINS

Vitamin A

The failure of rats to grow on certain experimental diets led to the discovery of "fat soluble A" in 1913. Since then, hundreds of papers have been published on the chemistry, physiology, and occurrence of the substance. The references at the end of this chapter contain many of these journal references, but our space is limited to a brief review of the results. In 1920, the name became vitamin A.

In addition to failure of the rats to grow, they developed an eye infection, called *xerophthalmia,* and finally blindness. So-called night blindness in humans was discovered in 1925. It consists of inability to see in a dim light or to adapt to changes in the intensity of light.

In the absence of vitamin A, the epithelial tissue of the respiratory and gastrointestinal tracts deteriorate, nerves degenerate and the bones of the young do not form properly.

Requirement.—The International Unit (IU) of vitamin A is 0.300 micrograms (μg). The Food and Nutrition Board's (FNB) recommended allowances of the vitamin start with 1,400 IU for infants under 6 months and gradually increase to 5,000 IU for adults with an extra 1000 for pregnancy and an extra 3000 for lactation.

The daily allowance is complicated by several factors: the vitamin is stored in the liver and there are four other substances in foods that are converted into the vitamin. They are alpha, beta and gamma carotene and cryptoxanthin, which are converted to vitamin A during absorption through the intestinal wall. Beta carotene is the most important of these four substances because it occurs most abundantly in leafy and other vegetables and also yields more vitamin than the others. One unit of carotene is equivalent to 0.6 unit of vitamin A.

Excess.—Students of nutrition have been concerned about possible deleterious effects from excessive amounts of vitamins. With vitamin A, Sherman and Batchelder report considerable increase in vitality of rats from higher levels of vitamin A in the diet.

However, large doses, 20 to 30 times the RDA, taken over a long period of time, are toxic. Carotene is not toxic, but long continued ingestion of large doses will cause yellow skin. Either toxicity or skin coloration disappear with reduction of intake of the cause. None of the ordinary foods will cause adverse effects from either vitamin A or carotene, but some of the fish liver oils contain enough of the vitamin to be dangerous.

Sources.—The richest sources of vitamin A are the fish liver oils. Some samples of oil from the livers of halibut or tuna have been found to contain 100,000 IU per gm, herring liver oils contain about 5,000 IU and cod liver oil, 1,000. The richness of the fish liver oils is readily realized when we note that these values are in IU per gm, while the usual reports of food content are in IU per 100 gm.

Vitamin A occurs only in animal foods: beef liver 43,900 IU per 100 gm, calf liver 22,500, chicken liver 12,500, margarine 4,800, butter 3,300, cream 1,540, American cheese 1,200, eggs 1,180, and milk 140.

Vegetables and fruits contain some of the carotenes, which are provitamins. The vitamin A equivalent of some foods of plant origin are: dried apricots 14,100 IU/100 gm, carrots 11,000, sweet potatoes 8,800, spinach 8,100, turnip greens 7,600, cantaloupe 3,400, broccoli 2,500, dried prunes 2,170, green onions 2,000, pumpkin 1,600, yellow peaches 1,330, sour red cherries 1,000, tomatoes 900, green peas 640, okra 520, sweet peppers 420, cabbage 130, and apples 90.

In addition to the vitamin content of the natural foods, baby foods, cereals, and some other processed foods have vitamin A added.

In the 1920s, cod liver oil was fed to infants but some of them did not tolerate so much fat, then manufacturers began to make more concentrated vitamin A products. Before World War II, vitamin A was obtained from halibut liver oil, but with the advent of war, the government prohibited the fishing boats going to sea in the Pacific, whereupon the vitamin was synthesized and soon in commercial production.

Chemistry.—The chemical name of vitamin A is *retinol,* which doesn't tell us anything except that it is an alcohol. The following formulas indicate the chemical nature of vitamin A and its relation to beta carotene, which is the most important carotenoid.

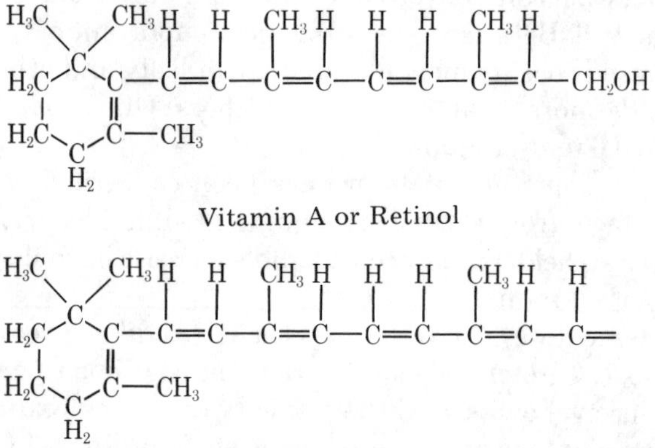

Vitamin A or Retinol

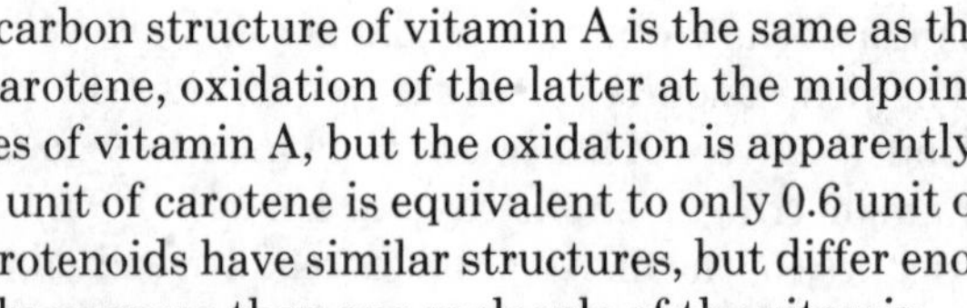

Half of β-Carotene

Since the carbon structure of vitamin A is the same as that of half that of the beta carotene, oxidation of the latter at the midpoint will produce two molecules of vitamin A, but the oxidation is apparently not very efficient since 1 unit of carotene is equivalent to only 0.6 unit of the vitamin. The other carotenoids have similar structures, but differ enough that none of them produces more than one molecule of the vitamin.

All four of the double bonds in vitamin A are *trans.* The vitamin is a pale yellow substance that melts at 62° to 64°C. It is soluble in alcohol, fats, and fat solvents. It undergoes the reactions indicated by the double bonds and the primary alcohol group.

Stability.—Vitamin A is unstable to light, particularly light of short wavelength and the ultraviolet. It is also unstable in solutions that are strongly acid or alkaline and rapidly destroyed when heated in the presence of oxygen. Oxidation seems to be the chief manner of destruction; the peroxides of rancid fat destroy it rapidly.

The chemist cannot take the vitamin A content of a food for granted from the values reported in the published tables. If the food is a natural plant product, its carotene content differs with variety and maturity; if it is an animal product, such as eggs and dairy products, it differs with the diet of the producing animal; if it is a processed food, it differs not only with the carotene or vitamin content of the original food, but also to the extent to which it has been exposed to air and high temperatures.

Determination.—The Official Methods (Chap. 43, AOAC) for the determination of vitamin A involve extraction from plant material with acetone, separation from fat and other substances by chromatography and determination by spectro-photometric methods.

There are also biological methods that may be useful in some circumstances; they use rats and measure the vitamin content of a food by measuring the rate of growth of the rats or the rate of cure of xerophthalmia.

Vitamin D

During early experiments with fat soluble A, it was discovered that the vitamin also cured rickets. This led to the discovery that fat soluble A was a mixture of two separate vitamins, and the discovery of vitamin D was announced in 1922. Since then, hundreds of papers have been published on the various aspects of this vitamin and its importance in nutrition makes it a matter of concern to the food chemist.

Rickets is a disease of mineral metabolism that results in improper calcification of the bones and teeth. The bones are composed largely of a calcium phosphate, but the diet can contain ample calcium and phosphorus and still they will not deposit in the bones unless the diet also contains vitamin D; a deficiency of any one of the three nutrients causes rickets.

Rickets is a disease of childhood, characterized by bowlegs, knock knees or other bone malformation. It has been recognized for centuries, but its cause was not discovered until 1918 when an English Professor of physiology found it to be a deficiency disease. It has been estimated that 90% of the children in Europe in the year 1900 had rickets and some died.

Requirement.—The International Unit of vitamin D is 0.025 μg. Vitamin D is essential at all ages for the maintenance of bones and teeth; it is not stored in the body and must be ingested daily. Of course, the requirement is greater during the growing years. The exact minimum requirement after infancy is impossible to determine because of the differences in the exposure in sunlight. The FNB recommends 400 IU a day for all ages. This amount should not be exceeded in infants, but children and adults can tolerate more. However, continued daily intake of 2,000 units or more may lead to calcification in the kidneys, heart, lungs, blood vessels and other tissuès. Moreover, the condition does not seem to be corrected by lowering the intake.

Sources.—The occurrence of Vitamin D is greatly limited, which was the cause of the enormous number of cases of rickets. Cod liver oil is a rich source, but few of the common foods contain it; egg yolk and the fat of milk are the principal food sources. Fish liver oils vary enormously in the vitamin content with the variety of fish; the liver oils from the tuna, bass and swordfish contain from 7,000 to 50,000 IU per gm, cod liver oil contains 60–300, sturgeon none. Egg yolk contains 1.5–5.0 IU/gm and butter 0.1–1.0. The amount of the vitamin in eggs and milk fat varies with the exposure of the animal to sunlight.

The scarcity of vitamin D in foods has led to its addition to several types of food. Fortunately, sunlight converts certain sterols, which the body contains, into the vitamin, but unfortunately it supplies very little vitamin to the human because of lack of exposure.

Chemistry.—The D vitamins are sterols, and as many as ten have been reported to have antirachitic activity. Only two are of importance to the food chemist: D_2 or ergocalciferol, and D_3 or cholecalciferol.

Ergosterol

Vitamin D-2

The formulas show the structure of vitamin D and the provitamins from which it is produced by irradiation.

Ergosterol occurs in many fungi and is obtained from yeast for the commercial production of vitamin D_2; 7-dehydrocholesterol occurs in animal skins and their appendages, hair, fur, etc. Pig skin has been found to contain 4.6%, but human skin only 0.015% to 0.42%. Vitamin D_3 is the one that occurs in animal products.

The various D vitamins are white crystalline substances that melt from 115° to 118°C. The fraction of sunlight that converts the sterols to the vitamins lies in the ultraviolet, the wavelengths between 275 and 300 mμ; 280 is the most effective.

Stability.—Vitamin D is stable in acid solutions but unstable in alkali and it is destroyed by oxidation. Cooking losses may amount to 40%. Eggs have been found to retain all their vitamin D during spray drying and 9 months storage at a temperature below freezing.

Determination.—A chemical method for the determination of vitamin D is based on the fact that the D vitamins give a yellow color with antimony trichloride, $SbCl_3$. The method cannot be used for the determination of the vitamin in foods, but is the official method of the U.S. Pharmacopeia. The Official Method of the AOAC is a biological method using rats.

Vitamin E

An enormous amount of research over the past 35 yr has failed to provide a clear picture of the importance of vitamin E in human nutrition. It is often deficient in newborn infants and its requirement for adults seems to vary with the amount of unsaturated fat in the diet. The problem of the requirement is complicated by the findings that its effects differ with the species of animal under investigation. Pronounced deficiencies in rats, for example, have effects that do not occur in man at all. With our present state of nutritional knowledge of vitamin E it is unlikely to be a problem for the food chemist. The FNB does not think that there is likely to be a deficiency in the adult, and deficiency in infants is a problem for the physician. The AOAC, however, has an Official Method for the analysis of pharmaceutical preparations.

The effects of vitamin E are those of several tocopherols of which alpha tocopherol is the most important. The IU is 1 mg of synthetic alpha tocopherol acetate.

Vitamin E occurs in many foods. Wheat germ oil is particularly rich in it and smaller amounts are present in the common food oils.

Vitamin K

This fat soluble vitamin has been known since 1929. It is required in the clotting of blood, which accounts for skipping the letters from E to K in naming it—the discoverer was a Dane, and "koagulation" is the Danish version of our coagulation.

No allowance has been recommended by the FNB. The vitamin occurs in most green leaves and green vegetables and a diet that includes them will contain an adequate amount of the vitamin. Foods of animal origin are poor sources.

The AOAC has no Official Method for the determination of vitamin K, but it can be determined by both spectographic and biological methods.

Pharmaceutical preparations of this vitamin are available for the treatment of hemorrhages.

Like the other fat soluble vitamins, vitamin K activity is shown by several compounds that are closely related chemically. Vitamin K_1 is the one present in plants. Its formula is

Vitamin K-1

WATER SOLUBLE VITAMINS

Water soluble B, which was announced in 1915, turned out to be more complex than fat soluble A. The FNB recognizes 10 different B vitamins. The A, B, C nomenclature had been proposed in 1920 and vitamins C and D were already named; consequently, the nomenclature of the B vitamins became a problem. When the second one was discovered, the Americans called it vitamin G and the British, B_2. The British system prevailed and today we have B_1, B_2, B_6 and B_{12}; the others are known by their chemical names. The system, as it has developed, is confusing to students. What it amounts to is that scientists are inclined to use the chemical name if it is not too complicated, and the vitamin and letter name if it is. A glance at the formula of vitamin K or B_{12} will explain why they are referred to as vitamins and not chemical compounds.

Thiamine, B_1

This is the antineuritic or antiberiberi vitamin; chemically, it is called *thiamine*, which is simply an empirical name.

Beriberi was a severe problem in the Orient; about 1883, one third of the sailors in the Japanese navy suffered from the disease and many of them died of it. Their diet was mostly polished rice and the medical director noticed that British sailors did not get the disease, and guessed that diet made the difference. He changed the diet on an experimental cruise and proved that his guess was correct. But it was not until 1901 that a Dutch doctor, working in the East Indies (Indonesia) found that the outer layer of the rice grain contained a substance that could be extracted with alcohol and would prevent or cure beriberi. Many investigations followed and in 1936, the substance, which had been named thiamine, had been synthesized. McCollum (1957) stated that between 1906 and 1941, there were 1617 papers published on this nutrient. These, and more recent papers show that thiamine has several other functions in nutrition besides the prevention of beriberi.

Requirement.—The FNB recommends from 1.0 to 1.5 mg daily.

Sources.—Thiamine occurs in foods from both plant and animal products. A few of the richer foods contain the following numbers of milligrams of the vitamin in 100 gm of the food: peanuts 1.14, Canadian bacon

0.92, oatmeal 0.7, lean pork 0.63, whole wheat 0.57, brown rice 0.3, green peas 0.28, beef liver 0.26, lima beans 0.24, oysters 0.14, avocados 0.11, eggs 0.10 and potatoes 0.1.

Chemistry.—The formula shows that thiamine differs profoundly in chemical nature from any of the fat soluble vitamins.

Thiamine chloride

Thiamine is a white crystalline compound soluble in water and alcohol. It is now manufactured by the ton.

Stability.—Thiamine is stable to heat and to oxidation in acid solution of pH 5.0 or less, but is sensitive to both in neutral or alkaline solutions. Aside from losses by decomposition and oxidation, thiamine is very soluble in water and, in some processes, 50% or more of it may be lost. The loss is often due to all three of the factors. In one experiment, cabbage was boiled for 2 hr and 68% of the vitamin was in the water. Adding soda to boiling vegetables may keep them green, but it increases the loss of thiamine. In another experiment, toasting white bread for 1 min destroyed 27% of the thiamine. Losses in cooked food, held on a steam table for 2 hr or more may amount to as much as 70%. Even vegetables held in cold storage for a month lost 20% of the vitamin.

Determination.—Several methods have been developed for estimating the thiamine content of foods. Two of them are Official Methods. The chemical method oxidizes the thiamine to thiochrome, a fluorescent blue pigment, with potassium ferricyanide, $K_3Fe(CN)_6$, and then measures the thiochrome with a fluorometer.

The other AOAC method is a bioassay. The various biological methods use the growth of yeasts, molds, bacteria and other microorganisms for the determination of thiamine.

Riboflavin, B_2

In 1879, a yellow pigment was extracted from milk and named *lacto-chrome*, but nothing was done with it until 1925; then, its chemistry was investigated and it was named *lactoflavin.* Meanwhile, water soluble B was under investigation and in 1919, it was definitely found to be a mixture. That one of the components was essential to the growth of rats was established in 1926. Later, the name riboflavin was adopted by three scientific societies and the FDA.

Riboflavin is thought by some public health scientists to be the vitamin most likely to be deficient in the American diet. It is most notable as a growth vitamin. In fact, the rate of growth of rats was the first method for its determination. It occurs in all the tissues of the body, however, and has several metabolic functions besides growth.

Requirement.—The FNB recommends a daily allowance of 0.4 mg to 0.6 mg for infants under 1 year old, 0.8 to 1.2 for children 1 to 10, and values rising to 1.8 in the 15–18 age group and descending to 1.1 after the age 51.

Sources.—The occurrence of riboflavin in mg/100 gm of food is: torula yeast 5.06, raw beef liver 3.26, raw calf liver 2.72, raw chicken liver 2.49, almonds 0.92, cheddar cheese 0.46, eggs 0.30, bread 0.22, ham 0.22, broccoli 0.20, asparagus 0.18, and many other vegetables contain from 0.1 mg to 0.2 mg. Fruits are very poor sources of riboflavin.

Chemistry.—Riboflavin is a yellow solid with a green fluorescence. By 1935, its formula was found to be:

Riboflavin
6,7-Dimethyl-9-(1′-D-ribotyl)-isoalloxazine

The chain attached to the N of the second ring is D-ribose. Some of the riboflavin molecules that occur in food are attached to molecules of protein, which causes some difficulty in analysis and also accounts for the strange phenomenon of more riboflavin in some cooked foods than in the same foods raw: the cooking releases the vitamin from the protein.

Stability.—Riboflavin is stable to heat, acids, and oxidation, but not to alkali or the ultraviolet. Milk in a glass bottle, exposed to sunlight for 2 hr, lost half its riboflavin.

Riboflavin is classified as a water soluble vitamin, but it is much less soluble than thiamine; 100 ml of water dissolves 10 mg at 25°C and 235 mg at 100°C.

Blanching vegetables destroys riboflavin and the loss differs with the time and the vegetable. One experiment, in eight tests found that lima beans lost from none to 41% of the vitamin and spinach in 37 tests lost none to 22%. There is little lost during storage of canned vegetables. Dehydration of vegetables causes about 10% loss and the pasteurization of milk destroys only about 5%. Eggs in storage near 0°C lost only 5% in 3 months. It is not lost in any great quantity in the preparation of food for the table.

Determination.—There are two Official Methods for the estimation of riboflavin: a fluorometric method and a biological method. The latter uses *Lactobacillus casei* as the organism.

Niacin

Pellagra was first described by a Spanish physician in 1730. The cause of the disease was soon traced to the diet, but the doctors thought that there was some component of the food responsible, rather than that the cause was the lack of some substance in the diet; consequently, it took nearly 200 yr to arrive at the cause and cure of the disease. Shortly after 1905, there was an outbreak of pellagra in the southern United States and, by 1914, it was definitely shown to be a dietary deficiency disease.

The disease begins with a sore mouth, then attacks the mucous membranes of the digestive tract, followed by a dermatitis and then injury to the brain and nerves; many cases have resulted in death.

Research on pellagra is so complicated that it is impossible to say when the antipellagric vitamin was discovered. Although Dr. Joseph Goldberger had shown pellagra to be a deficiency disease as early as 1914, a book on the subject of pellagra published in 1919 still called it an infectious disease and thereby added to the developing controversy. But by 1938 the matter was settled.

Requirement.—The vitamin is nicotinic acid, which had been made from nicotine in 1867, but it was not tested physiologically until after it was isolated from liver in 1937. Here another problem arose. Nicotinic acid sounded more like a poison than a vitamin. It is nontoxic, but the biologists changed its name to *niacin.* Moreover, nicotinamide was also found to have antipellagric activity and so it became *niacinamide.* Then to complicate matters further, it was discovered that the body converts tryptophane into niacin at the inefficient ratio of 60 mg tryptophane to 1 mg niacin. Therefore, the nature and amount of protein in the diet affects the niacin requirement.

The FNB recommends 5–8 mg niacin for infants under 1 yr, 9–16 mg for children under 11, rising to 20 mg at age 15 and declining from 20 to 12 mg at ages 22 to 51. For females from 10 to 18, 15 mg are required, and 13 mg thereafter.

Sources.—The richest sources of niacin are the yeasts; torula contains 44 mg in 100 gm, bakers yeast 11 mg, peanuts 17 mg, the various kinds of liver in the market from 11 to 16 mg, chicken 11 mg, turkey 8 mg, smoked mackerel 8.9 mg, canned tuna 11 mg, mushrooms 4.2 mg, whole wheat 4.3 mg, fresh vegetables contain under 3 mg and fresh fruits usually contain less than 1 mg in 100 gm.

Chemistry.—Niacin is one of the simplest vitamins in chemical structure. Both the acid and the amide are nutritionally active.

$$\text{—CONH}_2$$

Niacinamide

$$\text{—COOH}$$

Niacin

Niacin is a white crystalline solid that melts at 232°C. It has been found to constitute part of a coenzyme that seems to appear in all living cells if they are active in tissue respiration.

Stability.—Niacin is stable to heat; 20 min in an autoclave at 120°C did not destroy any of it. It is also stable in both acid and alkaline solutions. It is liable to extraction because of its solubility in water; 18 gm dissolve in 60 ml of water at 25°C and it is much more soluble at higher temperatures. Cooking losses may reach 70%, and blanching removes 5 to 25% from vegetables, but dehydration losses are slight.

Determination.—Niacin and niacinamide react with cyanogen bromide to produce a greenish-yellow color. The Official Methods use this reaction as the basis of a colorimetric method for the determination of the vitamin.

The Official Methods also include a microbiological method using the organism *Lactobacillus plantarum.*

Folacin

Another of the B vitamins is folic acid or *folacin,* and here again there are several substances with the same nutritional activity. The absence of the vitamin results in a sore tongue, diarrhea and then one type of anemia. This vitamin was not discovered until 1938 and so it has been less explored than thiamine and riboflavin, but considerable literature on it has been accumulated.

Requirement.—Establishing the requirement is difficult. It is synthesized by bacteria in the intestines and several different compounds are active. Furthermore, an experimental deficiency is hard to produce.

The FNB recommends 0.05 mg daily for infants up to 1 yr, 0.1–0.3 mg for children 1 to 10, and 0.4 for adults.

Sources.—Not much quantitative information is available for folacin. The best sources are liver, yeast, green vegetables, dry beans, lentils, bananas, mushrooms, and whole wheat. The synthetic product is available in the drug trade.

Chemistry.—Folic acid is a yellow solid that crystallizes in platelets.

It is only slightly soluble in water; 10 mg per liter at 0°C and 500 mg at 100°C. The chemical name is pterylglutamic acid and its structure is

Pteryl-L-glutamic acid
(Folic acid)

Stability.—Folic acid is easily destroyed. It is unstable in both acid and alkaline solutions and is destroyed by oxidation, light and heat. Cooking losses may reach 100%.

Determination.—Three types of method have been devised for its estimation: chemical, biological and microbiological. The biological method, which uses chicks as the experimental animal, is slow and expensive, but it does measure the total vitamin activity.

The microbiological methods have used various organisms. The Official Method for folic acid is a microbiological method using *Streptococcus faecalis.*

The chemical method is based on the amount of aromatic amine produced by reduction of the acid. It is more rapid than the above methods, but any substance that may be present and reducible to an aromatic amine interferes. The method is most useful for the assay of pharmaceutical preparations in which interfering substances are unlikely to be present. Another chemical method that is sometimes useful is based on the fluorescence of an oxidation product.

Vitamin B_6

Like some of the other B vitamins, B_6 consists of more than one substance, in this case three. The chemical name most used is *pyridoxine,* the other two are *pyridoxal* and *pyridoxamine,* both simple derivatives of pyridoxine. The relationships are shown by the formulas.

Requirement.—A deficiency of vitamin B_6 results in skin lesions, anemia, and convulsions. The effects of vitamin B_6 deficiency in humans develop slowly, and so the absolute requirements at the various ages are not known. The FNB, after consulting all the available evidence, recommends an increasing amount for infants from birth to 6 months of 0.3 mg, then 0.4 mg to 1 yr. Children 1 to 10 require 0.6–1.2 mg, and adults, 2.0.

Sources.—Vitamin B_6 occurs in many foods. The best sources are liver, meat, fish, whole wheat, oatmeal, yeast, milk, eggs, several fruits, vegetables, and nuts.

The vitamin is most important for infant feeding, and any processed food designed for that purpose should contain it in adequate amounts. An outbreak of severe deficiency symptoms occurred in 1951 because of the destruction of the vitamin in the processing of milk for feeding infants.

Chemistry.—The chemical nature of the three forms of vitamin B_6 is indicated by the following formulas:

$$CH_2OH \quad\quad CHO \quad\quad CH_2NH_2$$

Pyridoxine	Pyridoxal	Pyridoxamine

Pyridoxin is a white, crystalline compound that melts at 160°C. The nitrogen atom in the pyridine ring renders the compound alkaline, and it reacts with HCl to form a hydrochloride that melts at 206°C.

Stability.—Vitamin B_6 is soluble in water and alcohol, and cooking losses by extraction and oxidation may amount to 40%. It is moderately stable to heat, but destroyed by light.

Determination.—Vitamin B_6 can be estimated by either biological or chemical methods. The one Official Method is microbiological using *Saccharomyces uvarum* as the organism. The procedure is very involved.

Both chicks and rats have been used for the biological assay method.

A chemical method that determines a metabolic product of the vitamin in urine has been useful in nutritional studies, but no chemical method has been devised that is satisfactory for the determination of the vitamin in foods.

Vitamin B_{12}

This vitamin, which is also called *cobalamine* is a late arrival in the vitamin family. In 1926, two Boston physicians discovered that liver will cure pernicious anemia—a discovery that earned them the Nobel Prize; but it was not until 1948 that chemists succeeded in isolating the curative substance. Its chemical structure was worked out in 1956, but it was not synthesized until 1972.

Requirement.—The amount of vitamin B_{12} required to cure or prevent pernicious anemia has been well established, but the vitamin has other functions and the requirement for general health has not. However, the FNB has considered all the experimental evidence and recommends 0.3 micrograms (μg) daily for infants under 1 yr, 1–2 for children 1 to 10, and 3.0 for adults.

Vitamin B_{12} occurs in all the active cells of the body, particularly those of the bone marrow, the nervous system and the digestive tract. It has many functions in nutrition, and its absence finally causes failure of the bone marrow to produce red corpuscles, resulting in pernicious anemia.

Sources.—Not many foods have been examined for vitamin B_{12}, but it does not appear to occur in foods of botanical origin. Liver, lean meat and other foods of animal origin are the principal natural sources; the actual amount of the vitamin in animal tissues varies with the amount of cobalt in the animal's diet.

Chemistry.—Vitamin B_{12} crystallizes in dark red needles. It is the only vitamin discovered to date that contains a metal in its constitution; of its 167 atoms, 1 is cobalt. Its empirical formula is

$$C_{63}H_{88}O_{14}PCo$$

Its molecular weight varies from 1360 to 1576 according to its state of hydration. Compounds of high molecular weight are difficult to purify, and the inherent nature of the methods results in some lack of precision in the results of the molecular weight determination.

Stability.—Vitamin B_{12} is not very soluble in water and cooking losses are slight, usually under 10%. It is sensitive to light and oxidation, but stable to heat. Milk was found to lose 20% by roller drying and 35% by spray drying. Both heat and oxygen were factors in the loss, but the processed milks are still good sources of the vitamin; evaporated milk has been found to contain 1.9 μg per kg, sweetened condensed milk 6.3 μg, dried whole milk 28 μg and fresh, raw milk 32 μg.

Determination.—The Official Method is microbiological, using *Lactobacillus leichmannii* as the organism.

Pantothenic Acid

This vitamin was discovered in 1933 and isolated in 1940. It has many functions in nutrition. Its deficiency results in fatigue, abdominal distress, nausea and other symptoms. It is thought that gray hair is one of them.

Requirement.—The FNB estimates that the average American diet contains from 6 to 20 mg daily. It does not make any recommendation, but estimates 5 to 10 mg daily to be enough.

Source.—It occurs in yeast, animal organs, such as liver, heart, brains and pancreas, and less in muscles. Eggs and milk are sources as are some vegetables, among them, potatoes, tomatoes, broccoli, and also whole wheat. The amount in cheese increases as the cheese ripens.

Chemistry.—Pantothenic acid is a pale yellow oil, but it is marketed

$$\text{HOCH}_2\text{—}\overset{\displaystyle \text{CH}_3}{\underset{\displaystyle \text{CH}_3}{\text{C}}}\text{—CHOH} \cdot \text{CO} \cdot \text{NH}_2 \cdot \text{CH}_2 \cdot \text{CH}_2\text{COOH}$$

Pantothenic acid

as the calcium salt, which is a white solid. It is one of the simpler vitamins in structure. Its formula was established in 1940.

Stability.—The vitamin is unstable to acid and heat; cooking losses may reach 50%. The activity is well retained in cold storage. Eggs stored for 12 months at 0°C retained 92% of the vitamin activity.

Determination.—The AOAC has an Official Method for calcium pantothenate. It is a chemical method that hydrolyzes the vitamin with HCl, chlorinates the beta alanine with NaOCl and destroys the excess NaOCl with phenol. Then KI is added and the free iodine determined with a spectroscope. There is a microbiological method for the acid using *Lactobacillus plantarum.*

OTHER B VITAMINS

Several substances, other than those just described, have been included in the list of B vitamins. Some of them affect the nutrition of experimental animals, but do not appear to be required by man. Some of long standing have been abandoned, others are recent discoveries about which little information has been obtained. Two deserve brief mention: biotin and choline.

Biotin

The discovery of biotin began in 1916 with the observation that raw egg white is toxic to rats; cooked egg white is not. Investigation of this strange phenomenon revealed a vitamin that was originally called *biose* and *vitamin H,* and then *biotin,* which superceded the other names when its chemical nature was discovered in 1940.

Human deficiency causes lassitude, loss of appetite, followed by nausea, depression, muscle pains, dermatitis, anemia, and changes in the electrocardiogram. Such deficiency effects are rare, however, because the vitamin occurs widely in foods and is also synthesized in the intestines. The FNB does not recommend a daily allowance.

Biotin is present in liver, molasses, yeast, milk, and egg yolk and to a lesser extent in vegetables and grains. It is stable to heat up to 100°C.

Egg white contains a glycoprotein called *avidin,* which has a molecular weight of 70,000–80,000 and is strongly basic. Since biotin is an acid, the avidin combines with it to form an insoluble compound that is not absorbed from the intestine, hence the egg-white injury of the early experimenters. The avidin is very sensitive to heat and is destroyed by cooking. The chemical formula of biotin is

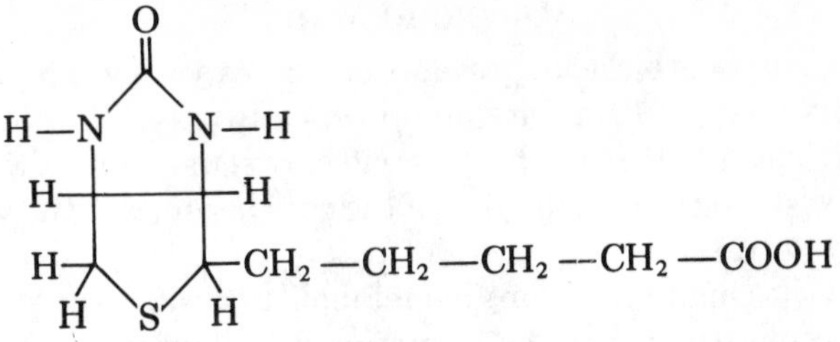

$$O$$

$$H-N \quad N-H$$

Biotin

Biotin is a white, crystalline solid that melts at 232°C. It is optically active with a specific rotation of +91°. It is only slightly soluble in cold water, but very soluble in hot water, which contributes to the cooking loss when water is used. Strong acids and oxidizing agents destroy it. Cooking loss of as much as 60% has been reported. There is no Official Method for the determination of biotin.

Choline

A choline deficiency has been demonstrated in experimental animals but not in man, although there is evidence that its deficiency may cause the deposit of fat in the liver. Nevertheless, choline has several functions in metabolism. It supplies methyl groups for the synthesis of compounds that are essential, and it constitutes part of the phospholipids that are essential to normal metabolism and also part of acetyl choline, which is essential to nerve function.

The FNB reports that the daily content of choline in the average mixed diet of Americans is 500 to 900 mg. Fresh eggs contain 1.7% choline, beef liver 0.6%, legumes 0.2% to 0.35% and other vegetables contain smaller amounts. There is little cooking loss of choline so that the diet is probably adequate in the content of this vitamin, with the exception of some special diets.

Choline is synthesized in the body but not in sufficient quantity to meet the nutritional requirements.

Choline is a very hygroscopic solid and very soluble in water; the solution is strongly alkaline. Chemically, it is a quaternary ammonium derivative of ethyl alcohol

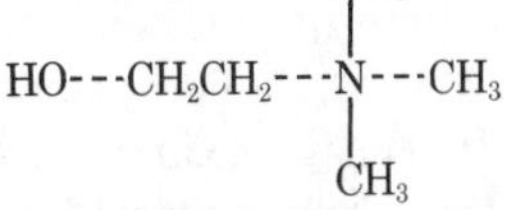

$$HO\text{-}\text{-}\cdot CH_2CH_2\text{-}\text{-}\text{-}N\text{-}\text{-}\cdot CH_3$$

Choline

There are chemical methods for the determination of choline, but the AOAC has not adopted an Official Method.

ASCORBIC ACID

Vitamin C, or ascorbic acid, prevents and cures scurvy. The vitamin has a long history (Garard 1974; McCollum 1957). Briefly, in 1753, an English physician discovered that fresh fruits will cure scurvy, but it was not until 1907 that a systematic investigation of the cure began, and the vitamin was not isolated until 1931.

Vitamin C has many functions in metabolism besides the prevention of scurvy. It is concerned with the formation of collagen and the healing of wounds. It is present in the adrenal glands and is involved in the formation of the dentine of the teeth and the growth of bones. It affects the blood vessels and the heart action.

Requirement.—As little as 10 mg of ascorbic acid has been found to prevent scurvy, but more than that is required because of its many other functions; an average healthy man uses over 20 mg daily. Furthermore, some people require more than others; therefore, the FNB recommends more than the actual requirement. For infants under 1 yr the recommendation is 35 mg daily, children 1 to 10, 40 mg and adults, 45 (pregnant and lactating women, 60).

In 1970, Professor Linus Pauling claimed that 200 to 5,000 mg daily will prevent most cases of the common cold, but the recommendation has become controversial both as to the effect on colds and the possible side effects of such a massive intake of the vitamin. More recently, it has been reported that the vitamin lowers the cholesterol content of the blood. Some infectious diseases, however, destroy it.

Sources.—Some of the richer sources of ascorbic acid in mg per 100 gm of food are: hot peppers 300, parsley 172, orange juice 158, sweet peppers 125, broccoli 90, brussels sprouts 87, turnip greens 69, strawberries 59, lemons 53, cabbage 47, grapefruit 38, cantaloupe 33, lime juice 32, green onions 32, spinach 28, asparagus 26, green peas 20, lettuce 18, and tomato juice 16.

There is a variation in the amount of vitamin C among different varieties of the same fruit or vegetable, and in any variety, the amount increases with the amount of sunshine in which the food was grown. Sunshine on the human or other animal, however, has no effect on the vitamin C requirement.

Chemistry.—Ascorbic acid was synthesized shortly after it was discovered and is now available in quantity in pure form. It is a white crystalline solid that melts at 190°C; its specific rotation is +21. It is soluble in water and and alcohol. It is easily oxidized to dehydroascorbic acid; the reaction even occurs in stored raw vegetables. Further oxidation occurs readily and is catalyzed by the presence of copper and some other metals.

The vitamin is the levo form and is readily prepared from glucose. The formulas for the two forms are

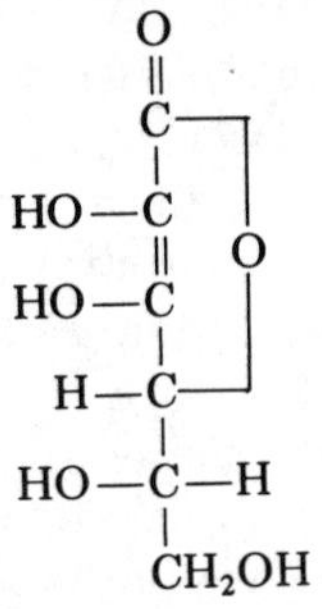

L-Ascorbic acid
(Vitamin C)

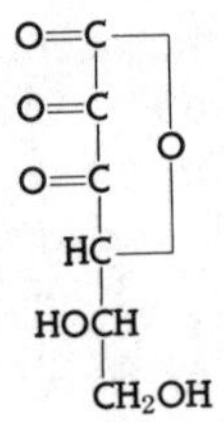

L-Dehydroascorbic acid

Stability.—Ascorbic acid is the least stable of the vitamins. It is extremely soluble in water and as much as 100% may be lost in cooking, which is not all due to extraction; the vitamin is destroyed by heat, light, oxidation, and alkalinity. It is more stable in acid solution than in either neutral or alkaline media. Rancid fats destroy it, apparently because of oxidation by the peroxides.

There is a vast literature on the loss of ascorbic acid under various conditions, but our space permits the mention of only a few examples. Asparagus stored 24 hr at 35°C lost 3% but at 66–78°C the loss was 40%. Broccoli lost 80% in 96 hr at room temperature. Spinach lost 78% in 48 hr at room temperature. Acid foods, such as tomatoes, retain the vitamin much longer than foods of less acidity.

Since the loss of ascorbic acid in stored foods is largely caused by oxidation, the manner of storage affects the results.

Processing is also a factor: drying riced potatoes caused a loss of 16 to 37%, the loss varying with time, temperature and method. Peaches, blanched, sulfured, and sun dried, lost 46%.

There is so much variation in the destruction of ascorbic acid by storage, processing, and storage after processing, that the chemist must do considerable research before making any claims for the vitamin C content of his product.

Determination.—There are two Official Methods for the determination of ascorbic acid. The first method is titration with a standard solution of 2.6-dichloroindophemol. The ascorbic acid decolorizes the dye, and the endpoint is the appearance of a faint pink color.

The dye solution is standardized against a standard solution of ascorbic acid made by direct weighing of a sample of ascorbic acid of specified purity.

The method determines only the reduced form of the acid and gives a result that is too high if there is Fe^{++}, Sn^{++}, Cu^+, SO_2, SO_3^{--} or $S_2O_3^{--}$ in the solution. The presence of any of these except Sn^{++} can be detected by adding 2 drops of a 0.05% aqueous solution of methylene blue—the color disappears in 5–10 sec. The presence of Sn^{++} can be determined in a similar manner with indigo carmine.

To prepare the sample, dry material is ground and extracted with a dilute mixture of acetic and hypophosphoric, HPO_3, acids.

In the second method, which measures both forms of the acid, the prepared solution is clarified with Norit and reacted with O-phenylenediamine to form a fluorescent substance, which is then measured with a fluoromicrophotometer and compared with a standard solution of the acid treated similarly.

The literature contains several other methods for the determination of ascorbic acid. The manuals of food analysis describe some of the more common ones, and the AOAC book of methods cites many references.

USE OF FNB ALLOWANCES

The FNB has considered all the evidence of nutritive requirements critically, and for most of them has made recommendations in detail for both sexes and all ages. The food chemist will be concerned with these details if he deals with food for infants, children in an orphanage or in school, or the inmates of a home for the elderly. They are of the utmost importance to dietitians and nutritionists.

However, most processed food is eaten by families that may consist of both sexes and ages from 1 to 90; consequently the food chemist will generally be concerned with the maximum allowance only for each nutrient. Any special case will require study of the bulletin and other references, but labeling requirements may make a knowledge of all recommended allowances imperative.

BIBLIOGRAPHY

ANON. 1965. Landmarks of a half century of nutrition research (Symposium). J. of Nutrition *91*, No. 2, Supplement 1, Part 2.

ASSOC. OFFIC. ANAL. CHEMISTS. 1975. Official Methods of Analysis, 12th Edition. Assoc. Offic. Anal. Chemists, Washington, D.C.

AURAND, L. W., and WOODS, A. E. 1973. Food Chemistry. Avi Publishing Co., Westport, Conn.

AXELROD, A. E., and MARTIN, C. J. 1961. Water soluble vitamins, Part 1. Ascorbic acid, biotin, inositol, nicotinic acid, pyroioxine group. Ann. Rev. Biochem. *30,* 383–408.

BEATON, G. H., and McHENRY, E. W. 1964. Nutrition: a Comprehensive Treatise, Vol. II. Academic Press, New York.

DE LUCA, H. F., and SUTTIE, J. W. 1969. The Fat-soluble Vitamins. Univ. of Wisconsin Press, Madison.

FREED, M. 1966. Methods of Vitamin Assay, 3rd Edition. John Wiley & Sons, New York.

GARARD, I. D. 1974. The Story of Food. Avi Publishing Co., Westport, Conn.

GOLDBLITH, S. A., and JOSLYN, M. A, 1964. Milestones in Nutrition, Vol. 2. Avi Publishing Co., Westport, Conn.

HARRIS, R. S., and VON LOESECKE, H. 1960. Nutritional Evaluation of Food Processing. John Wiley & Sons, New York.

McCOLLUM, E. V. 1957. A History of Nutrition. Houghton Mifflin Co., Boston.

SHERMAN, H. C., and BATCHELDER, E. L. 1931. Further investigation of quantitative measurement of vitamin A values. J. Biol. Chem. *91,* 505–511.

WILLIAMS, R. R. Toward the Conquest of Beriberi. Harvard Univ. Press, Cambridge, Mass.

CHAPTER 7

Food Analysis

In the discussion of the various nutrients, I have given a brief outline of one or more methods by which the amount of the nutrient in a natural or processed food can be determined. Operating details of the method must be obtained from a manual of food analysis or from the Official Methods of the AOAC, the Methods of the American Association of Cereal Chemists or those of the American Oil Chemists Society. Also, some of the methods of the United States Pharmacopeia or the National Formulary may prove useful. Besides these compiled lists of methods, which have official standing within their respective spheres, there are analytical methods in several journals both domestic and foreign. The only way to be up to date on analytical methods is to make a card index of methods that are found by scanning current issues of the journals, particularly *Chemical Abstracts.*

This chapter is not concerned with how to analyze a food, but with what analytical determinations are significant.

REASONS FOR ANALYSIS

There are three general reasons for analyzing a food: to learn whether or not it meets a state or federal standard; to assess its nutritive value; and to control processing. There are also minor reasons for analysis; for example, a chemist may come by a tropical fruit for which no analysis has been published. Another reason that frequently appears is to satisfy the curiosity of someone. I once knew a man who sent his product to two consulting laboratories just to see if they could find out what was in it; he, of course, knew its composition. Such analysis is the most difficult of all.

Standards

At least ten branches of the United States government, including the Congress itself, set standards for one or more food items. Gunderson *et al.* (1963) has listed the government agencies that set standards for food and explained the nature of the standards they set. In many cases, the standard specifies the method to be used in analysis and that method must be used. Whether or not the standard specifies the method, any analysis that involves a legal standard should be done by an official method if there is one.

Standards set by a federal agency may be obtained from that agency. Whether your company cans fruit or makes cheese, there is probably a legal standard for the product. State standards are usually set by the state Dept.

of Agriculture or the Board of Health, and each of the 50 states may have food standards; if any two of them are the same, it is coincidental. Only the big companies sell in all the states. Before a small company or a new one attempts to sell its product in a state, someone in the company must know what the standards for those products are in that state. Usually, a letter to the Secretary of State will produce the standard or, at least, the name of the agency that issues them.

Local standards may be obtained from the Board of Health, but local regulations generally refer to sanitary conditions rather than chemical composition or properties.

If food is to be exported, the standards and regulations of the country involved must be observed. Such regulations can usually be obtained from the Customs Service of the importing country or its consul in this country.

Sometimes the chemist needs to know standards that have not occurred to him. For example, the Defense Dept. has its own standards for food items that it buys. If your company does not sell to the Defense Dept., you may not bother about the standard, but perhaps some of your distributors bid on food contracts, and then the food must meet the standards.

The Food and Drug Administration (FDA) has established definitions and standards for 20 classes of food and more are likely. Current ones are reported in the Federal Register. In each class, there are several food items, with several hundred in all. Under cheese and cheese foods, 72 products are defined. The main classes that have been defined to date are:

Cacao Products.
Cereal Flours and Related Products.
Macaroni and Noodle Products.
Bakery Products.
Milk and Cream.
Cheeses, Processed Cheeses, Cheese Foods, Cheese Spreads and Related Foods.
Frozen Desserts.
Food Flavorings.
Dressings for Food.
Canned Fruit and Fruit Juices.
Fruit Pies.
Fruit Butters, Fruit Jellies, Fruit Preserves and Related Products.
Nonalcoholic Beverages.
Shellfish.
Fish.
Eggs and Egg Products.
Oleomargarine, Margarine.

Nut Products.

Canned Vegetables.

Tomato Products.

The definitions and standards are too long for inclusion here, but their general nature is indicated by the following items from the definition and standards of cheddar cheese:

> Cheddar cheese, cheese, is the food prepared from milk and other ingredients specified in this section by the procedures set forth in paragraph (b) . . . .
>
> It contains not more than 39% moisture and its solids contain not less than 50% milk fat as determined by the methods prescribed in paragraph (c) of this section.
>
> If the milk is not pasteurized, the cheese so made is cured at a temperature not less than 35°F for not less than 60 days.
>
> Harmless artificial coloring may be added . . . with or without purified calcium chloride in a quantity not more than 0.02% of the weight of the milk.
>
> (c) Determine moisture by drying in a vacuum oven at 100°C.
>
> Cheddar cheese in the form of slices or cuts in consumer-sized packages may contain an optional mold-inhibiting ingredient consisting of sorbic acid, potassium sorbate, sodium sorbate or any combination of two or more of these in any amount not to exceed 0.3% by weight calculated as sorbic acid.
>
> The word milk means cow's milk.
>
> Milk shall be deemed to have been pasteurized if it has been held at a temperature of not less than 142°F for a period not less than 30 minutes.
>
> Cheddar cheese shall be deemed not to have been made from pasteurized milk if 0.25 gm shows a phenol equivalent of more than 3 micrograms when tested by the method prescribed in paragraph (f) of this section. *Note:* Section (f) occupies 4 pages and describes the method for phenol equivalent in detail. It requires the use of a photometer.

The definition includes the procedure for manufacturing the cheese, for it is written as a guide for the manufacturer as well as for the benefit of the chemists concerned with legal control.

There are some hidden determinations besides those for moisture and fat; cow's milk and harmless coloring may need to be identified, which means that a Reichert-Meissl number and a Polenske number should be made to identify the fat.Elsewhere in the FDA regulations, the dyes legal in foods are described, which limits the dyes that must be identified, so that the identification of the color is not as difficult as it seems. However, if the dye is not one of those approved, it may be necessary to identify it. In that case, the chemist should test for those dyes that have been used in foods in the past and subsequently declared illegal.

Roquefort cheese is made from sheep's milk and its identification is very important for it looks like the blue cheese made from cow's milk. The fat of sheep's milk is white because the sheep does not store carotene in the fat as the cow does. Furthermore, the Polenske number of the sheep's milk fat is 3 or more while that of cow's milk is under 3.

Space does not permit a complete discussion of analysis to assess the agreement of food composition with all the federal standards. The chemist should consult the *Code of Federal Regulations 21*. He should also acquire copies of the standards of the states with which he is concerned if he is involved in food processing or in an independent consulting laboratory that does food work.

Nutritional Value

The nutritional value of food as reported in the tables is usually for 100 gm or 1 lb of the edible portion of the food. The general method for the determination of the energy value of a food is to determine the percentage of carbohydrates, fats, and proteins in the edible portion. If 100 gm portions are the goal, then the percentages are the number of grams in the portion. To get the caloric value, the number of grams of carbohydrate and protein are multiplied by 4, the number of grams of fat by 9, and the results are added.

The actual heat of combustion of proteins is about 5.75 kcal per gm, that of fats 9.3, and that of carbohydrates 4.2. These values are too high for the heat produced by oxidation in the body. In the calorimeter, the proteins burn to CO_2, N or NO_2 and water, whereas in the body, they are oxidized to urea, $CO(NH_2)_2$. The heat of combustion of urea is 1.25 kcal per gm, which brings the nutritional energy value down to 4.5 kcal/gm.

None of the major nutrients is 100% digested, absorbed, and metabolized. Different nutritionists have used percentages from 92% to 98% for utilization, which reduces the energy values to the familiar 4, 4 and 9 kcal/gm. These give values for the several foods that are probably within the variability of the individual foods. However, biochemical research has shown that the coefficients of digestibility of proteins vary from 70% to 90% with a few even below 70%. Fats vary only from 90% to 95% and carbohydrates from 90% to 98%. The Agriculture Handbook No. 8 has used the more exact heats of combustion and coefficients of digestibility for calculating the energy values in its tables.

Fat.—The principle involved in the determination of fat in a food is simple, but its practice is not. Fats are soluble in ether, petroleum ether, acetone, chloroform, carbon tetrachloride, and other chlorinated solvents, but anhydrous diethyl ether and petroleum ether, or a mixture of the two, are nearly always the solvents used in the quantitative determination of the fat content of a food. Both require special preparation before they are used. As purchased, diethyl ether contains water and alcohol and may contain peroxides, which are dangerous and may affect the results of the analysis.

The Official Method for the preparation of anhydrous ether washes the commercial product with water to remove alcohol. Solid NaOH or KOH is then added and allowed to stand until most of the water has combined with the alkali. Ether dissolves 1.5% water and most of it will be removed by the hydration of either of the alkalies. The ether is decanted from the alkalies and freshly cut pieces of sodium are added. Hydrogen is evolved and the NaOH and C_2H_5ONa formed adhere to the sodium. Since the ether must be stored in loosely stoppered bottles because of the hydrogen, it may be necessary to add fresh pieces of sodium if the storage time is long. The sodium prevents the formation of peroxides, but light encourages their formation and so ether should be stored in the dark.

Anhydrous ether is essential for the complete extraction of fat and therefore, the sample must be dried before the extraction. Moreover, if water were present it would be extracted along with the fat and would be difficult to remove from the fat, which forms a layer on top of any water present.

In addition to the bother of preparing and keeping anhydrous ether, the solvent is a hazard after the extraction. There are several devices and methods of extracting fats from foods, but in most cases, the solvent must be evaporated and the fat weighed. Ether vapor is 2.5 times as heavy as air, its ignition limits are 1.7% to 48% in an air mixture and its ignition temperature is 186°C. Consequently, it cannot be evaporated over a burner or a common electric hot plate because the vapor soon settles down around the apparatus and ignites. Even if the heat source is shielded, the vapor flows along a desk top like so much water and may ignite from a burner several feet away.

It is customary to remove the last traces of ether and water in an electric drying oven, but if the flasks still smell strongly of ether, they should not be put into such an oven. There is at least one case on record where this neglect resulted in an explosion that blew the oven door through a partition into the next room.

Petroleum ether is anhydrous as purchased because water is insoluble in hydrocarbons. Neither does it form dangerous peroxides, but since it may contain high boiling hydrocarbons, it must be distilled, and the portion boiling between 30° and 60° collected. It may be stored without special precautions except that in sunlight or near a heater it may blow the stopper and evaporate.

Petroleum ether in the 30°–60°C range is mostly pentane, C_5H_{12}, which boils at 36°C. Its vapor is about 2.5 times as heavy as air, its explosive limits 1.4 to 8.0% in an air mixture and its ignition temperature is 309°C; consequently, it is less of an explosive hazard than ethyl ether.

The chlorinated solvents, such as chloroform, $CHCl_3$ (boiling point 61.2°C), carbon tetrachloride, CCl_4 (boiling point 76.8°C) and methylene chloride, CH_2Cl_2 (boiling point 40.1°C) are not flammable and are easily purified, but they seldom extract all the fat from the sample. Even with ethyl ether, the sample may contain lipoproteins or other lipid compounds that are insoluble. In such cases, the sample must be hydrolyzed before the extraction.

There is one exception to the determination of fat in food by ether extraction: the analysis of milk by the Babcock method, which is an Official Method. The method specifies the addition of sulfuric acid to the milk in a special bottle with a long, slender neck with a scale etched on it. The bottle is then whirled in a special centrifuge. The acid has dissolved the protein, the fat gathers in the neck of the tube and the amount of it as read on the scale is the percentage of fat in the milk.

The Babcock method is very rapid and is widely used for routine work. It requires a special centrifuge, special bottles, and a special pipet. It also requires special attention to details, such as the size of the sample, the concentration and amount of sulfuric acid, speed of rotation, and temperature. Fat has a high coefficient of expansion and, since the final reading is a volume and the determination is fat by weight, a low temperature would mean a low result. The temperature is adjusted to 55° to 60°C before the reading.

Ether Extract.—The result of any of the extraction methods of analysis is usually referred to as percentage of fat, but to be exact, it is, and is often called the *ether extract*. The use of the term "fat" is traditional and appears in all the literature on the subject, but foods contain many other substances than fat that are soluble in the ethers. Many of these substances have already been mentioned. There are the vitamins A, D, E, and K, cholesterol, and other sterols, carotene and other caratenoid pigments. Some foods may contain waxes; honey for example. Processed food may contain esters, essential oils, or other additives. This list does not pretend to be complete; the chemist must be familiar with the composition of his foods to be able to interpret the result of a fat determination.

There are methods for the determination of unsaponifiable matter, which helps in the evaluation of a fat assay.

Protein.—The determination of the amount of protein in a food is made by some modification of the Kjeldahl method. The method determines nitrogen only and the result is converted to protein by multiplying the percentage of nitrogen by some factor. For mixtures of proteins or those of unknown composition, 6.25 is used. The Official Methods include special factors for proteins of known composition, such as the 5.7 for wheat protein and 6.38 for milk protein.

Protein analysis for nutritive value has become complicated in the past 40 years by the discovery that certain amino acids are essential. Therefore, the gross percentage of protein is no longer an adequate measure of its nutritive value.

The Official Methods contain procedures for the determination of the amino acids. The content of the eight or ten essential acids must be known before the nutritional value of a protein can be evaluated.

Carbohydrates.—In general, carbohydrates are determined by difference; add the percentages of water, fat, protein, and ash and subtract from 100%. This procedure includes cellulose, pentosans, and other polysaccharides, some of which are not digested by humans.

The determination of *crude fiber,* for which there are Official Methods, helps to solve this difficulty, but fails to separate all the indigestible carbohydrates. The usual method for crude fiber is to dry the sample, extract the fat if it exceeds 1%, hydrolyze the starch with 1.25% H_2SO_4, filter, wash the residue, dry, and weigh it.

If the analysis is being made to determine the nutritive value, the diastase method of starch hydrolysis is probably better. In this method, the sample is cooked to gelatinize the starch and then malt or taka diastase is added to convert the starch to soluble dextrins and maltose and, after the residue has been removed, the hydrolysis is carried on to glucose with acid and the latter determined by a copper reduction method. The starch is 0.9 the weight of the glucose,

$$C_6H_{10}O_5 + H_2O \rightarrow C_6H_{12}O_6$$

162	180
Starch	Glucose

Most carbohydrates have a heat of combustion of 4.20 kcal per gm. Sucrose, however, yields only 3.95 kcal and glucose only 3.75 kcal/gm. These two sugars are 98% utilized so they supply the diet with 3.87 and 3.68 kcal respectively. Mixtures of carbohydrates from various foods are utilized to the extent of 92% to 98% and supply from 3.60 to 4.16 kcal/gm.

Minerals.—The inorganic elements are determined in the ash of the food by one of the methods of inorganic analysis. The AOAC Methods specify the details to be followed in the ash determination in over 60 foods, although it specifies the same method for several of them. In general, the sample is heated to a specified temperature (500–550°C) in a muffle furnace until a white ash is obtained. A platinum dish is usually specified, but in some cases a quartz or porcelain dish may be used. Neither is as satisfactory as platinum.

Two precautions are necessary besides all those that are common to all analyses. Some ash constituents are volatile; among them are iodine, fluorine and some salts of sodium and potassium.

The other precaution arises when the ash is high in phosphate. Sodium and potassium phosphates fuse before the combustion is finished and coat particles of carbon so that the ash does not make contact with the air. This may require ignition at a lower temperature, but it can often be corrected by adding acetic acid, which dissolves the phosphate and the acid can then

be evaporated and the ignition resumed at a lower temperature. Sodium
and potassium are usually present in the ash as the carbonates or the
chlorides. Other metals are present as oxides, so the composition of the ash
is not altered by the acetic acid as the acetates decompose to the carbonates
or the oxides of the several metals. The usual metals in a food ash are sodium, potassium, calcium, and magnesium.

Trace Elements.—Those elements that occur in foods either as nutrients or pesticide residues require special methods of analysis because of
the small quantity involved. The Official Methods contain procedures for
all the elements that have been considered important, and will undoubtedly
devise others for any new ones that may appear. The methods are usually
spectrographic.

Vitamins.—The analysis of foods for vitamin content is a special field.
Some of the methods are purely chemical, but many of them are either biological or physical, which require the use of a spectrometer or fluorimet-
er.

Process Control

There are few cases in which process control requires Official Methods.
Speed and convenience are more important. However, food processors buy
foods from each other frequently, on specifications. A baker buys flour from
a miller and may specify the amount of moisture and the minimum amount
of protein. Both the buyer and the seller have chemists to make the necessary analysis. Suppose they do not agree! The chemists must communi-
cate and, if they are using different methods, they must try to agree on a
common method. Or one of them may be using the protein factor 5.70 and
the other 5.83.

In any event, the method of analysis, even for routine control, must be
a matter of record, and must be selected with some judgment. If the label
on a product specifies definite quantities, or if there are legal standards,
then the method must give the same results as the Official Method, or the
product may be seized as misbranded.

Filth.—One determination that is extremely important in process control
is filth in the final product, for the federal law specifies that a filthy product
is adulterated and most states have a similar law. Even the local Board of
Health may have a regulation that applies. Chapter 44 in the 12th Edition
of the AOAC Official Methods provides methods for the detection of *Extraneous Materials*. Included are glass, sand, soil, rust and other miscellaneous material, but the greater part of the section is devoted to filth which
is defined as: "Any objectionable matter contributed by animal contamination of product such as rodent, insect or bird matter; or any other objectionable matter contributed by insanitary conditions."

The various methods require special equipment, which is described in detail; directions are given for the manner of reporting the results. The more common items of filth are insects, dead or alive, in pupal, larval or adult form. Also included are insect fragments and excreta, rodent hairs and excreta, and unidentified foreign particles. In milk, for example, there may be dust of unknown origin; it may be clay, soil, seeds, spores, pollen, dry cow manure or other material.

The first problem of the chemist in a filth determination is the separation of the filth from the food. One of the simpler methods is the determination of sediment in milk. The milk is filtered through a white disk of specified diameter and material, and the color compared with that of a similar disk that has had a standard suspension of sediment of specified composition filtered through it. The composition of the standard sediment is a uniform mixture of the following materials that have been dried at 100°C and ground by hand with mortar and pestle:

	%
Cow manure, through No. 40 sieve	53
Cow manure through No. 20 sieve, retained by No. 40	2
Garden soil, through No. 40	27
Charcoal, through No. 40	14
Charcoal, through No. 20, retained by No. 40	4

Concentrated milks are reconstituted and then examined the same as fresh milks.

Butter and cheese are more of a problem, because when the fat is removed, the curd remains to hide the filth. The curd is made soluble by boiling with phosphoric acid, H_3PO_4 diluted 1 to 40.

In addition to filth, peanut butter is analyzed for rocks and glass.

Cereals, especially whole grain products, are very susceptible to insect contamination and the analytical methods differ with the product. Here the main problem is to destroy the starch, protein and pentosans without destroying the filth. In some cases, different methods must be used for different kinds of filth.

Each type of food is a different problem. The Official Methods recognize beverages (coffee, chocolate), dairy products, nuts, cereal foods, poultry, meat and seafoods, fruit and fruit products, sugar and sugar products, vegetables and vegetable products, spices and other condiments, hops (for aphids), and urine stains on food or containers.

The chemist is supposed to identify the filth if possible and report the amount of each kind. In most cases, this involves the use of a microscope and counting device.

GENERAL ANALYSES

A food chemist is sometimes given a sample and asked to analyze it. The request means nothing. It is possible to work for weeks and still not get the information the client (or boss) wants. The chemist must ask a lot of questions. What is this? Why do you want it analyzed? What do you want to know about it? It may be an oil and he wants to know whether it is pure olive oil. If so, the chemist knows what to do. Or it may be something of mixed composition that he wishes to duplicate. Or, he may think he has a secret composition and simply wants to know if a chemist can analyze it. The possibilities are infinite and many of them require ingenuity on the part of the chemist for he will not find a list of methods anywhere that are applicable. Some problems are impossible; for example, of what ingredients was a cake made? What kind of baking powder and how much of it? Is all the sugar sucrose and was it a syrup, powdered or crystalline? What kind of flour? Any eggs; if so, how many? Was a low-protein flour used or was it a high-protein flour and corn starch? What kind of shortening etc., etc.?

It is impossible to answer these questions because there is phosphate in the flour and in eggs and probably in the baking powder. The sugar is in solution and probably partly inverted; the starch is gelatinized. The cake can be analyzed for the nutrients of course, for dyes and preservatives, but to identify what the recipe called for is impossible. A good chemist, however, could produce a fair replica if that were the object of the analysis.

It is common practice for a processor to examine the products of his competitors, and problems of the composition of mixtures are common. Uncooked mixtures are easier, for a way generally can be found to identify the ingredients. A general knowledge of processing is often a help. In one example, most of the product dissolved in cold water and nearly all the remainder in ether. A microscope showed that the small residue was corn starch. Why this tiny amount of corn starch? It dawned on the chemist that processors put a little corn starch in powdered sugar, and a telephone call told him how much, which was quicker than buying some powdered sugar and assaying it. A sugar determination told him that all the sugar used was sucrose and the amount of starch told him that powdered sugar had been used.

The only general statement that can be made about analysis of this type is that the chemist must know the chemistry of all the food components and methods for their determination, or at least, where to find the methods. A wide knowledge of food processing will help. Beyond these simple requirements, the chemist is on his own.

SAMPLING

The student is seldom faced with a sampling problem because a small sample is handed to him for analysis. If the sample is a suspension, however, such as a bottle of milk, he must see that it is mixed to uniformity, before he removes his small analytical samples. The word "sample" has two meanings to the chemist. It refers to the entire supply the chemist has in the laboratory and also to the small portions that are taken for each determination. Except in the case of suspensions, the latter are no problem to the chemist; the former are the problem. If the sample is not representative of the lot it comes from, an analysis is worthless. The chemist may be confronted with a car load of flour, sugar, lard, cornmeal, apples or any other food, or he may be concerned with all the products his company's competitors make. For legal purposes, package sampling may not be important, for if one package is adulterated, the product is illegal, and it is unnecessary to have a representative sample. If it is his own company's product, however, he will need to know whether it is all adulterated.

The Official Methods sometimes include sampling procedures. For flour, they sample the square root of the number of bags in the lot, but not less than ten. Bags are selected on the basis of exposure: 4 from most exposed, 3 from next most exposed, etc. The dimensions of the trier are specified and cores are drawn from the edge of the surface to the center of the bag. The cores are then put into an airtight container, such as a pint fruit jar with a rubber gasket—a separate jar for each bag. Several other details are included.

The triers are metal tubes of definite specifications and are used for many products besides flour such as small seeds of any kind, whole or processed. Butter and other solid fats are sampled by these triers; they, of course, differ in dimensions with the purpose for which they are intended.

The care of the sample after it is obtained is important. Many products absorb moisture from the air, some lose moisture and some oxidize or otherwise change chemically because of the action of enzymes or organisms. Refrigeration is essential for some samples and desirable for most of them. Many samples require mixing each time a portion is removed for analysis because any changes that may occur during the storage of the sample are greatest at the surface.

Taking and handling samples requires experience and consideration of the lot to be sampled and the purpose of the analysis to be made.

TOLERANCES

Because of variation in the composition of both natural and processed foods, and also because of lack of great precision in many analytical methods, the FDA has established tolerances for many products. There is no tolerance for butter fat, however, because its standard was set by a special law of Congress in 1823 and contains the statement: ". . . contains not less than 80 per centum by weight of milk fat all tolerances being allowed for."

Tolerances are not disclosed by the FDA for it does not wish to encourage food processors or distributors to market products that are substandard.

Some tolerances, however, are disclosed during a court action, but of course they may be changed afterwards.

In trying to hew to the line in food processing for economic reasons, some batches may be low in some component, the 50% fat requirement in cheese for example. If a single batch is low, there is not much cause for worry, but if several are low, the reputation and purse of the company may become involved.

BIBLIOGRAPHY

ANON. 1974. Nutrition labeling. Food Tech. *28,* No. 7, 42–49.

ASSOC. OFFIC. ANAL. CHEMISTS. 1975. Official Methods of Analysis, 12th Edition. Assoc. Offic. Anal. Chemists, Washington, D.C.

BOBBIT, J. M. 1963. Thin Layer Chromatography. Reinhold Publishing Corp., New York.

FDA. 1975. Code of Federal Regulations *21,* Food and Drugs, parts 10–199. U.S. Govt. Printing Office, Washington, D.C.

GUNDERSON, F. L., GUNDERSON, H. W., and FERGUSON, E. R. 1963. Food Standards and Definitions in the United States. Academic Press, New York.

HEFTMANN, E. 1961. Chromatography. Reinhold Publishing Corp., New York.

JAMIESON, G. S. 1943. Vegetable Fats and Oils. Reinhold Publishing Corp., New York.

KRAMER, A., and TWIGG, B. A. 1970. Quality Control for the Food Industry. Avi Publishing Co., Westport, Conn.

MACLEOD, A. J. 1973. Instrumental Methods of Food Analysis. John Wiley & Sons, New York.

MELOAN, C. E., and POMERANZ, Y. 1973. Food Analysis Laboratory Experiments. Avi Publishing Co., Westport, Conn.

POMERANZ, Y., and MELOAN, C. E. 1971. Food Analysis: Theory and Practice. Avi Publishing Co., Westport, Conn.

SAND, R. E. 1974. Rapid method for calculating values of food components. Food Tech. *28,* No. 7, 29–33, 40.

WINTON, A. L., and WINTON, K. B. 1945. The Analysis of Foods. John Wiley & Sons, New York.

CHAPTER 8

Colloids

In 1861, Thomas Graham, an English chemist, published a paper reporting the results of his experiments on the rate of diffusion of several substances in aqueous solution. He found that amorphous substances, such as alumina, gum arabic, albumin, and gelatin diffused very slowly, but that crystalline substances such as salt and sugar diffused rapidly. Furthermore, the crystalline substances diffused readily through a parchment membrane, but the amorphous substances did not pass through at all.

Glue was one of the noncrystalline substances of Graham's list and so he named them *colloids* from the Greek work *kolla,* which means glue. As a corresponding term, he called the crystalline substances *crystalloids.*

Since 1861, hosts of chemists have developed the field of colloid chemistry and some knowledge of it is essential to the food chemist, because gums, proteins, polysaccharides and several other food components are colloidal.

The term "colloid" does not indicate a *kind* of matter, but a *state* of matter: a dispersed state. A sugar solution is a dispersion of the sugar molecules in water.

We are concerned here mainly with dispersions in liquid media. Such dispersions are of three kinds: suspensions, solutions, and colloidal dispersions, or *sols.*

Suspensions are temporary dispersions in which the dispersed phase settles out when the system remains undisturbed. Mud in a flowing river and raw starch in water are suspensions; the dispersed particles are large and insoluble in water.

Solutions are permanent dispersions in which the dispersed particles are either molecules (sugar) or ions (salt).

The sols contain particles, or *micelles,* intermediate in size between those of the solution and those of the suspension. The size of the micelles differs with the substance, but in general, the molecule or ion in solution is less than 1 nanometer (nm), or 1 millimicron (mμ). The micelle of a sol is between 1 and 200 nm and the suspended particles are larger than 200 nm in diameter. The resolving power of the best microscope will not detect the dispersed particles in either a solution or a sol.

Aqueous sols are of two kinds: *suspensoids* and *emulsoids.* The micelles of the former are mostly inorganic; metals (gold, silver, platinum) or insoluble oxides and salts (Fe_2O_3, As_2O_3, CdS, and many others).

In food chemistry, we are concerned primarily with organic substances and therefore with three types of colloidal dispersions: emulsoid sols, liquid in liquid dispersions, or *emulsions,* and gas in liquid dispersions, or *foams.*

EMULSOIDS

The most common emulsoids in food science are dispersions of gums, pectins, starch, dextrins, or proteins in water.

Preparation

The preparation of an emulsoid sol is usually a simple procedure. Dehydrated nonfat milk solids can be added to water with a minimum of stirring and in a few seconds a colloidal dispersion is produced; fluid nonfat milk. Gelatin may be dispersed in cold water in a similar manner, but agar requires hot water and agitation.

Properties

The properties of sols are very different from those of solutions. A mole of a solute depresses the freezing point of water 1.86°C and increases the boiling point 0.52° at a concentration of 1 mole per liter in both cases. A mole in 1,000 gm of water has an osmotic pressure of 23.6 atmospheres, or 347 psi. The micelles of a colloidal dispersion have little effect on these three properties; a 1% dispersion of egg albumin depresses the freezing point of the water only 0.0008°, which is too small to measure with any useful precision. These three properties of solutions have been widely used to determine molecular weights of organic compounds up to values of a few hundred, but the freezing point depression and the boiling point elevation are useless for colloidal dispersions. The extraordinarily high osmotic pressure has been used for proteins and other colloids, but any slight impurities of low molecular weight may affect the pressure more than the compound under investigation.

There are some properties that are affected more by colloids than by solutes.

Viscosity.—An outstanding property of emulsoid sols is their viscosity, or the resistance of a fluid to flow. It is generally measured by timing the flow through a capillary tube and comparing the rate of flow with that of a standard, usually water. The capillary tube viscosimeter is most useful for measuring low viscosities; there are several other types of instrument for the high values.

The high viscosities of emulsoid dispersions is indicated by comparison with the viscosity of a solution; a 1% sucrose solution has a viscosity only 3–4% higher than that of water, a 1% dispersion of starch is 50% higher and one of agar is 500% higher.

Electrical Behavior.—Emulsoid colloids may or may not migrate in an electric field. Proteins are amphoteric and, at their isoelectric point, they have no charge and do not migrate. On the acid side of neutral they have a negative charge and migrate to the anode, while on the alkaline side they are positive and go to the cathode.

Carbohydrates have no electric charge unless they adsorb one of the ions of some electrolyte that happens to be present. Consequently, they do not migrate in a pure dispersion.

Diffusion.—Emulsoids diffuse very slowly and, as Graham discovered, they do not diffuse through a membrane or into a jelly. This indicates that their micelles are very large. Whether the micelle is a large molecule or an aggregate of smaller ones is seldom known. However, it is generally thought that in the common emulsoid sols the micelles are molecules. Molecular weights of organic compounds ascend gradually from the smallest, CH_4, which is 12, to compounds of many millions, and there is no sharp dividing line between solutions and colloidal dispersions. Some of the latter may contain both molecules and aggregates.

An amorphous solid that will form an emulsoid sol usually absorbs water, which indicates that the remarkable stability of these sols is at least partly due to the hydration of the molecules, and so they are often called *hydrosols*.

Dialysis.—There are three methods for removing solutes from colloidal dispersions: ultrafiltration, dialysis, and electrodialysis. The simplest of them is dialysis, which depends on the principle discovered by Graham; compounds of low molecular weight pass through a semipermeable membrane but colloids do not. The most common semipermeable membranes used are parchment, collodion, cellulose acetate, and cellophane. These membranes appear to be porous and operate like a bolting cloth that allows the flour to pass but not the coarser particles of the grain.

Dialyzers are of different designs; a simple one consists of a bell-shaped glass part, much like the top half of a large bottle. A membrane is stretched across the large end and tied tightly in place. The dialyzer is suspended in another vessel containing distilled water—the membrane well below the surface of the water. The dispersion is poured into the dialyzer. The salts, sugars, and other compounds of low molecular weight diffuse through the membrane until equilibrium is reached, then the water is changed and the system allowed to reach equilibrium again.

It is impossible to remove all the solute from a colloidal dispersion by dialysis. In order to keep the arithmetic simple, let us suppose that 1 liter of the sol contains 1 gm of sugar and is dialyzed against 1 liter of water, then:

	Sugar in sol	Sugar in water
1	0.5 gm	0.5 gm
2	0.25	0.25
3	0.125	0.125

After three changes of water, there is still ⅛ gm of sugar left in the sol, and it is obvious that it can never be reduced to zero. It can be reduced more rapidly by using a larger volume of water each time. If 9 liters of water are used in the first dialysis, there will be 0.1 gm of solute left at the first equilibrium and 0.01 gm after the second.

The preparation of collodion membranes requires no special apparatus but considerable skill. Select a flask of suitable size and a long neck, such as a Kjeldahl flask. The flask must be perfectly smooth inside and quite clean and dry. Pour a few milliliters of collodion gently down the side of the neck of the flask to avoid bubbles. Rotate the flask gently so the collodion wets the entire inside surface of the flask including the neck. Pour the excess collodion back into the bottle. Support the flask in an inverted position on a ring stand until the odor of ether has disappeared. Fill the flask with tap water and allow it to stand for a few minutes. Pour out the water, cut the film free from the mouth of the flask with a sharp knife and work the film loose from the sides of the flask. You now have a collodion bag the size and shape of the flask, which can be tested for pinholes by filling it with water. Any bubbles, grease or rough spots on the inside or the flask will cause holes in the collodion.

A collodion bag, such as that just described, requires no special apparatus. The sol can be poured into it, the mouth tied around a bent glass tube and the bag immersed in a vessel of water with the tube hooked over the edge of the vessel. The mouth of the bag cannot be closed because water diffuses into the bag and the pressure will burst it.

By varying the procedure, bags of different permeabilities can be prepared (Thomas 1934).

I have compared a semipermeable membrane to a screen, and in many cases penetration is purely mechanical, but in other cases penetration may be chemical; that is, the solute combines with the membrane on one side and is released on the other.

The walls of the intestines are semipermeable membranes that permit the passage of the low molecular weight compounds, the sugars, alcohol and the amino acids, but not cellulose, starch or proteins. Biological membranes are much more complicated than those of the laboratory. One problem is the absorption of fat. Fats are digested in the intestines by lipases to glycerol and fatty acids. The glycerol has a low molecular weight (92) and is soluble in water, but the fatty acids, sterols, and fat soluble vitamins are insoluble in water and do not diffuse from an aqueous solution through the common semipermeable membranes of the laboratory; yet they are absorbed from the intestines into the blood stream. Several theories have been advanced to explain this anomoly. The present view is that the fatty acids are emulsified by the bile acids or by monoglycerides and are absorbed as emulsions (Beaton and McHenry 1964).

Coagulation.—Emulsoid sols are easily prepared but they are not so easily coagulated. Proteins often precipitate at the isoelectric point and so changing the pH may cause coagulation. Changing the pH may also be effective with dispersions that are largely carbohydrate if they are acidic, such as the citric wastes from a fruit processing plant. The stability may be caused by the adsorption of the H^+. In some cases, a positive ion from a salt or base may have an effect on stability; $Ca(OH)_2$ is often more effective in causing coagulation than NaOH.

Many proteins can be precipitated by the addition of a salt. Although the relative value of salts for salting out proteins may vary with the conditions, Pauli in 1903 established the following series of effectiveness of anions in descending order: F^-, SO_4^{--}, PO_4^{---}, citrate, tartrate, acetate, Cl^-, NO_3^-, ClO_2^-, Br^-, I^-, CNS^- (Thomas 1934).

The cations differed in their effect much less than the anions, but a few appear effective in descending order: Li^+, Na^+, K^+, NH_4^+, Mg^{++}.

The stability of an emulsoid dispersion also differs with the dispersing medium. They are most stable in water, therefore, the addition of a miscible liquid to a water dispersion will often cause coagulation. Ethyl alcohol, isopropyl alcohol, and acetone have been found effective.

GELS

Many emulsoid sols set to a gel on cooling or simply on standing undisturbed. A 1% gelatin dispersion will not set at room temperature in several hours; a 2% dispersion will set in ½ hr at 15°C. The formation of a gel occurs in less time with decreasing temperature and with increasing concentration.

The food chemist encounters gels of starch, gelatin, pectin, carragenan, and other proteins or polysaccharides. The same substance as an emulsoid dispersion may serve as a thickener for a soup and, as a gel, add firmness to a pudding or a pie filling.

Gels have been investigated for more than a century. Gelatin has come in for more investigation than any of the others. Its preparation is simple; add water to the dry gelatin and wait. The dispersion of gelatin is quicker, however, in hot water and with agitation. Once a clear dispersion is obtained, cool it.

Properties

The vapor pressure of an aqueous gel, its refractive index, and its electrical conductivity are not different from those of water itself. However, as the water evaporates from a gel, the vapor pressure decreases as dryness is approached.

Solutes diffuse through a gel just as they do in water, but the rate is slightly retarded even in gels of low concentration, such as 2% agar or gelatin.

Some sols viewed in an ultramicroscope show Brownian movement, which was discovered by Robert Brown, a botanist, in 1827 and was named for him. It is a zigzag vibration of the dispersed particles. Metal sols, in particular, show a rapid vibration of micelles, but it is harder to observe the phenomenon in emulsoid dispersions, because of the near transparency of the micelles. Some have been observed, however, and when the sol sets to a gel, the movement ceases.

A gel is elastic, that is, it will return to its shape after a mild distortion, but it has a *yield point,* a pressure at which the elasticity fails and the gel begins to flow.

All gels can be destroyed by mechanical action, but some of them will set again after the agitation stops. Such gels are said to be *thixotropic.*

Many amorphous substances absorb water and swell. Gelatin exerts an enormous pressure as the dry powder, or a concentrated gel, imbibes water. The phenomenon is very common; drawers swell and stick in humid weather and raisins, prunes, or dried beans all swell if they are covered with water, and exert an enormous pressure in doing so.

Although most gels absorb water, some of them release water. This is called *syneresis* and is familiar in fruit jellies, custard puddings and other food gels.

Structure

The properties of gels indicate their structure. The polysaccharides and many proteins are large, linear molecules, and the properties of their gels indicate that the molecules attach to each other to form a rigid framework, the strength of which depends on the nature of the material and the amount of it. In some cases, the attachment may be by primary valence. The proteins have free $-NH_2$ and $-COOH$ groups that react on contact to form a union similar to the result of the simple reaction

$$NH_3 + HCl \rightarrow NH_4Cl$$

In the case of polysaccharides, the union may be of the hydrogen bond type.

Gels have been examined by X rays and the result indicates a definite structure; the molecules are not simply piled together like so much brush.

Food Gels

The federal standard for fruit jelly in the FDA code is long and detailed. In brief, the definition is: a food which is made from a mixture composed of not less than 45% by weight of one or any combination of two, three, four, or five of the fruit juice ingredients specified in paragraph (c) to each 55 parts by weight of a saccharine ingredient specified in (d) (as well as one or more of eight optional ingredients).

Paragraph (c) lists 28 fruits; it contains the common fruits, but restricts the jelly maker's efforts to introduce new ones—acerola, black currants, elderberries, gooseberries, lemon, lime, nectarines, pear and prune plum are among the fruits missing from the list.

The saccharine ingredients of (d) are sugar, invert sugar syrup, dextrose, corn syrup and honey and several combinations of them. The optional ingredients are:

1. Spice.
2. Vinegar, lemon juice, lime juice, citric acid, tartaric acid, fumaric acid or any combination of them.
3. Pectins.
4. Sodium citrate, sodium potassium tartrate or both, but not over 3 oz per 100 lb of the sugar used.
5. Sodium benzoate or benzoic acid, limited to 0.1% by the GRAS list.
6. Mint flavoring and artificial green coloring.
7. Cinnamon flavoring.
8. The antifoaming agents—butter, oleomargarine, lard, corn oil, coconut oil, cottonseed oil, mono- and diglycerides—in a quantity not greater than reasonably required to inhibit foaming.

For generations, the housewife has made jelly by cooking fruit, pouring it into a cloth bag to drain, adding sugar to the clear juice, cooking it again, pouring it into glasses and letting it stand undisturbed until it cooled and set. Then she tied a paper around the top of the glass to protect the jelly. Making jelly was an art, and if she were skillful, the jelly would keep for months although mold sometimes grew on the surface.

The housewife's success was the result of trial and error. Experience was passed on to her daughters and her neighbors. The jelly was always edible even though it might be the consistency of molasses, or that of a solid with a limpid liquid oozing out of it. The jellies were put into glasses wider at the top, and a good jelly would slide out into a dish, retain the shape of the glass, give to slight pressure and sustain a cut surface without weeping or collapsing. It was plastic so that it could be spread evenly on bread.

The commercial jelly maker seeks the same goal as the housewife, but the shape of the container has been altered for greater convenience in closing so that it can no longer be turned out into a dish, but must be served in the glass or dipped out.

All the technical knowledge of the chemist has been obtained in the same manner as that of the housewife but with better tools and control. Pectin, acid, and sugar are the requisites. Commercial jellies contain over 50% sugar, and some, 65–70%.

The housewife knew that in order to gel, the mixture must be acid; very sour apples or crabapples were the easiest fruit to make jelly from. Grapes and blackberries had to be picked before they were quite ripe. Chemists investigating this problem found that the kind of acid did not matter, hence the long list of permissible acids in the standard. But whatever the acid, the pH of the jelly has to be between 3.2 and 3.5. Many fruits are not that sour and acid must be added. If the jelly is too acid, it develops syneresis.

The pectins vary in composition and properties and must be extracted from the fruit with the least change possible.

Apples, crabapples, grapes, and some other fruits contain both enough pectin and acid to make a good jelly. Strawberries contain enough acid but are low in pectin; sweet cherries, quinces and several other fruits have enough pectin but are too low in acid content. With all fruits, the stage of ripening is a factor in both pectin and acid content change. The chemist will find procedures in the Official Methods for the determination of acid, pectin, and pH.

The viscosity of the juice helps to determine its setting qualities and may prove to be a better guide than the amount of pectin present.

Since clarity is one of the essential characteristics of jelly, the housewife knew that squeezing the jelly bag produced a cloudy juice by forcing insoluble material through the bag, but it gave a greater yield of jelly. The commercial jelly maker is faced with the same problem. He cooks the fruit with water. Solid fruits, such as apples and quinces require more water than the softer grapes and berries. After the cooking has converted the protopectin to pectin and softened the connective tissue, the fruit is drained in cloth bags or pressed. In the case of apples, the press cake may contain enough pectin to warrant breaking it up, adding water and pressing again.

If the fruit has been pressed, the juice is clarified, which may be done with a centrifuge, or it may be necessary to add a filter aid to adsorb the suspended matter and then remove it with a filter press.

The clarified juice is tested and adjusted by the addition of acid or pectin and transferred to the boiling kettle. Sugar is added in an amount that depends on the quality of the pectin; too much sugar produces a soft, weak gel and too little makes it tough and rubber-like. Furthermore, the sugar is the preservative unless sodium benzoate is added.

After the composition is established, the juice is boiled. During the boiling, fruit protein is coagulated and gathers on the surface as a scum that must be skimmed off or it will cloud the jelly.

The purpose of this second boiling is to evaporate water to a concentration at which the gel will set at room temperature. The boiling should not be continued longer than necessary because it removes the more volatile flavors, depolymerizes the pectin and may affect the color. Since the sugar is the main solute present, the proper concentration is reached when the sugar content reaches 65%, which can be determined by measuring the refractive index, the specific gravity, or the boiling point. According to Raoult's laws, sucrose, dextrose, invert sugar, and corn syrup solids of the same concentration should have different boiling points, but apparently a solution with the physical properties of a 65% sucrose solution is all that is required for proper gelation.

The beginner in jelly making or the expert faced with an unfamiliar fruit will need to make some laboratory tests, for a jelly failure is difficult or impossible to correct. The housewife tested the finishing point by cooling a little of the mixture in a spoon, a test that no chemist should disdain.

Pectin

For both home and commercial jelly making, there are two pectin preparations available if extra pectin is needed. Apple pectin is a byproduct of the cider or vinegar industry; it is separated from the pomace and purified. The product sold to the public is a fluid containing 4–5% pectin and 2–3% lactic acid. It usually has a strong apple flavor. For commercial use, the pectin is more highly purified and dried to a powder.

The other commercial sources of pectin are the citrus industries, where tons of orange or grapefruit peel are available from the juice and canning plants. The pectin is released from the protopectin of the rind with sulfurous acid, purified and dried.

The dried pectins are sometimes marketed as jelly powders that contain about 4% pectin, 8% citric acid and dextrose.

Artificially Sweetened Fruit Jelly

There is a standard for jellies made without added sugar, and since sugar is essential to the setting of pectin gels, the standard contains substances not permitted in normal jellies.

The gel is produced by pectin, agar-agar, carob bean gum, karaya gum, gum tragacanth, algin, sodium carboxymethyl cellulose, carrageenan, or any combination of them.

The gums may require the calcium ion to form a gel. Therefore, six calcium salts are optional, as well as several sodium or potassium salts, ascorbic acid and its salts, benzoic acid, and two other preservatives. The sweetener is saccharin.

There is a similar standard for artificially sweetened jams and preserves.

The labeling requirements for these artificially sweetened products are specific and detailed.

EMULSIONS

An emulsion is a colloidal dispersion of one liquid in another in which it is insoluble. Three substances are required to produce an emulsion: two immiscible liquids and an emulsifying agent. A simple illustration is an emulsion of oil in water with soap as the agent. To make it, pour any oil (kerosine, corn oil) into a bottle, add water and shake the bottle violently; the oil will become dispersed in droplets throughout the water, but if it is allowed to remain undisturbed for a few minutes, the dispersion separates into two layers because of the difference in densities. Now add soap and agitate it again and the oil remains dispersed in the water—the soap is the emulsifying agent. An emulsion is a two-phase system. In this case, the oil is the *internal* or dispersed phase, the water the *external* or continuous phase.

Commercial soaps are mixtures of the sodium salts of the acids of the fats from which they are made. Since oleic acid is common to all fats, its salt, $C_{17}H_{33}COONa$, will serve to illustrate one theory of emulsification. There are several theories, but the *oriented-wedge* theory seems the most likely one in the above illustration.

The sodium oleate consists of a hydrocarbon radical, $C_{17}H_{33}$-, which is soluble in the oil and the -COONa radical, soluble in the water, therefore, the hydrocarbon dissolves in the oil but leaves the salt radical in the water. The salt radical ionizes to $-COO^-$ and Na^+. This leaves a negative charge surrounding the drop of oil and so the droplets repel each other and do not coalesce.

In the food field, unfortunately, the emulsions are not as simple as the one just described unless we include the action of detergents in cleaning the equipment.

The most common emulsion in the food industry is milk; the fat is the internal phase and the water the external. There is no simple polar compound, such as soap, present in milk; the protein appears to be the emulsifying agent. Other food emulsions are the fat in egg yolk, the oil in olives and avocados and many composite foods of which the salad dressings are the most obvious.

There are two types of emulsions: oil-in-water and water-in-oil, that is, either liquid may be the internal phase. There are two ways to distinguish between the two types. If water is added to an oil-in-water emulsion, they mix completely, whereas a water-in-oil emulsion will form a layer on the surface of the water. The other method is to add a dye that is soluble in one phase and not in the other, then examine the emulsion with a microscope. An oil-soluble dye added to an oil-in-water emulsion will produce colored dots, but with a water-in-oil emulsion, the color will be continuous.

Most of the research on emulsions has been done on model systems, such as benzene or other hydrocarbon and water with a soap or other known compound as emulsifier. Food emulsions are much more complex and require more study.

Emulsoid colloids, such as the proteins and gums, which are common emulsifiers in the food industries, are invariably mixtures of several compounds. Likewise, the oil may be a pure fat, but fats are mixtures, and the aqueous phase generally contains solutes that may affect the emulsion—sugars may, salts do.

The interfacial tension between the two liquids is important: the less tension, the easier it is to emulsify the liquids. This gives rise to the *surface tension* theory of emulsification, which supposes that the emulsifying agent lowers the surface tension between the two liquids and concentrates at the interface. A 2% solution of a soap was found to lower the surface tension of water from 72 to 33.5 dynes/cm and the interfacial tension between water and a petroleum oil was only 8 dynes/cm. Salts raise the surface tension of water slightly and make emulsification more difficult.

The interfacial tension between water and an immiscible liquid is always less than that between water and air. Table 8.1 illustrates this general statement.

TABLE 8.1

SURFACE AND INTERFACIAL TENSIONS

Liquid	Surface Tension Dynes/cm at 20°C	Liquid	Surface Tension Dynes/cm at 20°C
Benzene	28.88	Ethyl alcohol	22.2
n-Butryic acid	26.8	Ethyl bromide	24.2
Chloroform	27.14	Glycerol	63
Carbon Tetrachloride	26.77	Water	72.75
Ethyl acetate	24.3		

Interfacial Tensions vs Water

Liquid	Surface Tension Dynes/cm at 20°C	Liquid	Surface Tension Dynes/cm at 20°C
Benzene	35	Carbon disulfide	48.36
Chloroform	32.8	Ethyl bromide	31.20

The surface tension theory explains the emulsions that can also be explained by the oriented wedge theory. The two theories are not contradictory because the wedge theory simply details the arrangement of the molecules of the emulsifier at the interface.

A third theory assumes that the emulsifying agent forms a film around the dispersed droplet and thus prevents coalescence, surface tension not entering into the explanation. This may be a correct explanation because emulsoid colloids are readily adsorbed and do have protective action for metal suspensoids such as those of gold and silver. The drug Argyrol is a colloidal dispersion of metallic silver protected from coagulation by a derived protein. Emulsions may owe their stability to similar protective action.

Making and Breaking Emulsions

In making an emulsion, the emulsifier, which may be less than 1% of the weight of the finished emulsion, is added to the dispersing liquid, usually water or an aqueous solution, such as vinegar. Then the liquid to be dispersed is added and the mixture agitated. No general rules apply, but work with several systems indicates that if the second liquid is added gradually, the time required to form the emulsion will be less. Also, the greater the proportion of the dispersed liquid the greater the time required for emulsification.

The chemist does not always have a choice of either the materials or the amount of them in making an emulsion. The official standard for mayonnaise specifies not less than 65% vegetable oil in vinegar, not less than 2½% acetic acid or lemon or lime juice and not less than 2½% citric acid. The emulsifier is egg yolk. Several flavors, colors or preservatives are permitted, which may affect the emulsion.

The standard for French dressing requires only 35% oil. The concentration of acid is not specified but it may be either vinegar, lemon juice or lime juice. More choice is allowed in emulsifiers, which may be egg yolk, gum acacia, carob bean gum, guar gum, gum karaya, gum tragacanth, extract of Irish moss, propylene glycol ester of alginic acid, xanthan gum, sodium carboxymethyl cellulose, methyl cellulose or any mixture of any two or more of them. Various flavors, colors and preservatives are permitted.

The standard for salad dressing is much the same as that for French dressing except that it may contain a starch paste made from a food starch or from tapioca, wheat or rye flour or any two or more of them.

The method and violence of agitation are important factors. In the emulsion of benzene and water, Thomas (1934) reported that it took less than 1 min to process a 30% by volume emulsion, but 3 min for a 50% emulsion and 40 min for a 95% emulsion, all by the same manner of agitation. Intermittent shaking with rest periods was more effective with the benzene–water emulsion. Whereas, an 80% emulsion required 15 min of continuous shaking, but only 3.5 min when shaken by hand with 30 sec rest periods.

The manner and violence of the agitation affect both the stability of the emulsion and the time required to make it. Violent agitation breaks up the continuous phase as well as the dispersed phase and may prevent emulsification altogether. The smaller the droplets, the more stable the emulsion, but it does not follow that stability increases with the violence of agitation.

Small scale emulsions can be made with some systems by grinding the oil with the emulsifier, such as gum arabic, in a mortar until an intimate mixture is obtained and then adding the water and stirring gently.

Since the ease of emulsification is a function of the interfacial tension, it increases with increase in temperature. It has been claimed that the exact temperature is important in the formation of some emulsions.

A chemist who proposes to make a new salad dressing must first make a laboratory study of the best conditions for producing a stable emulsion of the materials he plans to use. The gums, the proteins, the lecithins, and the mono- and diglycerides are the principal emulsifiers used in the food industries. The GRAS list contains a few others so the food chemist has several emulsifiers to choose from.

Not all emulsions are desirable; some are a nuisance. The petroleum chemist frequently encounters emulsions of salt water and petroleum. Engine condenser water may contain a low percentage of emulsified oil, which renders it unfit for boiler water. Where steam distillations are used, as in the manufacture of aniline or turpentine, the liquids sometimes emulsify and are difficult to separate.

Breaking an emulsion varies from very easy to almost impossible. Gentle agitation may break an emulsion; mayonnaise sometimes separates when subjected to rough handling.

The manufacture of butter is an example of breaking an emulsion on an industrial scale. The milk fat has been concentrated in the cream by standing or by the use of a centrifuge, because the density of the fat is less than that of the water. When the cream is agitated, the fat globules coalesce into huge clumps of butter. The agitation of a churn is much less than that of an emulsifier, such as a milk homogenizer.

A factor in the stability of some emulsions is the pH of the continuous phase. Oil-in-water emulsions are most stable around pH 4.2 if made with gum arabic, but at pH 2.5 with gum tragacanth. The effect of pH on the making and breaking of emulsions has not been studied thoroughly, but it has been known for centuries that the sourness of cream affects the ease of churning. Food systems that include vinegar or fruit acids need individual study for the preparation of emulsions.

In case the emulsifying agent is unknown, which is likely in unwanted emulsions, trial and error is the only possible procedure. One may try:

1. Changing the pH.
2. Gentle agitation.
3. Centrifuging.
4. Freezing.
5. Heating.
6. Adding a liquid in which both phases are soluble.

7. Filtration through a fine powder.
8. Destruction of the emulsifying agent.

Analysis seldom helps because of the small amount of the emulsifier in most cases and the difficulty in recovering any of it.

When all else fails, just let the emulsion stand exposed to the atmosphere and hope for the best.

FOAMS

Colloidal dispersions of gases in liquids are called *foams*. Like the emulsions, some are desirable (meringues, marshmallows) and others are a nuisance.

Pure liquids do not foam; boiling water brings bubbles to the surface but they break instantly, and every cook knows that boiling vegetables foam and overflow the pot.

Whipped cream is a foam that lasts for hours. Obviously the stability of a foam depends on the composition of the system. A liquid containing an emulsoid colloid forms the most stable foams.

A type of fire extinguisher that produces a foam is useful where tanks of oil or other flammable liquid are a fire hazard. The system consists of water, saponin, sodium bicarbonate and an aluminum salt in an acid solution. The acid and bicarbonate generate CO_2, which produces a foam that is stabilized by the saponin and the aluminum ion so that an atmosphere of CO_2 floats on the surface of the liquid and excludes the air, whereas water, which is more dense than most flammable liquids, simply sinks to the bottom of the tank so the liquid overflows and spreads the fire.

Marshmallow is a foam stabilized by gelatin, which produces the most foam at pH 5 where gelatin is the least soluble; its isoelectric point is between 4.4 and 5.6.

Meringues are foams stabilized by egg albumin, which has an isoelectric point between 4.6 and 5.0, and in the film, the heat denatures the albumin and stabilizes the foam.

The foam on champagne or soft drinks lasts only long enough to indicate the presence of CO_2, but the foam on beer is more important than a chemist would suspect, the importance being psychological, of course. The foam is caused by the escape of bubbles of CO_2 from a solution containing dextrins and a gum extracted from the hops. I once saw a man at a rest station on a golf course pour beer into a waxed paper cup. No foam formed and so he threw the beer away, but later drank the same brand from a glass in the club house.

Several liquids will destroy foams; ether, toluene, the aliphatic alcohols, C_6–C_9, and some oils break most foams.

Some foams, such as that of boiler water, are stabilized by solid particles and can be prevented by filtering the water.

ADSORPTION

Some substances adhere to solid surfaces much more firmly than others, even though there is no apparent chemical action between the solid and the adhering substance. In many cases, the force of attachment is very strong and apparently the result of a secondary valence of some kind.

This adhesion is called *adsorption*. Since a solid in the form of a fine powder has a large surface, most of the research on adsorption has been done with such powders.

The idea is not new, for boneblack has been used to remove the colored substances from raw sugar since the beginning of the 19th century.

Research on adsorption was stimulated by the demand for gas masks after the Germans introduced chlorine into warfare on April 22, 1915. The masks that were soon developed contained activated carbon and soda lime. The latter removed acids, but the carbon adsorbed the various war gases that were used. Research soon disclosed that charcoals differ enormously in their ability to adsorb. Charcoals made from peach seeds or coconut hulls were the most effective, but further research developed a method for activating the common wood charcoals. Such products are now referred to as *activated* carbons. They became available to industry shortly after World War I.

The use of animal charcoal, or boneblack, in sugar refining and of active carbon in gas masks indicate that adsorption is selective; sugar is not adsorbed by the former and the gases of the air were not absorbed by the latter.

When butter takes up the odor of onions, the odorous substance penetrates the butter; such action is called *absorption*. In the case of fine powders, it is difficult to tell whether the substance involved adheres to the surface or penetrates the particle, consequently, some authors evade the question by calling the process *sorption* without any prefix.

When substances are adsorbed from solution, the pH of the solution affects the degree of adsorption and the effect differs with the adsorbent. In an exhaustive study of the effect of pH on adsorption, Thomas (1934) reported results with silica, alumina, iron oxide, and fuller's earth. One striking result was that silica adsorbed 94.5% of the piperidine (pH 11.5) from an aqueous solution but no aspartic acid (pH 3.1), while alumina adsorbed only 5.7% of the piperidine but 84.5% of the aspartic acid.

The food chemist is usually concerned with the adsorption of emulsoid colloids to clarify or decolorize solutions. There does not seem to be any general rule to cover such operations. The adsorption of proteins is greatest near their isoelectric point. For gums, there is no rule; boneblack adsorbs little or no gum arabic from solution but hydrated alumina adsorbs it readily. Trial and error is the only procedure.

The adsorption of compounds of high molecular weight has not been thoroughly investigated and any available information is the result of trial and error experiment. In general, adsorbants differ in their ability to adsorb different substances. The pH of the system affects the action of some of the adsorbents more than others, and the least soluble substances are the most readily adsorbed. Each clarification is a separate problem.

Chemists may have occasion to use gas masks and the mask must be adapted to the hazard involved.

DUSTS

Smoke and dusts are colloidal dispersions of solids in gases. Being solids, the particles do not vibrate nearly as rapidly as molecules of a gas, and so they are usually not completely stopped by gas masks, for some of the solid particles pass through the layers of adsorbent without touching it. Neither are they readily washed out of the gas by water even though they are soluble. A water shower may remove all the sulfur dioxide or chlorine from the air, but it will not remove all the smoke or dust because some of the particles never touch the drops of water. As a result of this lack of vibration of solid particles, the removal of industrial dusts and smokes is a difficult problem.

BIBLIOGRAPHY

ADAMSON, A. W. 1967. The Physical Chemistry of Surfaces. John Wiley & Sons, New York.

ANON. 1973. Carageenan. J. Food Sci. 28, 367–368.

ASSOC. OFFIC. ANAL. CHEMISTS. 1975. Official Methods of Analysis, 12th Edition. Assoc. Offic. Anal. Chemists, Washington, D.C.

BEATON, G. H., and McHENRY, E. W. 1964. Nutrition, Vol. 1. Academic Press, New York.

BECHER, P. 1965. Emulsions: Theory and Practice. Reinhold Publishing Corp., New York.

DEFAY, R., PRIGOGINE, I., and BELLEMANS, A. 1966. Surface Tension and Absorption. John Wiley & Sons, New York.

GLICKSMANN, M. 1962. Utilization of natural polysaccharide gums in the food industry. In Advances in Food Research, No. 11, E. M. Mark and G. F. Stewart (Editors). Academic Press, New York.

GRAHAM, H. D. 1971. Determination of carboxymethyl cellulose in food products. J. Food Sci. 36, 1052–1055.

LAWRENCE, A. A. 1973. Edible Gums and Related Substances. Noyes Data Corp., Park Ridge, N.J.

LEWIS, W. K., SQUIERS, L., and BROUGHTON, G. 1943. Industrial Chemistry of Colloidal and Amorphous Materials. Macmillan Co., New York.

MATZ, S. A. 1962. Food Texture. Avi Publishing Co. Westport, Conn.

OSIPOW, L. I. 1962. Surface Chemistry: Theory and Industrial Applications. Reinhold Publishing Corp., New York.

SMIT, C. J. B., and BRYANT, E. F. 1968. Ester content and jelly pH influences on the grade of pectin. J. Food Sci. 33, 262–265.

THOMAS, A. W. 1934. Colloid Chemistry. McGraw-Hill Book Co., New York.

CHAPTER 9

Food Additives

Additives have been recognized as a problem in this country since about 1900 when Wiley began to investigate them. The history of this early problem has appeared elsewhere (Garard, 1974).

Under the law of 1906 and also that of 1938, the FDA had to prove that an additive was harmful in any amount in order to be considered an adulterant. In 1958, the law was amended in several profound respects. The burden of proving safety was transferred to the person who proposed the use of the additive. The *per se* doctrine (harmful in any amount) was eliminated, and a toxic substance can be used provided it is safe in the amount in which it is contained in the food. A third provision, added by the Delaney amendment in 1960, however, prohibits the use in any amount if the substance has been found to cause cancer in any animal species.

Toxicological tests are animal experiments, may take a year or more, and are very costly, so food processors were in a quandary. Here were salt, pepper and a host of other substances that had been used for centuries and were essential components of many foods on the market, but they were additives and were subject to proof of safety.

The law provided that a substance could be used without testing if it had been in use before January 1, 1958, and was considered safe by experts qualified by training and experience to evaluate its safety. The FDA selected 800 "experts" who drew up a list of substances generally recognized as safe (GRAS). By March 1961, the list numbered 768; it keeps changing by addition and subtraction of substances but it is now much longer.

OFFICIAL DEFINITION

Food additives include all substances not excepted by Section 401 of this Act (1958), the intended use of which results, or reasonably may be expected to result, directly or indirectly either in their becoming a component of food or otherwise affecting the characteristics of the food.

Space does not permit the presentation of several pages of definitions and regulations. The chemist who is concerned with additives should acquire a copy of the *Code of Federal Regulations,* which contains all regulations, a list of GRAS substances and detailed regulations of many of these substances.

Current changes in regulations and proposed changes are published in the Federal Register. Also, over 100 reports on the various items under investigation are available and may be bought from the FDA for present prices from $2.75 to $5.45. A few of the titles in this list are: butylated hy-droxyanisole, ascorbates, corn sugar, gelatin and sorbitol.

GRAS—1973

The list of substances generally recognized as safe is too long for complete inclusion here, but the descriptions of the groups indicate the nature of the substances that are excluded from testing.

Anticaking Agents

There are seven of these and the limitations are specified for each; for example, table salt may contain up to 2% aluminum calcium silicate, and baking powder may contain up to 5% calcium silicate.

Chemical Preservatives

This list contains 31 substances, but not all of them are for general use. Caprylic acid is limited to cheese wraps and the sulfites cannot be used on meats. Sodium benzoate is a general preservative but is limited to 0.1% of the weight of the food. Ascorbic acid has no limitations.

Emulsifying Agent

There are 9 of these and so the choice of emulsifiers is not very great. Moreover, 5 of the list are limited to use in dried egg white. The list includes mono- and diglycerides for general use.

Nonnutritional Sweeteners

The GRAS list of 1961 contained 8 nonnutritional sweeteners. Four of these were cyclamates and the other four were saccharin and three of its derivatives. The cyclamates were removed from the list in 1969 and sac-charin is now under investigation. Saccharin has been a controversial ad-ditive since the days of Wiley and the Poison Squad.

In August 1974 the FDA approved a new sweetener called *aspartame*, which is said to be 180 times as sweet as sucrose. The compound is a di-peptide of aspartic acid and phenylalanine and as such, has the caloric value of a protein. The chemical name is L-aspartyl-L-phenylalanine methyl ester.

Nutrients: Dietary Supplements

This group contains 81 items, mainly amino acids, vitamins, salts of calcium, copper, iron, magnesium, manganese, potassium, sodium, zinc, and a few miscellaneous substances. Mannitol (5%) and sorbitol (2%) are permitted in special dietary foods. Cuprous iodide and potassium iodide, each to the extent of 0.01%, are limited to use in table salt.

Sequestrants

There are 29 sequestrants, which are mostly calcium or sodium salts of organic acids. Calcium chloride and two calcium phosphates are included as well as tartaric acid, citric acid, and two of its esters.

Sodium thiosulfate is limited to 0.1% in salt. Isopropyl citrate is limited to 0.2% and stearyl citrate to 0.15% of the food product.

Sequestrants are used to suppress the activity of some ingredient that is objectionable but practically unavoidable. For example, in the manufacture of soft drinks, iron in the water may react with one of the flavor ingredients to form an insoluble compound that will make the product cloudy.

In some cases, the sequestrant renders the substance insoluble, in others it forms complex soluble compounds. The citrates form soluble complexes with iron. Each sequestration is a special problem to be solved. The chemist must know what is to be sequestered and which of the GRAS list will be the most effective and economical.

Stabilizers

There are 12 stabilizers. They are gums or mucilages, such as agar agar, gum arabic and sodium alginate. The purpose of stabilizers is to prevent the separation of a mixture. These gums are protective colloids and serve to keep oils dispersed in an aqueous medium.

The classification of additives, like all other classifications, is arbitrary and simply a matter of convenience. The line between emulsifiers and stabilizers is a fine one and many of the latter also serve as emulsifiers. However, those substances listed as stabilizers serve purposes other than emulsification. Such mixtures as ice cream and other frozen desserts, gravies, cottage cheese, and the like tend to separate and a stabilizer, such as egg white, gelatin or a gum will prevent it.

The food chemist must be aware that although the GRAS list may place no limits on the amount of an additive that can be used, the food standards often do.

General Purpose Additives

There are 81 additives that are used for miscellaneous purposes that defy classification. Some of them have quantitative limits. Caffeine may be used only in cola drinks and there the amount may not exceed 0.02%. Ethyl formate may be used to fumigate cashew nuts, but only 0.0015% may remain on the product. Nitrous oxide may be used as a propellant for certain dairy and vegetable-fat top dressings. For various uses, nitrogen, monoammonium glutamate, papain, rennin, and baking powder chemicals are all members of this group.

Spices and Other Natural Seasonings and Flavorings

There are 90 of these seasonings including allspice, caraway, cinnamon, cloves, and all the other common ones. Then there are several uncommon ones, such as alfalfa seed, chamomile, fenugreek, marigold and rue. If you can name a spice, it is probably in this list.

Essential Oils and Natural Extractives

There are 172 essential oils and extracts. Many of them are the flavorings from the spices of the preceding group, although there are several items in each of these two lists that do not occur in the other. In some cases, there are several products with approximately the same flavor. For example, there is the oil of cinnamon bark from Ceylon, China, and Saigon and of cinnamon leaves from the same three sources.

Miscellaneous

These are 17 substances that are sometimes used with flavors or seasonings, such as algae, kelp, quince seed, ambergris, and musk.

Trace Minerals Added to Animal Feeds

The animal's diet is not neglected. There are 45 salts of cobalt, copper, iodine, iron, manganese, and zinc that may be used to add those trace elements to animal feed.

Synthetic Flavors

There are 26 of these. Vanillin is the most familiar member of the group.

Components of Packaging Materials

These are in two subgroups: (a) substances that migrate from paper and paper board products, and (b) those that migrate from cotton fabrics used in packaging. There are 67 substances in (a) and 41 in (b). This is a heterogenous list of substances some of which have been included in previous lists. Among these are calcium chloride, cornstarch, glycerine, sodium chloride and sucrose. Many of the items will be a great surprise to many a food chemist because they would not be suspected in packaging material. However, many food chemists are concerned with packaging their products, and the law covers contamination of food from the container. Wooden containers deserve investigation for the wood may be treated with a preservative against decay, and also some woods impart flavors from their natural components, turpentine and rosin, for example.

Plastic packaging material may contain objectionable compounds that may migrate into the foods in which they are wrapped.

LABELING

Section 201 of the Federal Food, Drug and Cosmetic Act of 1966 refers to misbranding, which does not appear to be the concern of the chemist, but when the section is read, it proves to be of vital concern. In part, it reads:

> ... in determining whether the labeling is misleading there shall be taken into account (among other things) not only representations made or suggested by statement, word, design, device, or any combination thereof, but also the extent to which the labeling fails to reveal facts material in the light of such representation of material with respect to consequences which may result from the use of the article to which the labeling relates under the conditions of use prescribed in the labeling thereof or under such conditions of use as are customary or usual.

All this seems vague, but the FDA is empowered to make regulations for its enforcement, and those regulations are specific and detailed. For example, in dietary foods, any statement of the quantity of Vitamin A must be biologically measured.

An infant is under 12 months old, a child is between 12 months and 10 years and an adult is over 11 years old.

There are long and involved regulations for vitamins A, B, C, D, riboflavin, and niacin. There are also regulations of calcium, phosphorus, iron, and iodine. Infant foods also come in for many specific regulations.

Good Manufacturing Practices

The law and the regulations refer in several places to good manufacturing practices and the criteria as set forth applies mainly to sanitation of plants where food is prepared, packaged or held. The grounds around the plant must be free from improperly stored equipment, litter, waste, refuse, weeds, or grass that may attract or harbor insects, rodents, or other pests. No dusty roads, yards or parking lots, and no inadequate drainage are permitted.

Water supply, sewage, plumbing, toilet facilities, hand washing facilities, rubbish and general maintenance of equipment and the health of the personnel along with the food and packaging are all considered and subject to inspection by federal inspectors and probably by state and local inspectors also.

In addition to the general regulations, some apply only to certain industries, for example, fish and seafood products and thermally processed low-acid foods packaged in hermetically sealed containers.

The FDA has a large and active staff that continually studies all features of food processing, handling, display and sale. The law specifies four ways of adulterating food, eleven ways of misbranding it and provides for the establishment of standards and the control of additives. As new information warrants change, the new regulation is published in the Federal Register, which means that the duties of the food chemist are subject to sudden change.

ANALYSIS

The long list of additives created a sudden and enormous problem for the analytical chemists, but the AOAC has made a good beginning in the devising of methods of analysis for unusual substances. There is a polarographic method for fumaric acid and spectrophotometric methods for boric and sorbic acids. Qualitative tests have been devised for the antioxidants, propyl gallate, butylated hydroxyanisole, butylated hydroxytoluene and nordihydroguairetic acid, although the last of these has now been removed from the GRAS list.

Benzoic acid and sodium benzoate have been used as preservatives for a century or more and the AOAC has three methods for their determination. In the chemical method, the acid is extracted with chloroform, the solvent evaporated and the residue titrated with standard NaOH solution. The methods for benzoic acid are very important, for the preservative is widely used in foods and its use is limited to 0.1%.

There are qualitative or quantitative methods for all the traditional preservatives both legal and illegal, such as borates, dehydroacetic acid, formaldehyde, formic acid, hydrogen peroxide, monochloroacetic acid, nitrites, propionates, quaternary ammonium compounds, sulfites, and thiourea.

Chapter 20 of the 12th Edition of the Official Methods contains both qualitative and quantitative methods for the more common additives. Unfortunately, additives, including the vitamins, do not belong to a single class of chemical substances. Therefore, there is no possible test for *an additive;* each one must be tested for separately. This may be a long and difficult task; there are 31 preservatives on the GRAS list for instance, along with a host of others that are not on the list. More than 150 preservatives for use in food have been patented, although many of them were never in commercial use, but many have been used and some of them are still available.

Exposure of a food sample under the same conditions as one of known composition may help the chemist to locate a preservative if the difference in the rate of spoilage is very great, but it is not always possible to arrange such a test. Milk can be so tested by collecting a sample from a milk plant personally. Soft drinks and prepared mixes can be duplicated, but canned and many bottled products are difficult.

It is very difficult to find rate of spoilage data in the food literature that is applicable to the problem the chemist may have. Unless finding the preservative is imperative, the chemist can only test for the most likely ones.

If he wishes to add a preservative to his own product, he has the GRAS list to select from.

The appearance of a method for a new additive in the Official Methods supplement is a good indication that the additive is coming into general use.

In addition to the general analytical methods for food additives, methods for many of them are among the methods for the specific foods. For example, in the methods for the analysis of flour, there are procedures for the determination of bromates and iodates; consequently, the first place to look for an analytical method for an additive is among the methods for the analysis of the food in question.

PESTICIDE RESIDUES

The chemist concerned with fruits and vegetables faces the problem of pesticide residues. Chapter 6 of the 12th Edition of the Official Methods contains procedures for the analysis of pesticide formulations, and Chapter 29 for the determination of such residues in foods.

Washing fruits and vegetables does not remove all the pesticide, and if the food is blanched prior to further processing, more is removed in the blanching, but not all of it. One investigation showed that washing spinach removed 17% to 48%. Blanching still left 40% to 62% of the original DDT and 29% to 51% of the parathion on the spinach. With green beans, blanching removed 50% of the DDT and 71% of the malathion present on the raw product.

Washing and blanching processes vary and the pesticides differ in solubility and toxicity; consequently, the chemist needs to examine the pesticide content of the raw product and the effectiveness of his processes in bringing the final product within the legal tolerances.

COLOR

The addition of a coloring agent to foods is as old as history. The first coal tar dye was manufactured in 1857 and the colors used thereafter were either natural dyes, synthetic dyes or inorganic substances. Investigation that preceded the passage of our first food law in 1906, disclosed copper salts used to color green peas. All through the 19th century, inorganic salts were common coloring agents. Cheese was colored with vermillion, HgS, and red lead, Pb_3O_4; and the yellow lead chromate, $PbCrO_4$, was common in candy—the inorganic salts were cheaper than the natural dyes.

SYNTHETIC COLORS

After 1857, the number of coal tar dyes increased rapidly, and because of their great tinctorial power they became cheap sources of food color. But many of them are toxic, and regulations were imperative. The National

Academy of Sciences reported in 1971 that about 65 synthetic dyes are known to be in use: 33 in Denmark, 25 in Great Britain, 22 in Japan, 22 in the European community countries, 12 in the United States, 10 in Canada, 5 in Chile, 3 in the Soviet Union and none in Greece.

If the chemist is concerned with imports or exports of food that is or may be colored, he must be familiar with the color regulations of the country of origin or destination. Even with our smaller list, some of the dyes permitted in the United States may be banned in the country of export.

The regulation of synthetic dyes in foods in this country began about 1900. At that time, there were 695 of these dyes on the market. A long investigation reduced the list for foods to 7 dyes, which were reported in the first Food Inspection Decision issued July 13, 1907. From that day to the present, the list has increased and decreased as new information on the various dyes became available. The process still goes on. As of 1976, there were 7 synthetic dyes permitted in foods:

Blue No. 1	Red No. 3
Blue No. 2	Violet No. 1
Green No. 3	Yellow No. 5
	Yellow No. 6

Four more were given provisional listing. They were:

Green No. 1	Orange B
Green No. 2	Red No. 4

One, Citrus Red No. 2, is permitted for coloring the skins of oranges only.

Originally, these dyes were known by names assigned to them in the color industry, such as amaranth (Red No. 2), erythrosine (Red No. 3), tartrazine (Yellow No. 5) and sunset yellow (Yellow No. 6). After the passage of the Food, Drug and Cosmetic Act, the FDA assigned the names and numbers to them that are now in use. Nobody uses the true chemical name except the manufacturer and the FDA who use it for identification. The reason is obvious when we note that Red. No. 3 is the disodium salt of 9-o-carboxyphemyl-6-hydroxy-2,4,5,7-tetraiodo-3-isoxanthone.

Yellow No. 5

These dyes are made in batches and each batch is analyzed and certified, for although the dye may be harmless, it may contain harmful intermediates from the synthesis that have not been removed.

Specifications and uses are provided by the FDA. For example, the information on Yellow No. 5 in somewhat condensed form is:

The identity is indicated by the chemical name. Color additive mixtures for food may contain only those diluents listed in subsection D (*Note:* list of 17, including fruit juices, dried algae meal and annatto extract).

Specifications.—Volatile matter (At 135°C) not more than 6%.
Chlorides and sulfates (calculated as the sodium salts) not more than 7%.
Water insoluble matter, not more than 0.2%.
Phenylhydrazine-p-sulfonic acid, not more than 0.1%.
Other uncombined intermediates not more than 0.2% each.
Subsidiary dyes, not more than 1%.
Lead as Pb not more than 10 parts per million.
Arsenic, (As As_2O_3) not more than 1 part per million.
Pure color, not less than 87%.

Uses and Restrictions.—May be used for coloring foods generally, subject to the following restrictions: (1) not over 300 ppm on weight of food. (2) may not be used to color foods that have a standard of identity unless it is included in the standard.

Labeling Requirements.—These are elaborate and relate to the dye itself.

Certification.—All batches of FD & C Yellow No. 5 shall be certified in accordance with regulations in Subpart A of this part.

Other Dyes

In addition to the list of dyes certified for use in food, there are other synthetic dyes that may be used in drugs, cosmetics and animal feeds. There are also over 20 dyes, mostly natural colors that do not require certification. Some are restricted to certain uses; for example, ultramarine blue is restricted to 0.5% in salt intended for animal feed. Natural colors that may be used in human food are: annatto extract, dehydrated beet powder, caramel, B-carotene, carrot oil, cochineal extract (carmine), cottonseed flour (partially defatted, cooked and toasted), fruit juice, paprika, riboflavin, saffron, titanium dioxide (not to exceed 0.1%), turmeric and vegetable juice.

Use of Color

To the layman, the mere addition of color to food is adulteration. There is one case where it is, and that is when it is used to conceal inferiority or to make the product appear better than it is. For example, the addition of a yellow dye to a pasta would make it appear to contain eggs.

There are several legitimate reasons for adding color to foods. During processing, the color of many fruits and vegetables is largely destroyed and needs to be restored. Furthermore, some foods that are naturally white are unattractive. Margarine is naturally white or nearly so and butter may be a very pale yellow. Any housewife would consider either to be an inferior product and wouldn't dare serve it to guests.

Aside from restoring the traditional color of foods, dyes serve to identify many processed foods; cheddar cheese must be orange; lime candy, drink or pie filling, green; cola drinks must be brown; wintergreen candy must be pink and so on. Imagine an orange drink that is not an orange color! Many people think that an orange that is not a deep orange color is not fit to eat, although many oranges are fully ripe while the skin is partly or entirely green; valencia oranges after they are fully ripe and have been orange in color turn green again.

Each synthetic dye and some natural ones have limits on the amount added, but if a processor adds more than enough to produce the desired color effect, it is because of the carelessness of some employee, for too much color defeats its purpose—imagine bright red wintergreen lozenges or an orange colored butter. Moreover, the food colors are expensive—$40 to $50 a pound perhaps—and economic considerations are as effective for control as legal restrictions.

ANALYTICAL METHODS

Colors have been added to foods ever since there was a food law and before, and so their detection has been a problem in this country since 1906. Chapter 34 of the 12th Edition of the Official Methods contains procedures for the identification of synthetic dyes in foods.

Natural Colors

Since there are few quantitative requirements for the natural colors, there are no Official Methods for their quantitative determination, but there are methods for their separation and identification. The color is extracted from the food. Ether extracts carotene, xanthophylls, and chlorophyll from neutral solutions and alkanet, annatto, turmeric and some others from acid solutions. Amyl alcohol extracts logwood, archil, saffron, and cochineal.

The solvent is evaporated, the residue dissolved in water and various reagents added to portions to produce color changes. In addition to the color tests, there are special tests for chlorophyll, annatto, turmeric, archil, and caramel, but these also are color tests.

The tests for the natural pigments are not satisfactory, but the color of the dye itself limits the possibilities and the use of more than one dye in a processed food is unlikely.

Synthetic Colors

Some of the synthetic colors are fat soluble and some are water soluble. The former are extracted with 90% alcohol; the latter with water. In both cases, the solutions are purified by treatment with various solvents. There are specific tests for some of the dyes, but the general procedure is to dissolve the fat soluble dyes in chloroform and the water soluble dyes in water and then separate any dyes that may be in a mixture by chromatography and identify the dyes with a spectroscope.

The food dyes as purchased are from certified batches, and each dye has rigid specifications; consequently, there are Official Methods for the determination of volatile matter, water-insoluble matter, ether extract, sodium salts, mixed oxides, pure dye and any other item in a particular specification. There are Official Methods for over 50 dyes because those used in drugs, cosmetics and feeds are included. In each case, the dye is identified by both the FDA number and the number from the Color Index of the Society of Dyers and Colorists.

The FDA and color manufacturers are doing research constantly on the safety and suitability of dyes for use in foods, drugs, and cosmetics and the Official Analytical Chemists are doing their best to supply the necessary analytical methods, and so the chemist must follow the changes in both certification and analysis if he is concerned with the use of official colors.

ENRICHMENT

Although vitamins had been known for years, a survey of the state of nutrition in this country made in the 1930's disclosed a woeful lack of them in diets in general. Information was available, but through ignorance, disbelief, or carelessness, people were not getting vitamins, calcium, and iron in sufficient amount. Vitamin D was one of the greatest deficiencies; rickets was common. Babies were given vitamin D preparations, but the earlier fad of sun suits had passed and older children failed to get enough vitamin D.

Milk

By 1930, a dairy in New Jersey that produced certified milk increased the vitamin D content of the milk by feeding the cows irradiated yeast and, by 1932, irradiated milk was on sale in New York. In 1934, the evaporated milk industry was adding vitamin D to its products.

In 1974, evaporated milk could have 25 USP units of vitamin D per oz added to it. Plain condensed milk may be raised to the same level as the evaporated milk and nonfat dried milk may be fortified with both vitamins A and D so that when the milk is reconstituted according to directions, it contains 500 USP units of A and 100 of D in 8 oz of the reconstituted milk.

Vitamin D fresh milk is now marketed in all the states, and the usual requirement is 400 units per qt. The labels on the various milk products must inform the consumer that the product is fortified and when a product is so labeled, it must contain the legally specified amount of the vitamin. In some cases, a small excess is permitted so that the product will not fall below its standard during its shelf life.

The FNB specified allowances in International Units (IU), and the legal standards are in USP units. However, several years ago, the pharmacopoeia adopted the International Units so that now the two are the same.

Margarine

Aside from the milks, the only other food to be fortified with vitamin D is margarine. Both A and D are optional ingredients, but if A is added, the finished margarine must contain not less than 15,000 USP units per lb. Carotene is a permissable color for margarine and, since it is provitamin A, it may be included in the 15,000 units.

There is no specification of the amount of vitamin D that must be added, but manufacturers generally add 2,000 units per lb.

Other Enrichment

Nutritionists and public health officials have continually stated that any normal person can get all the essential nutrients from common foods, but it is no less a fact that most of the people in the world do not. Many prosperous Americans, even the well schooled, fail to be properly nourished. The poor are in the worst condition because they cannot afford a great variety of foods.

As soon as vitamin A was available, a New York baker made a vitamin A bread but it was not a commercial success and was soon abandoned. Then came a campaign to increase the use of whole wheat bread. Few liked the bread and the bakers were not enthusiastic about it because of insect infestation in the flour, and so that campaign failed.

The U.S. Public Health Service made a survey in the 1930s and found over 200,000 cases of pellagra and an uncounted number of beriberi and other deficiency symptoms, especially deficiencies of the B complex vitamins. They estimated that 75% of the diets in the United States were not good. The Surgeon General called a conference of nutritionists and others to try to decide what to do about the poor state of our nutrition, but it was not until May 27, 1941 that the government decided on a policy—they decided to enrich the flour. Although the consumption of flour had dropped to 199 lb per capita in 1940, white bread was still the major food of most families. Within a year, 70% of the flour and bread were enriched, but enrichment increased the price of the flour and so the project met with resistance and was about to go the way of the whole wheat campaign; then came the war.

In 1942, South Carolina passed a law requiring enrichment and in January 1943, War Food Order No. 1 required the enrichment of all white flour and bread. The Order was repealed in 1946, but by then, 19 states had passed laws requiring enrichment.

Although there is now no federal law requiring bread to be enriched, standards have been set up for eight enriched cereal products. The requirements are shown in Table 9.1.

TABLE 9.1

FEDERAL STANDARDS FOR ENRICHED CEREAL PRODUCTS

Cereal	B_1 Min (Mg)	Max	B_2 Min (Mg)	Max	Niacin[1] Min (Mg)	Max	D_2 Min (USP)	Max	Iron Min (Mg)	Max	Calcium[2] Min (Mg)	Max	Wheat Germ[2]
Bread	1.1	1.8	0.7	1.6	10	15	150	750	8.0	12.5	300	800	5%
Flour	2.0	2.5	1.2	1.5	16	20	250	1000	13.0	16.5	500	625	5%
Farina	2.0	2.5	1.2	1.5	16	20	250	—	13	—	500	—	8%
Macaroni	4.0	5.0	1.7	2.2	27	34	250	1000	13	16.5	500	625	5%
Noodles	4.0	5.0	1.7	2.2	27	34	250	1000	13	16.5	500	625	5%
Cornmeal	2.0	3.0	1.2	1.8	16	24	250	1000	13	26	250	700	—
Corn grits	2.0	3.0	1.2	1.8	16	24	250	1000	13	26	500	750	Dried yeast 1.5%
Rice	2.0	4.0	1.2	2.4	16	32	250	1000	13	26	500	1000	—

[1] Niacinamide may be used.
[2] Optional.

The enrichment of the wheat products attempted to restore the nutrient content to that of whole wheat.

In 1974 there was a proposal by the FNB to add vitamin A, vitamin B_6, folic acid, magnesium and zinc to the cereal enrichment list, and the FDA is proposing the enrichment of ready-to-eat breakfast cereals.

Addition of these nutrients is not a simple matter and more research is needed into methods and the stability of the nutrients in the products.

BIBLIOGRAPHY

ANON. 1968. Use of FD & C colors in food; guidelines for good manufacturing practice. Food Tech. *22*, No. 8, 946–949.

ANON. 1975. Code of Federal Regulations *21;* Food and drugs, parts 10–199. U.S. Govt. Printing Office, Washington, D.C.

ANON. 1975. Symposium on food colors. Food Tech. *29,* No. 5, 38–54.

ASSOC. OFFIC. ANAL. CHEMISTS. 1975. Official Methods of Analysis, 12th Edition. Assoc. Offic. Anal. Chemists, Washington, D.C.

FDA. 1968. Color Additives. Regulations under the Federal Food, Drug and Cosmetic Act, part 8, title 21. FDA, Washington, D.C.

FDA. 1975. Federal Food, Drug and Cosmetic Act, as amended. FDA, Washington, D.C. January.

GARARD, I. D. 1974. The Story of Food. Avi Publishing Co., Westport, Conn.

GRAHAM, H. D. 1968. The Safety of Foods. Avi Publishing Co., Westport, Conn.

JOSLYN, M. A. 1950. Methods in Food Analysis. Academic Press, New York.

MFG. CHEMISTS' ASSOC. 1960. Food additives. Mfg. Chemists' Assoc., Washington, D.C.

MERORY, J. 1968. Food Flavorings: Composition, Manufacture and Use, 2nd Edition. Avi Publishing Co., Westport, Conn.

NATL. RES. COUNCIL, NATL. ACAD. SCI. 1971. Food Colors. Natl. Res. Council, Natl. Acad. Sci., Washington, D.C.

OSER, B. L. 1960. Food additives: new law causes many problems. Chem. and Eng. News *38,* 108–111.

PINTAURO, N. D. 1974. Food Additives to Extend Shelf Life. Noyes Data Corp., Park Ridge, N.J.

POMERANZ, Y., and MELOAN, C. E. 1971. Food Analysis: Theory and Practice. Avi Publishing Co., Westport, Conn.

WILLIAMS, R. R. 1964. What has enrichment accomplished? Food Tech. *18,* No. 12, 63–64.

CHAPTER 10

Food Preservation

The preservation of food is the most important task of the food chemist. Foods of plant origin are seasonal and those of animal origin are usually obtained in large batches, such as a cargo of fish.

There is no definition of spoilage that can serve as a guide to the operations of the chemist. A food need not be inedible or poisonous to be spoiled; any deterioration in the organoleptic properties, such as staleness in bread or sourness in milk, is spoilage, although the food is not harmful nor have its nutritive properties deteriorated.

CAUSES OF SPOILAGE

It is difficult to separate the biological and chemical processes of spoilage. The changes are chemical, but in most cases, they are the result of biological activity.

Probably all cases of food spoilage may be attributed to microbiological action, enzyme activity, oxidation by air, or the absorption or evaporation of water. In fact, water has a role in all spoilage. Since all spoiling is physical or chemical activity, it increases in speed with the temperature, roughly double for each 10°C.

CURES

All types of spoilage are retarded by lowering the temperature. No keeping time for any food can be established because there are so many types of spoilage in operation—wilting, rotting, loss of vitamins, enzyme activity. Each item is a separate problem; apples may keep for many months but peaches for only a few weeks.

Refrigeration

The milkhouse built around a spring or other arrangement for cooling milk as quickly as possible after milking, the underground cave and the 19th century icehouse are examples of early American refrigeration.

In 1851, Dr. Gorrie, a physician in Appalachicola, Florida, invented a machine to make ice, but such a machine was not commercialized until 1875, and the expansion of the use of it for everything from cooling an automobile to freezing food, is common knowledge. However, expansion was slow. The cold storage warehouse appeared in 1865 and railroad refrigeration in the 1880s so that by 1890, commercial refrigeration was well under way.

With modern air conditioning equipment, it is possible to maintain any temperature down to $-80°C$ or possibly lower. Since lowering the temperature retards all reactions, it might seem that the lower the temperature the better. But lowering the temperature of the air is expensive; low temperatures are unpleasant for employees and, most important of all, the idea of refrigerating food is to keep it in its original condition by transporting and storing it above its freezing point.

Fresh Fruits and Vegetables.—Both fruits and vegetables are undergoing respiration and therefore emit heat. The USDA reports that apples stored at 32°F emit 660–880 Btu per ton; lettuce, 638 Btu and pears, 660–880 Btu. It is obviously difficult to maintain a definite temperature in a storage room because of the activity in it, but it must not fall below the prescribed temperature.

Trucks and railroad cars are also a difficult problem because of exposure to sun and to great variations in atmospheric temperature.

The ideal storage temperature for fresh fruits and vegetables is either 31°–32°F or 32°–33°F; for example, apples are stored at the former temperature and lettuce at the latter. The idea is to store the food at a temperature as low as possible without freezing it. According to Raoult's Law, a mole of solute in 1000 gm of water depresses the freezing point 3.35°F. In the case of glucose, a mole is 180 gm and in 1000 gm of water that is a 15.2% solution and freezes at 28.65°F. The aqueous phase of fruits and vegetables varies in the amount of solute. Apples contain sugars, soluble acids and some minerals, but potatoes and leafy vegetables contain very little soluble matter in the liquid portion.

Respiration requires oxygen and produces carbon dioxide, water, and heat. Furthermore, fruits and most vegetables have odors, which means that they are emitting organic compounds. These substances raise the question of the composition of the atmosphere. The products with the most active respiration require the circulation of air and the humidity should be high in order to retard the evaporation of water. Shriveling and wilting are spoilage and the loss of water is an economic one. If a ton of fruit loses 1% water, only 1980 lb remain for sale.

Under the best conditions, some foods keep only a few days in storage; berries are available only during the harvest season except where they are transported from one area to another. Peaches, green beans and a few other foods keep for a month, and apples keep for 10 months or more. Of course, the keeping time of any commodity depends on the definition of spoilage. If it is defined as unchanged from the original, the time is short, but if it means still edible, the time is much longer. Refrigeration only retards spoilage. Enzymes continue to ripen some fruits and to cause changes in many vegetables. Potatoes may begin to sprout and the enzymes convert starch to sugar and proteins to amino acids, which render them unfit for either potato chips or french fries because the heat causes a reaction between the acids and the sugars to produce a dark brown color and off taste.

Some fresh foods do not withstand low refrigeration; bananas, for example, lose their flavor and the skins turn brown at low temperatures.

Many fruits are very soft when ripe and are damaged by handling. Bananas, pears and persimmons are the only common fruits that ripen completely after they are removed from the parent plant. Some of the processes do continue; fruits may soften, develop color, lose some acids and produce sugar, but the characteristic flavor is not produced. Tomatoes, peaches, nectarines, plums and melons all suffer from failure to ripen in storage. Here is a field that deserves more research than it has had.

Some changes have been made in shipping, but they are not always to the advantage of the food; they may be simply an improvement in the convenience of handling. For over a century, bananas were shipped on the stem (bunch), but recently the hands have been removed and packed in boxes. This saves freight on the stems and is said to reduce bruising, but that is controversial. In France, peaches are packed in cotton, one layer to a wooden box, and in recent years, Florida shippers of gift fruit have packed each fruit individually within the container to reduce crushing. But such packing is expensive and not for bulk food supplies.

Milk.—The details of fresh milk storage will be discussed in a later chapter, but it is reduced to a temperature of 50°F or less as soon after milking as possible, because it is an excellent medium for the growth of spoilage organisms. According to work done at the University of Nebraska, milk held for 10 hr at 32°F contained 3000 bacteria per ml, at 50°F, 11,500 and at 80°F, over 2,000,000.

The most familiar milk spoilage is souring caused by *Streptococcus lactis,* which converts the lactose in the milk into lactic acid, $CH_3CHOHCOOH$. Since pasteurization does not kill all these organisms, the milk is cooled immediately after pasteurization, and if kept below 50°F it will keep for several days.

Butter.—Although the keeping qualities of butter are better than those of fresh milk, it contains over 15% water and 80% fat. Enzymes in the butter cause the hydrolysis of the fat and the low molecular weight acids, which are hydrolyzed first, have a disagreeable odor and taste.

Most butter is now made in creameries and the milk pasteurized, which increases the lifetime of the butter. The finished butter, however, still contains many organisms and some of them produce hydrolytic enzymes, which hastens spoilage. Ordinary handling of butter such as delivery to a retail store or a home kitchen may be done under ordinary refrigeration, but butter withstands freezing temperatures without damage, and for long storage, it is kept at a temperature of 0°F or below.

Refrigeration does not protect butter from spoilage by the absorption of odors from other foods, cleaning compounds, gasoline fumes or from its container.

Cheese.—In the matter of storage, the various kinds of cheese provide much the same problem. As an example, American cheddar cheese contains 37% water, 25% protein, 32% fat, 2.1% carbohydrates and minute quantities of vitamins. It may be made from pasteurized milk, but it still contains microorganisms.

Spoilage of cheese is not as common as that of milk or butter, but it can become rancid, dry out or become overripe and bitter. The ripening process depends on enzymes from organisms. Flavor develops during the ripening and the protein of the curd undergoes hydrolysis, which changes the consistency from rubberlike to the crumbly texture of sharp cheese. The development of a characteristic flavor and the change in texture are caused by different organisms. Ripening is carried out between 35° and 65°F. The higher the temperature, the faster the ripening, but high temperatures encourage molding and other spoilage. Once the cheese is cured enough for the market, it is stored at or near 0°F to retard further changes.

Meats.—The meat industry is too complex for discussion here. Cattle and hogs have been raised in the west or middle west and the slaughter centered in the large cities of the region. At one time, the USDA estimated that the average shipping distance for beef was 1,000 miles.

The problem is that the meat is at the body temperature of the animal at the time of slaughter and so it is cooled as quickly as possible to 32°–34°F, which is a safe temperature, for no meat freezes at or above 32°F. The reason for the quick cooling is that the surface of the meat accumulates bacteria and molds during the butchering, from the workers, the equipment or the air, and these organisms quickly attack the meat and cause rancidity and putrefaction. The meat is held at the low temperature for several days to become more tender. Also most meat is sold to the consumer in the fresh state. No keeping time can be specified because of the variation in handling and transportation. The meat must be transferred from cold room to truck or railroad car, then to the warehouse of a distributor, from there to another truck, to the retail store, to the display case and finally to the home refrigerator and, of course, it is removed from refrigeration at each transfer.

Poultry.—The raising and slaughtering of poultry is more widely dispersed than the production of beef and pork, consequently, the shipping distances and holding times are much shorter. Spoilage, however, is more rapid and the animals are cooled as quickly as possible after slaughter and either marketed immediately or frozen.

Fish.—Seafood is a special problem. Small boats come ashore daily and sell their cargo at the dock to restaurants, retail stores or consumers. The boat may or may not have ice.

Large boats may be out for days or weeks and may or may not be refrigerated. Seafood spoils the most quickly of all the meats, and so fish are partially cleaned on the boat and packed in ice. Dry ice (solid CO_2) and mechanical refrigeration of recent years have extended the shipping distance for fresh fish. Enzymes within the fish begin to hydrolyze the tissues soon after the death of the fish and bacteria attack the surface. The results are a soft and flabby texture and "fishy" odor, which is produced by conversion of some of the amino acids into amines.

Eggs.—The white of an egg contains 88% water, the yolk, 51%, and the two parts are separated by a semipermeable membrane, consequently, water slowly diffuses from the white into the yolk and weakens the membrane. This lowers the quality of the egg because the yolk may break when the consumer attempts to poach or fry the egg or to separate the yolk and the white for other cooking purposes. Also, the shell of an egg is porous so that water evaporates from the white and enlarges the air space at the end of the egg, which may increase from the ⅛ in. of a fresh egg to ¾ in. This does not change the nutritive value of the egg, but it does indicate the age of it or the way it has been handled. It also reduces the weight of the egg and might lower its classification as large, medium or small, which is based on weight.

Hydrolytic and other changes attack the protein with the evolution of ammonia and hydrogen sulfide and the pH of the white may increase from the 7.6 of a fresh egg to as much as 9.7.

Eggs are produced in every state. California leads the list with 8–9 billion annually, Nevada is at the bottom with 3–4 billion. The distribution of egg production, not only by states but within the state, makes the marketing of fresh eggs easy, but there are still problems. If a grocer can get fresh eggs daily from a nearby producer, atmospheric temperatures or mild refrigeration, such as an air conditioned store, may be sufficient to keep the eggs until they are consumed. But the larger cities cannot depend on receiving day-old eggs nor can the producer depend on selling his production daily unless his customer is a distributor. Furthermore, egg production is partly seasonal. The greatest production is in May and June, the least in the winter. Formerly, the price of eggs varied enormously from late spring to fall and winter, because all eggs were marketed fresh and the price varied with the supply.

After 1900, cold storage warehouses became common and both the price and the supply of eggs became more stable. Today, with the increase in refrigeration, the number of eggs in storage has decreased to about 12% of the total supply.

The cold storage of eggs is a peculiar problem. Water expands about 10% when it freezes. Consequently, if the temperature of storage drops enough to freeze the eggs, they are ruined—but egg white is a solution and its

freezing point is below that of water and so the eggs can be stored at 29°
to 31°F without freezing them.

Eggs remain in storage from 4 to 10 months, and the USDA estimates
that 75% of them are removed from storage within 7 months. Most states
have laws that limit the storage of eggs to a year or less.

Eggs absorb odors readily and so the atmosphere of the warehouse is
important; it must be free from odors, and ozone is sometimes used to keep
the air free from them. Humidity is also important; if the air is too dry, water
evaporates from the white more quickly and if it is too moist, molds or other
organisms grow on the shell, penetrate the egg and increase the rate of
spoilage. Molds form gray or black spots and bacteria cause decomposition.

Fortunately, most of the spoilage defects of eggs can be determined by
candling. Inspectors carry a box with a light bulb inside and a hole in the
top slightly smaller than an egg. An egg is translucent and, placed over a
bright light, will show embryo development, blood spots, molds, the size
of the airspace and the firmness of the yolk. Blood spots are original defects
in the egg, but the other visible defects are an indication of the age and
general quality of the egg. Decomposition and the absorption of odors
cannot be determined by candling, but they usually increase with the age
of the egg.

General.—Refrigeration is properly used to keep natural foods in the
fresh state from a few weeks to a year. Many processed foods are also refrigerated, canned hams weighing over 3 lb, for example.

Each food is a special problem and must be treated as such, for each has
its own lifetime and may have individual spoilage qualities.

Freezing

Most spoilage depends on the presence of water. Consequently, freezing
retards spoilage more than one might expect from the slight difference in
temperature between freezing and refrigeration, because the water is
converted into ice and is not available for the activity of either enzymes or
organisms. If all the water in the food were frozen, the food would probably
keep indefinitely, but some of it isn't.

However, frozen foods keep longer than the same kind of food under
refrigeration and so it may occur to the student to wonder why more food
is not frozen for preservation. There are two reasons. The first is the cost.
Water has a specific heat of 1 cal/gm, all other substances less. To freeze
1 gm of water requires the removal of 80 cal. Since 75% water is a fair average
for foods, 15 cal per gm are required to cool it from 20°C and 60 cal per gm
must be removed to freeze it, a total of 75 cal per gm of food. Then, frozen
food must be stored well below freezing, which is also expensive.

The second reason is of more concern to the chemist; the freezing causes physical changes in the food. When a frozen food is thawed, it has a different texture from that of the original food and spoils more quickly. The exact cause of this is somewhat controversial, but if the food is frozen slowly as by a climatic change, the damage is greater than if it is frozen quickly. The difference is due to the difference in the size of the ice crystals. It is a general rule of crystal formation from either liquids or solutions that the less time required for the crystal formation, the smaller the crystals. Precipitates formed in analytical chemistry are so fine that they pass through a filter paper, and it requires a microscope to show that they are crystalline, but crystals formed in a solution allowed to stand for several days in an open beaker are large.

Ice cream has been made commercially since 1851 and liquid eggs have been frozen since around 1900, but these foods do not have a cellular structure and so ice crystals were no great problem except for the grit they constituted in the early ice cream; that was corrected by the addition of a colloidal substance. Fruits, vegetables and meats were never frozen except as a matter of necessity until the 1930s. During World War I, sides of beef were frozen in Chicago and delivered to the soldiers at the front, still frozen too hard to cut with a knife. The meat was cooked and eaten immediately after it was thawed.

If the farmer allowed his store of apples to freeze it was ruined, and the same would be true of all his other fruits and vegetables. With the uninsulated railroad cars, shipping fresh food in winter was hazardous.

About 1930, Clarence Birdseye invented a process for freezing packages of food quickly and, since the mid-thirties, the practice has spread to most meats, fruits and vegetables and to many processed foods. Even quick freezing damages the food because there is some rupture of the cells by the expansion of the water, and there are other changes that require more study. Water is moved about as the food freezes and may not return to its original place when the food is thawed. Also, some of the water is bound to other molecules in the food and may be dislodged as the freezing proceeds. Whatever the correct explanation may be, the thawed food is of a different texture from that of the original and spoils more quickly. Microorganisms may be inactivated by freezing, but they are not killed, and once the food is thawed, they become active and there may be pathogens among them.

The keeping time of frozen foods differs with the product and its handling, but some foods will be in fair condition after two years or more; others in poor condition after a few months.

Frozen food should be stored and transported at 0°F, which is expensive. It requires mechanical refrigeration, and a zero temperature cannot be maintained at all times because of transfer from the freezing plant to a truck or freight car, them to a receiving warehouse, to another truck, to a retailer's display case and finally to a home refrigerator. Such transfers do not cause

FIG. 10.1.   LOADING A BIRDSEYE MULTIPLATE FROSTER

complete thawing because of the high specific heat of ice, 0.5 cal/gm, but as the product freezes, the cell fluids concentrate and the freezing point drops so that the final freezing point will be far below the 32°F of water and so slight warming may mean slight melting.

Each food requires the special attention of the chemist who must assess the nature and extent of the damage, the keeping time and even the cost. If he is receiving frozen food for processing, fruits for pies, for example, he must be aware of the merits of the food compared to fresh foods or those that have been otherwise preserved. Frozen food is generally closer to fresh food in quality than that preserved by any other method and sometimes it is even superior to the fresh product available.

Drying

The cereals have been known for centuries to keep well, although meats, fruits and vegetables spoil in a short time. That the difference is caused by the difference in water content was soon learned. If grain got wet, it soon spoiled. The water content of wheat is about 12% and it withstands further drying without loss of quality. Beans and lentils are slightly drier than the cereals, and nuts contain about half as much water; all these keep well,

FIG. 10.2. DRYING FISH IN THE SUN: NEWFOUNDLAND

except that nuts may turn rancid because of the high fat content. Meats, fruits and vegetables, by contrast, contain from 65% to 95% water and none of them will keep more than a few days at room temperature.

Sun drying has been used for centuries for fruits, fish and a few vegetables, and some food is still preserved by that process because of its low cost. In the tropics, coffee, cacao beans and nutmegs are dried in the sun; the drying is mostly superficial for the prevention of the growth of mold.

Climate in this country is the chief deterrent to sun drying. The driest states do not produce much food, and the humidity and rain in the states where fruits and vegetables are grown is too great in the harvest season; fish are dried commercially in New England and fruits in California.

Fish.—Fish are partly cleaned and packed in ice on the larger boats, but further preservation is essential as soon as possible. When they are unloaded, the cleaning is finished and the fish are packed in solid salt. Water diffuses out of the fish and forms a brine; at the same time, salt diffuses into the fish and coagulates some of the protein. Eventually, the salt content of the cells becomes the same as that in the brine. The process takes 2–3 days and the fish are then dried on slatted racks in the open air for several days (Fig. 10.2).

The principal fish dried is the cod, which contains only 0.3% fat. The fresh cod contains 81% water and the salting and drying reduces it to about 50%, but the high salt content helps with the preservation. Fat interferes with the drying process and also turns rancid. The salting and drying process inhibits enzyme action and stops the production of the odorous amines.

Smoking.—Some kinds of fish are salted, dried a few days in the air and then smoked. The usual method of smoking fish in this country is the cold smoking process by which the fish are supported in the smoke room and

FIG. 10.3. SUN DRYING GRAPES: CALIFORNIA

subjected to smoke from hardwood at a temperature of about 80°F. The water in smoked herring is reduced from 69% to 35% in hard smoked herring but only to 60% in kippered herring. Here the principal drying is the result of the salting and air drying; little moisture is lost during the smoking. The smoke adds flavor to the fish, but which component of it if any acts as the preservative is unknown.

Fruits and Vegetables.—The housewives of American farmers in the east have always dried apples, blackberries, raspberries, cherries, peaches and pears if they had a crop greater than they could use in the season. Small fruits dried well by exposure to the sun and withdrawal to cover at night, but fruits that had to be cut did not dry so well. Apples turned a dark brown and developed a flavor totally unrelated to that of the fresh apple, but the process did preserve some fruit for the lean winter months; there were no commercial drying plants. In California, the climate is more favorable for sun drying and that state is the chief source of our dried fruits.

Raisins.—In this country, raisins are produced only in California. Most of the grapes are dried in the sun, on trays placed between the rows in the vineyards. Towards the end of the drying, they are stacked in the shade where they continue to dry for two or three weeks. Although raisins are small, the drying is greatest near the surface. When the water content has been reduced below 16%, the raisins are stored in a bin to permit the even distribution of water throughout.

Raisins are dried to 10% to 12% water in which condition they will keep for years. They are hydrated to 18% moisture when they are packaged for the market. An average analysis is water 18%, protein 2.5%, fat 0.2% and carbohydrate 77.4%. The carbohydrate is mostly glucose and there is enough of it to preserve the interior of the raisin, but 18% moisture will permit mold growth on the surface, which is the most common form of spoilage.

Prunes.—Prunes are a variety of plums, that can be dried without fermenting. They are shaken from the trees, put into lug boxes and taken to the drying yard where they are dipped into very dilute, hot NaOH solution for a few seconds to check the skins. They are next spread on trays and dried in the sun to 18% moisture and then put into a bin for the equalization of moisture. They keep well at this moisture content, but are rehydrated to 28% for packaging. The average market product contains water 28%, protein 2.1%, fat 0.6% and carbohydrate 67.4%. The rehydration shortens the cooking time of the prunes and increases the weight available for sale, but it also promotes mold growth, which has induced many housewives to buy prunes in transparent packages.

Peaches and Apricots.—These fruits present problems not common to prunes and raisins; they must be cut for drying and the cut surface increases the chance of spoilage. Also, enzymes in the fruit darken the surface. The halves are placed on trays with the cut side up and taken to a room where they are exposed to the fumes of burning sulfur, which restores the natural color and kills many of the superficial organisms. The fruit absorbs the SO_2, which helps in its preservation.

For many years after 1906, the presence of SO_2 in food was controversial, and at one time the amount was limited, but the producers insisted that the fruit would not keep with the limited amount of sulfur and the regulation was abandoned. There is no legal limitation at present.

After sulfuring, the fruit is sun dried and handled in much the same manner as prunes and raisins. They are marketed at 25% water, 5–6% protein and 66–68% carbohydrate. If they have not been properly sulfured, they may turn dark in storage; the sugar content preserves the interior but mold may grow on the surface.

Dehydration

Aside from some fruits and fish, there was little dried food in this country until World War II. Dried eggs were imported from China for decades before the first commercial drying plant began operation in 1927. During World War I, the soldiers and sailors were fed fresh or canned food except for a few dried foods of the day, but as the number of men in service increased and transportation facilities became strained, an attempt was made to dry vegetables, but it was a complete failure.

World War II, with its armed forces scattered over the world, dropped a vital and urgent problem into the laps of the food technologists. Fruits, vegetables and meats contain an average of about 75% water, and spoilage is caused by enzymes in the food itself and by bacteria, yeasts and molds by contamination. Water promotes the activities of both enzymes and organisms. Drying food was the obvious answer, for it not only increased the keeping quality of the food but it also reduced the weight, which was imperative because of the shortage of shipping facilities by both land and sea. Sun drying was too slow and it only reduced the water to about 10%. Moreover, much of the food had to be shipped into the tropics and with 10% water, it would not keep.

Enzymatic Browning.—The dried potatoes of World War I turned brown and developed an unpleasant flavor, which is a common problem with many fruits and vegetables. The exact chemistry involved has not been clarified because of the large number of substances involved. In general, when the skin of the fruit or vegetable is broken and the interior exposed to oxygen, some phenolic compound in the food is oxidized to a quinone that polymerizes to form the brown pigment, which is called *melanin*. The reaction occurs in several steps and is catalyzed by several different enzymes. For example, tyrosine is a phenolic amino acid that occurs in many proteins, and one enzyme hydrolyzes the protein to free the tyrosine, then several reactions occur in succession; some of them are indicated by the following formulas:

Tyrosine   3,4-Dihydroxyphenylalanine

5,6-Dihydroxydihydroindol Carboxylic Acid

Red Pigment

Each of these changes is catalyzed by a different enzyme with the possible exception of the red pigment, which is unstable. There are many enzymes present in fruits; peroxidases, phenolases and others.

Some fruits do not contain these enzymes and do not darken on exposure of the interior to air. Oranges, lemons, grapefruit, strawberries and tomatoes seem to be free from these enzymes.

There is an enormous literature on enzymatic browning (Mrak and Stewart 1951), but all the details of the process are far from being understood. Fortunately, empirical procedures have provided methods of control. Blanching destroys the enzyme and prevents the browning. Numerous substances have been suggested as additives for the prevention of enzymatic browning and several of them have been patented. The two that are not patented and are most frequently used are sulfur dioxide and ascorbic acid. The latter has been used on cut fruits for freezing and on fruit salads in cafeterias. Either of them may be added to fruit juices that tend to darken. Both are strong reducing agents, so they consume oxygen and destroy the peroxidases.

With the enzymatic browning problem solved, a vast number of tunnel driers were hastily built, foods dehydrated to 2–5% water and the armed forces were fed (Garard 1974). After the war, some of the dehydrated products came into the civilian market and the process began to replace the sun drying of fruits. It is more expensive than sun drying but produces a better and cleaner product; the food is not exposed to the dust and insects of the field or the drying yard and more of the vitamin content is retained.

Dehydration is applied to more than fruits and vegetables. Over 1 billion lb of milk is dehydrated annually and 50 million lb of eggs. In these cases, the food is not only preserved, but it is also made more convenient for use. Bakers use dried eggs in preference to either fresh or frozen eggs. Some foods are dehydrated for convenience alone, among them are instant coffee, instant tea and dried soups.

Partial dehydration is also useful for both preservation and economy; concentrated milks, orange juice and soups are in this class. A problem with these products is the loss of flavor. Much of the flavor is volatile and the evaporation of the water is essentially a steam distillation, which carries off the volatile flavors.

The newer process of freezing the food and then evaporating the ice helps to solve the loss of flavor, but the freezing process is slower than spray drying and more expensive because of the extra process of freezing and the longer time required to obtain the product.

Nonenzymatic Browning.—The food dehydrator has not solved the color problem when he has blanched the fruit or vegetables; a color may develop during the heating that is not caused by enzymes. The effect is more apparent in other processes, such as toasting bread or roasting coffee, cacao beans or meat. Sometimes this reaction produces a desirable flavor; raw cacao beans taste much like any other raw beans, the chocolate flavor is the result of the browning. At other times, the reaction is ruinous—dark brown potato chips, for instance.

The reaction is known as the *melanoidin* or *Maillard* reaction, the latter name after the French chemist who studied the reaction from 1912 to 1917.

Foods are complex chemical systems and the Maillard reaction has never been completely explained. Enough has been learned, however, to enable the chemist to control the reaction or, in some instances, to prevent it.

The studies of Maillard, and many others since, have shown that the reaction occurs between amino acids and sugars. Work with model systems and also with proteins has shown that histidine, threonine, phenylalanine, tryptophane, and lysine are the most reactive acids but not the only ones. It has been shown that a casein–glucose mixture, held at 37°C for five days, lost 66% of its lysine and after 30 days it had lost 90% lysine, 70% arginine, 50% methionine and 30% tyrosine. The loss was measured by recovering the amino acids freed by enzymatic hydrolysis.

The initial reaction is thought to be between the aldehyde group of the sugar and the amino group of the acid. The simplest case would be

$$
\begin{array}{c}
H\text{---}C\text{=\!=}O \\
| \\
H\text{---}C\text{---}OH \\
|\\
H
\end{array}
\;+\;
\begin{array}{c}
H \\
| \\
H_2NC\text{---}COOH \\
| \\
H
\end{array}
\;\longrightarrow\;
\begin{array}{c}
H \\
| \\
H\text{---}C\text{=\!=}N\text{---}C\text{---}COOH \\
| \qquad\quad | \\
H\text{---}C\text{---}OH \;\; H \\
|
\end{array}
\;+\; H_2O
$$

In some reactions, if not all, CO_2 is evolved and so the second reaction would form

$$
\begin{array}{c}
H \\
| \\
H\text{---}C\text{=\!=}N\text{---}C\text{---}H \\
| \qquad\quad | \\
H\text{---}C\text{---}OH \;\; H \\
|
\end{array}
\;+\; CO_2
$$

From this simple illustration, it is obvious that the number of possible compounds is enormous, because all the aldose sugars undergo the reaction and the $-NH_2$ group of a protein or a polypeptide. In the case of caramel, the $-NH_2$ group may be from ammonia, NH_3.

The Maillard reaction seldom produces melanoidin at room temperature or in the presence of 75% water or more in natural foods, but as the temperature rises and the concentration of reactants increases, the brown color begins to develop. The conditions are different for different foods. The evaporation of milk is an operation in which it is difficult to prevent a brown color and a cooked taste.

Control of the Maillard reaction involves the removal of sugar. If that is impractical, then the control of temperature is the only remedy. In the case of concentrated milks, the removal of sugar is impossible, but the fermentation of eggs removes the slight sugar content and prevents browning. Immature potatoes contain sugar and, late in the storage of mature ones, they again develop sugar preparatory to sprouting. Such potatoes cannot be used for potato chips or French fries because of the deep brown color and off flavor. But here again, the sugar can be removed by fermentation of the cut pieces. On the other hand, a little sugar added to pie crust, biscuits or the surface of meats increases the browning.

Chemical Preservatives

The present GRAS list contains 31 preservatives exclusive of salt, sugar, vinegar, spices, and smoke.

A chemical preservative must be harmless when consumed and effective for the prevention of some type of spoilage. Among the 31 items on the GRAS list, several have limitations: caprylic acid may be used only on cheese wraps; gum guaiac to the extent of 0.1% in edible fats and oils; the various sulfites not in meats or in foods recognized as sources of vitamin B_1; and thiodipropionic acid and other antioxidants combined with it may not exceed 0.02% of the fat or oil (including essential oil) of the food.

Formaldehyde has been used in the past, but is now illegal. The most common chemical preservatives in use are SO_2 and the sulfites used mainly in fruits and fruit products and sodium benzoate used in many processed foods. These two are general preservatives intended to prevent the growth of any of the three spoilage organisms, bacteria, yeasts and molds. Other chemicals are intended to prevent specific kinds of spoilage, such as gum guaiac for the prevention of rancidity of fats, and calcium propionate to retard the molding of bread. The discovery and use of these preservatives was purely empirical and the mechanism of their operation is unknown.

Much research has been done in recent years on the harmlessness of preservatives, that is, harmlessness to both the food and the consumer.

It has been shown that the sulfites increase the ascorbic acid retention in fruits, but destroy some of the thiamine.

Salt and Sugar.—The chemical preservatives of the previous section are used in small fractions of 1% and therefore must attack the organisms, but salt, sugar and acetic acid used in these low concentrations have no preservative action at all. Vinegar is 5% acetic acid and salt and sugar are used in high concentration; they kill the organisms or prevent their growth by osmotic pressure. Cell walls are semipermeable membranes and the salt or sugar dehydrates the cell and the solutes penetrate until the concentration is the same inside and outside the cell. It takes much more sugar than salt to produce the same effect. The molecular weight of sucrose is 342 and that of salt is 58.5. The salt is almost completely ionized, so about 30 gm of salt will have the same osmotic effect as 342 gm of sugar. The amount of solute required varies with the product to be preserved and spores are seldom or never destroyed. In general, a brine of 10% or 15% is effective but a concentration of more than 50% sugar is essential.

Salt is seldom used for the preservation of foods for the market because salt is also a flavor and a 10% brine would be intolerable in a finished product. Olives contain about 6% salt, but they are heat treated in addition to the salt in order to avoid the possibility of *Clostridium botulinum,* which is an ever present hazard in the preservation of vegetables.

One of the problems of products preserved with sugar is the growth of molds. If the product is stored in cans, drums or tanks with a head space, water vapor collects in the head space and when the temperature drops, some of it condenses and drops on the surface of the product forming a dilute solution that will support the growth of molds.

Meats and fish intended for processing are packed in solid salt, which dehydrates organisms on the surface and thus preserves the product until further processing can be begun.

Smoking.—Smoking is one of the oldest methods of preserving meats and the one we know the least about. It is used mainly for high-protein foods—meats, fish, oysters and cheese.

Usually, the product is first salted. Fish and pork are packed in salt, which forms a heavy brine by dilution with water removed from the meat by osmosis. The product remains in the salt for several weeks and then is taken out and some of the superficial salt removed after which it is exposed to wood smoke. In this country, hickory, maple, beech and birch wood or sawdust are burned to produce the smoke. The composition of the smoke is very complex and differs with the kind of wood and its water content and also with the manner in which it is burned. Formaldehyde, furfural and other aldehydes, ketones, methanol and several acids, esters, phenols and hydrocarbons have been identified as components of the smokes. Rhee and Bratzler (1970) identified eight polycyclic hydrocarbons in the smoke from maple sawdust. From 4.5 kg of sawdust, they obtained 51.5 μg of phenanthrene, 7.0 of 1,2-benzanthrene, 5.7 fluoranthrene, 5.5 pyrene, 3.8 anthracene, 2.6 chrysene and 2.1 benzopyrenes. Some of these hydrocarbons are carcinogenic.

There has been much speculation as to which component of the smoke acts as the preservative. Wood smoke is a poor bactericide and the prevailing opinion at present is that preservative action is due to the salt and the drying of the surface of the meat. In the United States, meat is now smoked at a low temperature and is done mainly to add flavor.

So-called liquid smokes are now made by condensing and fractionating wood smoke. The liquid, or rather the solution, is either applied to the surface of the meat or injected into it. Gorbatov *et al.* (1971) report the composition of 9 commercial liquid smokes in millimoles per 100 ml as: Phenols 1.2–30, carbonyl compounds 1.4–18.1, furfural 0–2.3, acids 1.6–100, esters 0.8–59.5, methanol 0–15.3 and nonvolatiles 0.4–15.2.

The carcinogenic hydrocarbons have been removed from the commercial products, which differ in flavor as is evident from the enormous variation in composition. Meats treated with a liquid smoke must be refrigerated.

Canning

The methods of preservation discussed in the previous sections are all processes intended to preserve the food as nearly as possible in the natural, uncooked condition, and the success of the method can be measured by how nearly it provides the consumer with food in its natural color, texture and taste. All these methods of preservation impose limitations on methods of packaging, storing, handling and keeping.

Canning involves cooking the food. The process was invented by Nicolas Appert, a French confectioner, in 1795 and the results published in 1810. By then, the process was under experiment in England whence it was introduced to the United States in 1817, by William Underwood, who established the first American cannery near Boston.

Home canning soon became popular and farmer's wives preserved the excess fruits and some vegetables in earthenware (stone) jars, closed with a tin lid and sealing wax. Appert had used glass bottles and the English, cannisters of steel plated with tin, hence the name *tin can* from which the process has taken its name regardless of the type of container used.

The chemist's concern with canning may involve either the food or the container.

Aside from engineering details, the canning process consists in filling a can with food, sealing it hermetically and heating it to sterilize the food and inactivate the enzymes. Foods are mostly water, which is a poor conductor of heat. Consequently, it takes some time to raise the temperature of the interior of the food to the lethal temperature of the organisms. In the meantime, the food next to the wall of the can may become overcooked and its texture damaged. The inactivation of the enzymes occurs before the organisms are destroyed. The most dangerous organism is *Clostridium botulinum,* which is anaerobic and forms spores. It is a soil organism and likely to be in beets, green beans or other vegetables that grow in or near the soil. Acid foods, such as most fruits and tomatoes seldom contain it.

Botulism is primarily the problem of the bacteriologist, but another form of spoilage is the concern of the chemist; the presence of oxygen in the can, which destroys vitamin C and thiamine depending on the acidity of the food. In general, foods stored 1 yr at 65°F lose 5–15% of their vitamin C and at 80°F, 15–30%. The loss of thiamine is somewhat greater.

Foods that contain fat may become rancid by oxidation, which is a consumer problem with a bottle of oil or a can of coffee that lasts several weeks; every time the container is opened, the food gets a fresh supply of air. The only thing the processor can do to solve that problem is to supply smaller packages.

Oxygen is the most serious problem in dry powders, such as coffee, dried milk, dried eggs and cake or other prepared mixes. In these cases, the oxygen is mixed with the powder during the processing or filling and is difficult to remove for some of it may be adsorbed on the powder. Remedies consist in flushing the head space of the package with carbon dioxide or nitrogen or exhausting the package by vacuum, but neither process removes all the oxygen.

With some foods, such as concentrated soups, the oxygen may be largely removed by filling the can full. This is impossible with powders, which shake down to lower volume. Expansion of foods when they become warm may preclude filling the can full.

If more than one kind of container is suitable for canning a product, practical and economic considerations must be taken into account. Glass is heavy, breakable, a poor conductor of heat and exposes the food to light. But it also permits an effective display of stuffed olives, cherries, apricots and other attractive foods; the containers are also easier for the consumer to open. Metal is attacked by some foods, dents easily and conceals the contents, but it conducts heat well, is lighter than glass and can be made in larger sizes with greater safety. In the past, metal has been cheaper but that may no longer be true.

Tin plate is very thin. It was formerly made by a hot dip process in which sheets of steel were dipped into a vat of molten tin. This gave a thick, uneven coat and since our tin is imported from the Orient or from South America, the process was very expensive. During World War II, the government restricted the tin plate to 0.5% of the weight of the can, which required electrolytic plating to get a thin even coat.

TABLE 10.1

SUGAR CONTENT OF CANNED FRUIT SYRUPS

Syrup	Peaches (%)	Pears (%)	Apricots (%)	Cherries (%)
Light	14–19	14–18	16–21	16–20
Heavy	19–24	18–22	21–25	20–25
Extra heavy	24–35	22–35	25–40	25–35

There are cans made of bonderized steel coated with enamel and also cans of aluminum, but neither has come into general use in the food industry. Aluminum cans, however, have become popular for beverages.

Tin plate has tiny holes in the coating and the steel may be attacked by acid foods with the evolution of hydrogen causing the ends of the can to bulge. Since gas-forming organisms produce the same effect, the consumer rightly assumes that the food has spoiled. The selection of a container for a given food product is a problem for the chemist who must keep a constant shelf life test to measure his success.

Cans of fruits that leave large air spaces, such as cherries and apricots, are filled with a syrup to eliminate most of the air, add flavor and support the fruit against damage from handling. Foods with standards have the sugars that may be used and the concentration specified. The usual sugars are sucrose, glucose, invert syrup, and corn syrup. The sugar concentration of the syrup varies with the fruit as shown in Table 10.1. The syrup for various kinds of berries also varies and spices, ascorbic acid, vinegar and flavorings are among the additives, but differ with the fruit.

The composition of the brine for vegetables is more complex than the syrups added to fruits. Canned peas, for example, have 12 optional ingredients including salt, monosodium glutamate, sugar, dextrose, spice and 7 alkalies to bring the pH up to 8, as well as 7 natural flavorings including green or red peppers, mint leaves, onions, garlic, horseradish, lemon juice and butter. Canned green beans have much the same list of permitted additives.

The standards for canned foods are long and detailed and must be studied by the canner for each product that he processes.

PACKAGING

The use of cans and jars for canning food is a method of packaging that permits heat treatment in the package. It also prevents loss of volatile substances and contamination from the ambient atmosphere. Products that do not require heat treatment—soft drinks, syrups, vinegar—are likewise protected by glass bottles or cans of tin or aluminum.

Frozen foods are processed in the consumer package, but neither glass nor metal is required, because the contents remain solid until the product is used. Frozen food is not sterile and after it thaws, the organisms grow and the product spoils rapidly.

Since cost is an important factor in packaging, the chemist must first determine the adequacy of the package as a protection for the food, for that is the first consideration. Dehydrated foods must be protected from the absorption of water; foods containing even a low percentage of fat must be shielded from the air and all must be protected against the entry of dirt and organisms and must not absorb flavors from the package. Sometimes transparency is an important factor, not only for sales appeal, but also to enable the consumer to detect spoilage. Most cheese is over 30% water and subject to mold. In these products, a transparent package enables the grocer and the consumer to detect mold at once.

Packaging is an industry in itself affected by cost, sales appeal, and food protection. The student should go to a supermarket with a notebook, record the packaging of several items and then try to decide why each particular packaging was chosen.

In his own laboratory, the chemist conducts shelf life studies of his products. If they are dry mixtures, sugar, flour and fat will keep for months in a lined cardboard carton; add eggs and be prepared for returns.

BIBLIOGRAPHY

ANON. 1974. Shelf life of foods. Food Tech. *28*, No. 8, 45–48.

ASSOC. OFFIC. ANAL. CHEMISTS. 1975. Official Methods of Analysis, 12th Edition. Assoc. Offic. Anal. Chemists, Washington, D.C.

AYRES, J. C. 1958. Methods for depleting glucose from egg albumen before drying. Food Tech. *12*, 186–189.

DESROSIER, N. W. 1970. The Technology of Food Preservation. Avi Publishing Co., Westport, Conn.

GARARD, I. D. 1974. The Story of Food, Avi Publishing Co., Westport, Conn.

GILLIES, M. T. 1974. Dehydration of Natural and Simulated Dairy Products. Noyes Data Corp., Park Ridge, N.J.

GORBATOV, V. M. *et al.* 1971. Liquid smokes for use in cured meats. Food Tech. *25*, No. 1, 71–77.

GRIFFIN, R. C., JR., and SACHAROW, S. 1972. Principles of Package Development. Avi Publishing Co., Westport, Conn.

HARRIS, H. E. *et al.* 1974. Soluble coffee shelf life. J. Food Sci. *39*, 192–195.

HARRIS, R. S., and KARMAS, E. 1975. Nutritional Evaluation of Food Processing. Avi Publishing Co., Westport, Conn.

HEID, J. L. and JOSLYN, M. A. 1975. Fundamentals of Food Processing Operations. Avi Publishing Co., Westport, Conn.

KRAMER, A. 1973. Food and the Consumer. Avi Publishing Co., Westport, Conn.

MRAK, E. M., and STEWART, G. F. 1951. Advances in Food Research, Vol 3. Academic Press. New York.

POTTER, N. M. 1973. Food Science. Avi Publishing Co., Westport, Conn.

RHEE, K. S. 1968. Policyclic hydrocarbon composition of wood smoke. J. Food Sci. *33*, 626–632.

RHEE, K. S., and BRATZLER, L. J. 1970. Benzopyrene in smoked meat products. J. Food Sci. *35*, 146–149.

ROBINSON, J. F., and HILLS, C. H. 1959. Preservation of fruit products by sodium sorbate and mild heat. Food Tech. *13*, 251–253.

SACHAROW, S., and GRIFFIN, R. C. JR. 1970. Food Packaging. Avi Publishing Co., Westport, Conn.

SMITH, O. 1968. Potatoes: Production, Storing, Processing. Avi Publishing Co., Westport, Conn.

STADELMAN, W. J., and COTTERILL, O. J. 1973. Egg Science and Technology. Avi Publishing Co., Westport, Conn.

TORREY, H. 1974. Dehydration of Fruits and Vegetables. Noyes Data Corp., Park Ridge, N.J.

TRESSLER, D. K., and COPLEY, M. J. 1968. The Freezing Preservation of Foods, 4th Edition. Vols. 1–4. Avi Publishing Co., Westport, Conn.

VAN ARSDEL, W. B., COPLEY, M. J., and MORGAN, A. L. 1973. Food Dehydration, Vol. 2. Avi Publishing Co., Westport, Conn.

CHAPTER 11

Dairy Products

Fresh milk is the parent substance of all dairy products and, throughout the world, the milk of many species is used for food; principally the milk of the cow, goat, sheep, camel, mare and water buffalo. These milks differ somewhat in composition, but the main components of all of them are water, lactose, protein, fat, calcium phosphate and vitamin A.

In the United States, the dairy industry is almost exclusively concerned with cow's milk, and all the discussion in this chapter deals with that product unless the milk of another species is distinctly indicated.

FRESH MILK

The composition of milk is variously reported in the literature. The variation is explained by the fact that the composition differs with the breed of cow and also with her diet, the stage of her lactation and even with the time of day.

Sherman (1948) reported the milk from Jersey and Guernsey cows as containing 5.19% and 5.16% fat, and 9.68% and 9.53% solids not fat respectively; Ayershire and Holstein composition as 3.64% and 3.43% fat, and 9.09% and 8.53% solids not fat. Jerseys and Guernseys under special diet conditions may exceed 7% fat and Holstein milk may drop below 3.0%.

The composition of milk from different breeds of cows and that resulting from other variables is of concern mainly to dairymen and cattle breeders. Commercial production generally involves mixed herds to bring the composite milk to the standard that the market requires. The wholesale price of milk increases with the fat content, but the cows that yield milk of high fat content, produce less of it. Watt and Merrill (1963) reported commercial milk as: water 87.2%, protein 3.5%, fat 3.7% and carbohydrate 4.9%.

Legal Standards

Because of the variation in the composition of milk, the state governments have set minimum standards of composition. As of January 1, 1965, the USDA reported these standards, which may be changed at any meeting of a legislature. Five states—Hawaii, New Jersey, New York, Ohio, and Wisconsin—require a minimum of 3.0% fat and 8.25% solids not fat. Ohio requires 3.0% fat but a total of 12% solids. Colorado, Oregon, and Utah require 3.2% fat; 24 states require 3.25% fat; Massachusetts and New Hampshire 3.35%; Arizona, California, Delaware, Maryland, Michigan, Mississippi, Nevada, New Mexico, Tennessee, Vermont, Washington, and

West Virginia require 3.5%; Georgia 3.7%; Louisiana and South Carolina 3.8%. The total solids required vary from 11.0% to 12.3%.

It is generally illegal to add anything to milk except vitamins, which are permitted by most states and the vitamins must be declared on the label. The one vitamin permitted by all the states is D, to at least 400 units.

Sources

There are two sources of milk: the farmer who has a few cows and the large herd of a dairy. The farmer strains the evening milk into 5-gal. cans and cools it by water from a well or a spring. The morning milk goes through the same process and then the farmer delivers both morning and evening milk to a nearby milk plant, or the plant sends a truck around the countryside to collect it.

At the plant, each farmer's milk is filtered, weighed and a sample set aside for analysis for fat and dust. The milk is then run into a tank with all the other milks for further processing.

At the big dairies where the number of cows may be several hundred, the milk is handled variously according to the equipment and the nature of the business of the dairy. Dairies are necessarily in the country and those in or near a thickly populated area may process milk in the same manner as the milk plant and deliver it to the consumer, while those in more remote areas may simply strain and cool the milk below 50°F and ship it to a milk plant in a distant city.

Sanitation

Although the legal standards of composition must be met, the producer and processor is more concerned with sanitary matters. Milk is an excellent medium for the growth of microorganisms, some of which are in the milk before it is drawn from the cow and others are added from the cow, dust from the hay or the soil, or the hands and breath of the milk handlers. All these organisms grow faster at higher temperatures up to about 100°F which accounts for the requirement that the milk be cooled below 50°F as soon as possible after it is drawn, or pasteurized. In addition to the organisms such as *B. lactis,* which cause the milk to sour, there are the pathogens that may enter the milk from diseased cows or diseased personnel. In some states, the cows are required to have veterinary inspections and the personnel to have medical inspections at stated intervals.

The two most dangerous organisms from diseased cows are *Brucella abortus* and *M. tuberculosis.* The *B. abortus* is the organism of Bang's disease or contagious abortion in cows, and of undulant fever in humans. The latter is a lingering and fatal disease for which no cure was known until a remedy was announced a few years ago. How effective it is has not been publicized.

Tuberculosis in cows is readily transmitted by the milk and is also a lingering disease that was formerly incurable.

Contamination of milk from the milk handlers may be typhoid fever, scarlet fever or septic sore throat.

Pasteurization

To sterilize milk, it would be necessary to boil it, which would change all its physical properties. But over half a century ago, the pasteurization process was invented and has been widely used; it is now mandatory in many states.

Holding Process.—The older process of pasteurization is the result of much study of the effect of heat on both the milk and the organisms in it. The resulting conditions most widely used heat the milk to a minimum of 142°F for not less than 30 min. Every particle of the milk must be subjected to these conditions. The temperature may be allowed to rise to 145°F but not higher. These conditions destroy the pathogens and greatly reduce the population of the other organisms.

Flash Process.—A more recent process has come into use in which the milk is heated to 162°F for 10 sec. It is acceptable in most states, and the relative merits of the two processes are the concern of the milk processor.

Raw milk or milk that is insufficiently pasteurized can be detected by the phosphatase test. Raw milk contains an enzyme called *phosphatase*, which is an enzyme that hydrolyzes phosphate esters. Pasteurization destroys the enzyme and so the test is a measure of the amount of the enzyme in the milk and is said to be sensitive enough to indicate incomplete pasteurization.

The method uses a phenol phosphate ester, $Na_2C_6H_5PO_4$, which the enzyme hydrolyzes to phenol and Na_2HPO_4 and the amount of phenol is measured by a colorimetric method. The Official Methods give the details of the procedure.

Composition

Although pasteurization of milk frees it of pathogens, it may become infected again from milk handlers. However, if it is to be sold at retail, it is cooled and packaged immediately with little chance of contamination, but milk that is to be further processed (butter, cheese, dried milk) is susceptible to contamination. Although sanitation is a microbiological matter, the chemist frequently inherits the responsibility. Development of acidity and the phosphatase test, of course, are purely chemical matters.

Physical Nature of Milk

In addition to quantitative composition and sanitation, the physical condition of the components of milk is important in processing. Milk is about 87% water with some mineral salts, lactose and vitamins A, B_1, B_2, niacin and ascorbic acid dissolved in it. Pasteurization, however, destroys the ascorbic acid and about 20% of the thiamine (B_1).

The mineral content, which averages about 0.725%, is composed of calcium, potassium, phosphorus, chlorine, sodium, magnesium ions and traces of others of which the HCO_3- is the principal negative ion. Milk is such a dilute solution that its freezing point is only 31°F.

The soluble proteins are mainly albumin with a little globulin and traces of others although the solubility of the globulin is questionable. The principal protein in milk is casein; it has been found to vary from 1.79% to 6.29% with an average of 3.02%; the albumin from 0.25% to 1.40%, averaging 0.53%. The casein is in colloidal dispersion. The fat varies from 1.67% to 6.47% with an average of 3.64%. The fat is in the form of a rather coarse emulsion. The droplets range in size, but they are coarse enough to be clearly visible with a microscope and may be as large as 0.01 mm in diameter.

The specific gravity of milk fat is about 0.912 and that of whole milk is about 1.03. This difference causes the fat globules to rise to the top as the milk remains undisturbed. The housewife strained the milk into earthenware crocks, left it there overnight and then skimmed the cream off. Since this removed the richest part of the milk, the skimmed milk was looked upon as very poor and most of it was fed to the cats, dogs and pigs.

With the introduction of the cream separator into the industry, it became possible to separate a concentrated fat emulsion of different concentrations up to 94% and produce them at once.

This tendency of milk to fractionate creates problems. It separates somewhat in the udder of the cow. It is customary to milk twice a day, and in the interval, the fat concentrates at the top and so the last milk drawn is richer in fat than the first, which is why the legal definition of milk describes it as the "product obtained by the complete milking of one or more cows."

The rising cream created problems when milk was retailed in bulk; unless it was well stirred before a portion was withdrawn, the first portions were rich, the last were skim milk. The restaurant problem was solved by packaging the milk in drink size containers, but the housewife had a problem with the quart bottle. That problem has now been solved by passing the milk through a homogenizer, which breaks up the fat globules into extremely small ones of uniform size. This provides a much greater surface to the fat particles and results in a more stable emulsion. The milk does not form a cream line within its keeping time.

CREAM

The only difference between cream and milk is the fat content. During the standing or the centrifuging of the milk, the fat concentrates and thus produces a fraction of lower specific gravity. Since the separator will produce any fat concentration from that of the milk to over 90%, it is imperative that cream have a standard.

There are several adjectives used to describe creams of different fat content, but to be called *cream*, it must contain at least 18% fat, and if the fat is less than 30%, the cream may be homogenized. *Light cream, coffee cream, table cream* and *whipping cream* must contain 30% fat. *Light whipping cream* contains less than 36% fat and *heavy cream* or *heavy whipping cream* contains not less than 36% fat. These are the federal standards, but the standards of many states are the same with few exceptions. Some states do not define all the grades. Connecticut, Iowa and Massachusetts have a 16% minimum; California, Minnesota and Washington require 20% fat and Nevada 22%.

CONCENTRATED MILKS

There are both federal and state standards for several whole milk products that are concentrated for preservation and convenience in packaging.

Evaporated Milk

This product contains not less than 7.9% milk fat and not less than 25.9% of total milk solids. It may also contain disodium phosphate, sodium citrate, or calcium chloride; the total not to exceed 0.1% by weight and carrageenan not to exceed 0.015%. If the latter is used, there may be added sodium sulfosuccinate to the extent of half the weight of the carrageenan. The standard also contains details for the addition of vitamin D and for labeling. The product is heat processed.

Sweetened Condensed Milk

The limits for this product are a minimum of 8.5% milk fat and 28.0% total milk solids and sufficient sugar to prevent spoilage. The sugar may be sucrose, dextrose or both. There are no optional ingredients and the milk is not heat processed. However, it may be sweetened with corn syrup or corn syrup solids, but this variation must be stated on the label.

Some states permit slightly lower fat content in either evaporated or condensed milk or both.

OTHER DAIRY PRODUCTS

Many of the states recognize and set standards for fortified milk, 2% milk, cultured buttermilk, yogurt, chocolate flavored milk, half and half, cultured sour cream and whipped cream.

In *buttermilk,* solids, not fat are the only quantitative specification and 8.25% is the most common requirement; a few states require 8.0% and a few 8.5%.

Requirements for other products are too varied and too few to be listed here.

BUTTER

Butter has been the chief fat for the kitchen and the dining room for centuries and, although the consumption in the United States was down to 5 lb per capita in 1972, it is still one of the most important dairy products.

The chemical nature of butter is largely the result of the historical methods of production. The dairy maid or the housewife skimmed the cream off the milk that had stood for several hours, and the fat content of the cream depended on the skill of the skimmer. The cream was put into an earthenware jar to wait until enough accumulated for churning. This might be a week or more depending on the milk supply and so the cream invariably soured.

The churning was a mild agitation such as could be produced by pouring the cream from one vessel to another. Several types of churn were devised to furnish the proper degree of agitation. One type was a small barrel mounted so it could be turned end over end by means of a crank. Too violent agitation broke up the globules of fat and too little did not bring them together quickly enough to complete the churning in a reasonable length of time. The process was more rapid at higher temperatures up to about 65°F; the fat should be soft but not liquid. The melting range of milk fat is 68°–73°F (20°–23°C). If the fat is too cold, the globules will not adhere.

When the churning was finished large clumps of butter floated on the buttermilk and were removed with a wooden paddle. The butter was put into a large shallow wooden bowl and worked with the paddle until no more buttermilk could be drained off. Salt was then added and mixed evenly with the butter. Finally, the butter was formed into a roll about the shape of a loaf of bread and stored in a crock in the milkhouse for use or sale.

The cow concentrates carotene in the milk fat and the amount of it varies chiefly with the diet of the cow. On the farm, carotene was abundant in the summer when the cows were on pasture, but in the winter, it was scarce in the hay and grain of the feed. Consequently, the color of country butter varied from a pale cream color in winter to a deep yellow in summer. This did not bother the farm family, but the city dweller was inclined to think the white butter of inferior quality. It did, of course, supply less vitamin A.

During the latter half of the 19th century, creameries began to increase in number and by 1900, they were making 40% of the butter; by 1935, they were producing 75% of the market butter and today country butter in the groceries is a rare item.

With the transfer of butter making from the farm to the factory, Dairy Departments in the Colleges of Agriculture and Experiment Stations began research on several procedures in the butter-making process, and they have now been pretty well standardized, so that market butter is uniform.

Congress defined butter and set a standard for it in 1823, which reads in part: "Butter . . . is made exclusively from milk or cream or both, with or without common salt and with or without additional coloring matter and contains not less than 80 per centum by weight of milk fat all tolerances being allowed for."

The chemist who needs to analyze butter must determine not only the percentage of fat by an Official Method, but must also identify the fat. The Reichert-Meissl and Polenske Values are the most useful for the purpose.

In the manufacture of butter, the technologist has simply expedited and standardized the procedures of the dairy maid and added a few refinements.

Many creameries are owned by large corporations and operated by a manager, who is expected to produce all the butter that can be made from the milk he buys. The milk is weighed and the percentage of fat determined by a rapid method such as that of Babcock or Rose-Gottlieb. The milk is strained and passed through a separator and the fat removed as completely as possible. The cream is pasteurized. The milk is cooled to about 70°F and a culture of *Streptococcus lactis* or other acid producing organism added and the cream is held at that temperature until the acidity reaches 0.4%, calculated as lactic acid.

These procedures take less time that those of the domestic method and also remove pathogens and lessen the possibilities of "off flavors" caused by the absorption of odors or the action of organisms present in the raw milk.

The souring or *ripening* of the cream performs two functions: churning time is shortened and the characteristic flavor of butter develops. The mechanism of the effect of the ripening on churning has not been explained, but either the acid or the bacterial action weakens the film of protein that keeps the fat emulsified. Cultures of *S. citrovorus* and *S. paracitrovorus* are sometimes added to enhance the flavor. The principal flavor constituents are

$$CH_3CO-\underset{\underset{H}{|}}{\overset{\overset{CH_3}{|}}{C}}-OH \qquad\qquad CH_3-\overset{\overset{O}{\|}}{C}-\overset{\overset{O}{\|}}{C}-CH_3$$

Acetylmethyl-
carbinol

Diacetyl

The source of these compounds is uncertain, but there is evidence that they are produced by the action of enzymes on the citric acid, which is present in milk in small amounts.

The cream should be about 30% fat, 65°F and 0.4–0.5% acid for the best churning conditions, but the speed of the churn is also important.

When the churning is complete, the buttermilk is drawn off and the butter worked in the churn to remove more of the buttermilk. A weighed amount of salt is added, usually 1–2% of the weight of the butter. The grade of salt that can be used is limited because all of it must dissolve in the water remaining in the butter. Standard butter is 15% water, therefore a kilogram of butter contains 150 gm of water. At 0°C 100 gm water dissolves 35 gm salt and so the water in the butter will dissolve 52.5 gm salt or 5.25% of the weight of the butter. The actual solubility is somewhat less because of the other solutes present in the water and the storage of butter at 0°F which is −17.7°C. Salt is also a flavor, which places a further limit on the amount of salt used. Although 2% is probably enough for preservation, some samples contain as much as 3.5%.

A dye is added to produce a uniform color to all the butter of a given brand. Consumers, unfamiliar with country butter, may consider a pale colored butter as inferior or even spoiled. The dye used has varied over the years. As this is being written, there is no fat-soluble coal tar dye on the list of approved synthetic colors; annatto and carotene, both vegetable dyes, are the most likely colors now available.

The weight of the butter produced from milk of a given fat content is much greater than the weight of the fat because of the water and curd in the butter. The excess weight is called the *overrun* and amounts to about 22% under the best conditions.

Spoilage.—Off flavors in butter have many sources; they begin with the diet of the cow. A cow on pasture may eat wild onions or other odorous weeds in which the odor or flavor is fat soluble and not digested; therefore, it is concentrated in the milk fat. Exposure of the milk or the finished butter may cause absorption of odors from the environment. Also, the equipment used and the final packaging material are possible sources. Enzymes or organisms in the butter may cause a "tallowy" or "fishy" flavor and hydrolytic rancidity will develop if the butter is exposed to air and a warm temperature.

SKIM MILK

The separation of cream for butter making takes 10% to 12% of the volume of the milk, which leaves nearly 2.5 gal. of skim milk for each pound of butter made. The disposal of this milk has been a serious problem. It was used mostly to feed hogs until a few years ago, when processors began to dehydrate it because of its high nutritive value. The problem has been solved and in August 1974, the retail price of the dehydrated product was $1.28 a pound. Large amounts of it go into soups, salad dressings and other composite foods. It is marketed as *nonfat dried milk*. The federal standard for it limits the water content to 5% and provides that if vitamins A and D are added, an 8 oz glass of the reconstituted milk must contain 500 units of A and 100 units of D.

The nonfat dried milk has all the nutrient value of whole milk except the fat and vitamin A. The energy value is 350 kcal/100 gm, reduced from the 500 kcal/100 gm of dried whole milk. Dried to 4% water, it contains 36% protein, 51% lactose and 7.9% ash, the latter of which is rich in calcium, potassium and phosphorus. It also contains the water soluble vitamins thiamine and riboflavin.

CHEESE

The manufacture of butter has provided a luxury food for thousands of years; cheese is of similar lineage, but has been made largely as a means of preserving the nutrients of milk, because it keeps longer than milk or butter and preserves more of the milk than butter does.

It is impossible to say how many kinds of cheese there are; it has been made everywhere in the world that had an excess of milk and each product was named from the locality in which it was made; we have the familiar names of cheddar, Camembert and Roquefort along with many others. Then as local products became widely acceptable, they became types of cheese and were made outside the original geographic area. Both cheddar and Camembert are made in America and elsewhere, and Swiss is made in at least half a dozen countries. Our federal standards list 27 varieties, with one or more modifications of some of them. A USDA publication of 1969 describes over 400 varieties, but decides that there are only 18 distinct types of cheese.

Cheeses defy classification and so I shall describe only three distinct types: unripened cream, blue, and cheddar, for those three cover most of the known chemistry of the cheese process; the other varieties differ mostly in mechanical handling and in the ripening organisms.

Cream Cheese

The official standard defines cream cheese as "the soft uncured cheese prepared by the procedure set forth in paragraph (b) of this section [of the standard]. The finished cheese contains not less than 33% milk fat and not more than 55% moisture," and then designates the Official Method of analysis.

The procedure mentioned is given in detail in paragraph (b), but it covers several methods of procedure, and a brief description will indicate the chemistry involved. The cheese is made from cream and generally milk also, which may be fresh or processed just so the finished product meets the standard. The milk and cream mixture is pasteurized, cooled to about 80°F and a pure culture of lactic acid forming bacteria added. Sometimes rennet is added. The mixture is allowed to stand until a curd forms. It may take 12 hr for the mixture to reach the required 0.3–0.4% lactic acid. During this period, the fat begins to concentrate in the upper portion and make the final product nonuniform in fat content. To avoid this, the milk mixture is frequently homogenized before it is pasteurized.

Salt and any one or more of the usual food gums such as karaya, tragacanth, carrageenan and others may be added in an amount up to 0.5% of the weight of the finished cheese.

There are several ways of handling the curd after it has set. The simplest is to stir it and then transfer it to a cloth bag to drain.

The average cream cheese on the market exceeds the minimum standard; the water content is usually about 51% and the fat over 37%.

Cream cheese is unripened and must be kept under refrigeration until it is used. It is packaged in a cardboard or plastic container of consumer size.

Blue Cheese

This cheese is characterized by a blue-green mold throughout the cheese. It must not exceed 46% moisture and the dry solids must be at least 50% fat. It must be 60 days old before it is sold.

Blue cheese was not made in this country until the 1930s, but three famous ones were imported: Roquefort from France, Gorgonzola from Italy and Stilton from England.

Roquefort is made in southeastern France and Corsica and brought to the limestone caves in the city of Roquefort to be finished and ripened. It is made exclusively from sheep's milk and ripening has been carried on there since the days of Julius Caesar. It became the most popular of the blue cheeses both in Europe and in this country. Consequently, the blue cheese made from cow's milk in France, Denmark, the United States and elsewhere is all made to resemble Roquefort as closely as possible.

Blue cheese is made from cow's milk, which is much cheaper than ewe's milk. It is somewhat milder in flavor than Roquefort, but to call it Roquefort in a restaurant or a market is to invite prosecution.

The ewe's milk is not pasteurized. Ewes do not have Bang's disease and so the worst of the pathogens is absent from the milk; others do not survive the curing process. Ewes yield 500 to 1000 ml of milk daily for about 8 months. Sheep, unlike cows, do not breed at all seasons and the milk supply to the cheese factories arrives from February to September with the greatest supply in May. At one time, there were 301 creameries in five departments of southeastern France and Corsica that supplied cheese to the caves at

Roquefort. With improvement in roads and transportation, the number has been reduced.

The milk is collected from the shepherds in cans and taken to the cheese factory where it is measured and strained. In the smaller factories, the vats resemble large wash tubs; those in the larger plants are of the usual rectangular type. The milk has the following composition:

Fat	7.0–10%
Protein	5.0–6.0
Lactose	4.5–5.5

The milk is heated to 70°–80°F and rennet is added. In 1½ to 2 hr, the milk is coagulated and the curd is cut with a wire curd knife to facilitate drainage. A board is laid across the top of the vat and a hoop placed on it. The hoops are open-ended metal cylinders 7½ in. in diameter and 6 in. high. The wall is perforated. The curd is dipped into the hoop either by hand or a scoop and pressed slightly by hand. Whey drains back into the vat. The curd is placed in the hoop in three layers and dry mold spores added after the first and second layers. The curd is piled high above the top of the mold.

Over 90% of the Roquefort cheese is made by a cooperative that collects the cheese from the factories. It also sets standards of procedure and supplies the rennet and the mold, which is produced outside the organization; it is *Penicillium roqueforti,* a variety of *P. Glaucum.* It is grown on fresh bread, dried and ground to a powder, which is sifted on the layers of curd from a can that resembles a flour sifter.

After the hoops have been filled, they are inverted at regular intervals for several hours so the curd is well drained and well within the hoops. After five days, the cheese is very firm and a truck comes from Roquefort to take the cheese to the caves, which are limestone caverns in the mountainside on which Roquefort is built. The caves remain at 50°F and 95% humidity with slow circulation of air throughout the year. They are natural caves that have been altered only enough to make them workable—level floors, passages from one to another, etc.

The cheese is received in a large room in the cave where it is salted by rubbing dry salt on the surface, stacked three high and salted again after three or four days. The excess salt is then brushed off and each cheese perforated by a machine with about 50 steel needles each 3 mm in diameter. The cheese loaves then are placed on edge on a rack in a storage room with space between them for the circulation of air. Mold growth requires oxygen, which accounts for the perforation and the loose storage.

The ripening period is 2 to 5 months during which the enzymes from the mold partially hydrolyze both protein and fat. The fat contains a higher percentage of the C_6, C_8 and C_{10} acids and it is these free acids that constitute the greater part of the characteristic flavor although there are several other substances present in smaller amounts.

The curd is not weighed and so the cheeses differ in weight from 5 to 7 lb. At the end of the ripening period, which varies in length according to the demand of the various markets, the cheese is "put to sleep" as the French express it, that is, it is brushed to remove excess mold and salt, wrapped in tinfoil to exclude air and placed in storage at about 40°F.

The finished cheese is white except for the blue streaks of mold, because sheep do not store carotinoids in the milk fat. An average analysis is: moisture 40%, fat 32%, protein 21% and ash 6%, of which 40% is salt.

Blue cheese is made from cow's milk by the Roquefort process with a few modifications. Making cheese is largely an art and, although it would seem easy to adapt the Roquefort process to cow's milk and large scale production, many problems were encountered and nearly 20 years were required to bring the process to success in this country. Attempts to make it were begun about 1918, but no marketable cheese was produced until well into the 1930s. Several American companies now compete with the French, Danish and other European varieties.

The vats used are shallow rectangular metal structures with arrangements for heating or cooling the milk. The milk is weighed, inspected and analyzed as in the butter factories. The milk is then bleached with benzoyl peroxide, C_6H_5COOOH. The legal limit of the bleach is 0.002% of the weight of the milk. Vitamin A is usually added, for the bleaching destroys the carotene and thus reduces that vitamin. The U.S. standard also permits neutralizing the yellow color with a blue-green dye instead of bleaching.

Lactic acid bacteria and rennet are added at a temperature of 80° to 85°F and the curd sets in about 1½ hours. The curd is then cut and drained. The dry mold spores are mixed with it and it is transferred to a hoop in the same manner as the Roquefort curd is handled, and inverted at regular intervals for 24 hr.

The formed cheese is then dry-salted, perforated and aged for 60 days or longer in a room at 50°F and 95% humidity. In the early efforts to make the cheese, the importance of the atmospheric conditions for the aging were underestimated and, furthermore, the air conditioning equipment was rather primitive in the 1920s. A coal mine, an abandoned brewery cave and a limestone cavern were all tried as duplicates of the Roquefort caves, but in the end, air conditioning won.

The finished cheese is wrapped in tinfoil that carries the label. If the milk has been bleached with benzoyl peroxide, the label must say so. The size, shape and weight are the same as those of Roquefort. The standard limits the moisture to 46% and the dry solids must be 50% milk fat. An average analysis is: moisture 40%, fat 30.5%, protein 21.5%, carbohydrate 2.0% and ash 6%, 315 mg Ca, 339 mg P, 1240 units vitamin A and 368 kcal per 100 gm.

From Garard (1974)

FIG. 11.1. DANISH BLUE CHEESE

Cheddar

Cheddar cheese was named for the village of Cheddar in England where it was first made, probably in the 17th century. The process was brought to America by the colonists and its manufacture has become so widespread in this country that it is commonly called American cheese. It is now a big industry, for per capita consumption reached 13.1 lb in 1972.

For over two centuries, the production of cheddar cheese in the United States was a domestic process, but in 1851, the first cheese factory was established. Today, many cheese factories in the United States combine to produce 6.3 billion lb of cheese of which about two-thirds is cheddar.

Originally, cheddar cheese was a cream color, the natural color of the fat and the curd, and uncolored cheese is again appearing in the market, but most of it is the familiar orange color of annatto.

The milk is received at the factories and undergoes the usual treatment that it receives in creameries. It may also be homogenized and pasteurized; the latter is required by law if the cheese is to be sold under 60 days. The effects of pasteurization are controversial. Some cheese makers have claimed that a good cheese cannot be made from pasteurized milk and some claim that such a cheese is more uniform and of better quality. The latter would appear to be right for about 90% of the cheese made at present is from pasteurized milk.

The milk is heated or cooled to 85°–90°F and a lactic acid culture is added. The milk is stirred until the acidity reaches 0.2% lactic acid. The exact time depends on the amount of the culture used and the temperature. When the proper acidity is reached, the rennet and color (if any) are added and mixed thoroughly with the milk, which is then left undisturbed for the curd to gel. The time required for the gel to reach proper consistency varies from 15 min to an hour with the quality and amount of the rennet added.

When the gel is firm enough, it is cut into ½ in. cubes and whey is drained off from the bottom of the vat. The cut curd is stirred gently to facilitate drainage without breaking the cubes. In about 15 min, the curd ceases to drain and the temperature is slowly raised to 100°F which fuses the curd into a solid mass, and several chemical changes occur, although just what they are is not definitely known. Calcium combines with the casein and also some of the lactose becomes calcium lactate. The protein undergoes some change and some fat is hydrolyzed. The acid remaining in the curd enters into these changes.

After the draining and heating, the curd is 7–8 in. deep and is cut into strips 5–6 in. wide which are turned frequently until firm enough to be piled up. Then more whey drains off and the texture of the cheese begins to develop. This treatment is called *cheddaring*. The cheddared cheese is then cut into slabs, put through a curd mill, and spread on the bottom of the vat. Salt is added and the cheese is stirred. The salt dissolves and some whey drains away.

Finally, the cheese is pressed in metal hoops and placed in storage to age. The standard cheese is about 14½ in. in diameter, 12 in. thick and weighs 70 to 80 lb, but there are several other sizes and shapes. Unlike blue, Camembert and several others, the size of a cheddar cheese is unimportant in the ripening process.

The legal standard requires not more than 39% moisture and not less than 50% milk fat in the dried cheese. If the milk is not pasteurized, the cheese must be cured for 60 days at not less than 35°F.

An average analysis of cheddar cheese is: 37% moisture, 25% protein, 32.2% fat, 2.1% carbohydrate and 3.7% ash, which is mostly salt; also 750 mg calcium, 478 mg phosphorus, 1210 units vitamin A and 308 kcal per 100 gm.

Whey

In making any kind of cheese, the water of the milk is largely eliminated, and with it goes most of the lactose, soluble protein and minerals. An average analysis of whey is 90% water, 0.9% protein, 0.3% fat, 5.1% carbohydrate and 0.6% ash, which is mostly calcium phosphate. It also contains vitamin A, thiamine and riboflavin; it has an energy value of 26 kcal per 100 gm.

Despite its high food value, whey is a problem for the cheese manufacturer; most of it is run into the sewers where it overloads the disposal plant. The high water content makes the recovery of the food values of doubtful economic value. To evaporate the water would require over 600 cal per gm and to remove it by freezing would require over 80 cal per gm. Then, it would be necessary to educate the public to the use of whey solids, or the manufacturers to include it in composite foods.

At present, some of the food value of whey is retained by feeding it to hogs. Also small quantities are dehydrated and used in cheese spreads, salad dressings and a few other food products. It might be possible to convert it into a soft drink as a possible solution to the problem, but that would involve considerable chemical work to obtain palatability and it would also constitute a huge marketing problem.

Process Cheese

Most cheese continues to ripen during storage. Metal foil inhibits the ripening of blue cheese and a coating of paraffin protects several other kinds, but some organisms continue to grow and some of the ripening is enzyme action. Mild, sharp and extra sharp describe various states of ripening in cheddar cheese, but after extra sharp comes too sharp. In the United States and some other countries, there was an overproduction of cheese, and overripening was a problem. The solution lay in inactivating the organisms and the enzymes. As early as 1895, the Germans and Swiss were heating cheese and canning it. Then, in 1916, the first patent for processed cheese was issued in the United States.

The cheese is ground and run into a steam jacketed kettle. If it is too ripe, some mild cheese is mixed with it. One of several phosphates or citrates is added as an emulsifier and the batch is heated to 160°F and thoroughly mixed. The heating inactivates the ripening agents.

At first, the cheese was cooled and packaged in 5-lb blocks. The block was elongated with a 4 × 4 in. cross section which made it a convenient size for sandwiches. Some retailers even called it sandwich cheese.

Processing the cheese not only solved the overripening problem, but it created a very popular product, and now about a third of the cheese marketed in this country has been processed.

The requirements and permissions of the federal government for processed cheese are too long and too detailed for inclusion here, but a few of the specifications may be mentioned. The heating must be at least 150°F and the cheese held at that temperature for at least 30 sec. Moisture is limited to 43% and the fat content of the dried cheese must be 47%, except processed Swiss which must be 43% and gruyere, 45%.

Unripened cheese such as cottage, cream and Neufchatel, and a hard grating cheese cannot be used, but almost any kind of ripened cheese can be processed either alone or with other varieties. Where two or more varieties are used, each must be at least 25% of the total except that the limit for blue is 5% and for limburger, 3%.

There is considerable choice of emulsifier, but really only three: there are 9 sodium or potassium phosphates; sodium, potassium or calcium citrate; and sodium and sodium potassium tartrate.

The cheese may be acidified with vinegar or lactic, citric, acetic or phosphoric acid or any mixture of them, but the pH must not be below 5.3.

Water, salt, coloring, spices and flavorings (other than cheese flavoring) may be added. If the cheese is marketed in slices or in a consumer size package, it may contain 0.2% by weight of sorbic acid or its sodium or potassium salt, or 0.3% sodium or calcium propionate as a mold inhibitor.

$$CH_3—CH=CH—CH=CH_2COOH$$
Sorbic Acid

There are also definitions and standards for processed cheese with fruits, vegetables and meats. And for all of them, there are detailed labeling requirements.

Cheese Foods

About 50 years ago, one of the large cheese manufacturers added whey solids to cheese and made a product that closely resembled processed cheese, but it could not be called cheese because it did not meet the existing definition. So the company gave it a proprietary name; in Germany, in 1937, it was exhibited as a variety of American cheese. We now have definitions and standards for such products under the name of *cheese foods*.

In making cheese foods, the mixture must be heated to 150°F for not less than 30 sec and must pass a phosphatase test for adequate pasteurization.

The moisture content must not exceed 44% and the fat content must be at least 23%. Cheese must constitute 51% of the mixture. If two or more varieties of cheese are used, there are quantitative requirements similar to those for processed cheese. Hard cheese and the unripened varieties are excluded but the cheese food may be smoked.

Dairy ingredients other than cheese that may be used are: cream, milk, skim milk, buttermilk, whey and pure milk fat.

The emulsifying agents are the same as those used in processed cheese, as are water, coloring, spices, flavorings and mold inhibitors. There are also standards for cheese foods containing fruits, vegetables or meats. Labeling requirements are detailed.

Cheese Spreads

Cheese spread is essentially a soft cheese food, but it has its own standards. Among the differences are: the moisture content must be between 44% and 60% and the fat at least 20%. The product must spread at 70°F. Any of the usual food gums may be included but not to exceed 0.8% of the weight of the finished product. Sweetening agents may be sucrose, dextrose, corn sugar, corn syrup, corn syrup solids, glucose syrup, glucose, glucose syrup solids, maltose, malt syrup and hydrolyzed lactose in a quantity necessary for seasoning. A consumer package may contain 0.2% sorbic acid. There are also standards for cheese spreads containing fruits, vegetables and meats.

ICE CREAM

The federal definition and standard for ice cream is long and detailed. It begins: "Ice cream is the food prepared by freezing, while stirring, a pasteurized mix composed of one or more of the optional ingredients specified in paragraph . . .". Then follow paragraphs (b), (c), (d), (e) and (f) each of which contains options.

Briefly the options in (b) are:

1. Ground spice or vanilla beans, an infusion of coffee or tea or any natural food flavoring.
2. Artificial flavoring.
3. Chocolate or cocoa.
4. Mature fruit.
5. Nut meats, roasted, cooked in oil, preserved in syrup or salted.
6. Malted milk.
7. Confectionery.
8. Cooked cereal.
9. Alcoholic beverage not to exceed the amount necessary for flavoring.

(c) This paragraph lists 23 varieties of milk and cream. Among them are cream, butter, milk, buttermilk, concentrated milks and skim milk. The other 17 are variations of these.

(d) This paragraph contains 13 sweeteners, among them are sucrose, dextrose, invert sugar, corn syrup, maple syrup, honey, malt syrup, molasses, lactose and fructose.

(e) Ammonium, calcium, potassium and sodium caseinate.

(f) includes:

1. Liquid eggs or egg yolks, fresh, frozen or dried.
2. The usual gums.
3. Mono- or diglycerides.
4. Sorbates.
5. Propylene glycol alginate.
6. Micro-crystalline cellulose.
7. Dioctyl sodium sulfosuccinate.

8. Sodium citrate or phosphates. Calcium or magnesium oxide, hydroxide or carbonate.
9. Labeling requirements.

Many of the numerous optional ingredients have quantitative limits.

The finished ice cream weighs 4.5 lb to the gallon and contains not less than 1.6 lb of total solids; milk fat not less than 10% and total milk solids not less than 20% of the weight of the finished ice cream.

In addition to the federal standards, all the states have standards. Indiana, Vermont and Wisconsin adopt the federal standards. Texas and West Virginia set the fat content at 8%, Hawaii and Iowa at 12% and Nevada at 14%.

An average analysis of 10% ice cream is: water 63.2%, protein 4.5%, fat 10.6%, carbohydrate 20.8% and ash 0.9%; 100 gm contains Ca 146 mg, P 115 mg, Fe 0.1 mg, vitamin A 440 units, thiamine 0.04 mg, riboflavin 0.21 mg, niacin 0.1 mg, ascorbic acid 1.0 mg and 193 kcal.

OTHER FROZEN DESSERTS

There are federal standards for frozen custard, ice milk, fruit sherbets, nonfruit sherbets and both fruit and nonfruit water ices. The frozen custard or French ice cream is essentially the same as the ice cream above described except that egg yolk solids must constitute 1.4% of the weight of the finished product.

Ice Milk

The chief difference between ice cream and ice milk is in the fat content. Ice milk must contain 2% milk fat but not over 7% and total milk solids must be 11%; it must weight 4.5 lb to the gallon and a gallon must contain 1.3 lb of food solids. In some state standards, the milk fat minimum is as high as 4% and the maximum 10%. The weight of a gallon varies from 4.2 to 5 lb, but the 1.3 lb of food solids seems to be universal. An average analysis is: water 66.7%, protein 4.8%, fat 5.1%, carbohydrate 22.4% and ash 1.0%, and in 100 gm Ca 156 mg, P 124 mg, Fe 0.1 mg, vitamin A 210 units, thiamine 0.05 mg, riboflavin 2.2 mg, niacin 0.1 mg, ascorbic acid 1 mg and 152 kcal.

Sherbets

Fruit sherbet must contain not less than 1% and not more than 2% milk fat and total milk solids between 2% and 5% of the weight of the finished product. It must weigh 6 lb to the gallon. The milk must be pasteurized and the sherbet must have an acidity of 0.35% calculated as lactic acid; it may be acidified with citric, ascorbic or phosphoric acid.

The fruit is any mixture of fruit or fruit juices, either of which may be fresh, frozen, canned, concentrated or dried and the sherbet may be thickened with pectin or gums. With citrus fruits, the whole fruit may be added, including the peel but not the seeds. The fruit or fruit juice content must be 2% of the weight of the sherbet if it is citric, 6% with berries and 10% in sherbets made from other fruits.

The dairy products may be any of the forms of cream, milk or whey. The sweeteners are the same as those for ice cream.

Among the optional ingredients are: eggs, food gums, mono- and di-glycerides, sorbates, citric, tartaric, malic, lactic, ascorbic and phosphoric acids; casein, coloring, flavoring and microcrystalline cellulose. An average sherbet contains: water 67%, protein 0.9%, fat 1.2%, carbohydrate 30.8% and ash 0.1%, vitamin A 60 units and 134 kcal/100 gm.

The nonfruit sherbets have essentially the same composition as the fruit sherbets except that instead of fruit they are flavored with spices, coffee, tea, chocolate, cocoa, distilled alcoholic liquors and artificial flavors other than fruit flavors.

Water Ices

The ices do not contain dairy products and have no fat content specification. Otherwise they have the same composition as the sherbets.

ANALYTICAL PROCEDURES

The food standards for the processed dairy products serve as a guide to the chemist who wishes to develop a formula for one of these items. He has a definite list of permitted substances to choose from. The standards for the different kinds of cheese contain the directions for making them; that is, they contain the general directions. The special skills of the cheese-making art must be learned by experience or under the direction of an expert.

The analytical chemist is not so fortunate. The Official Methods contain procedures for a gross analysis, such as moisture, fat and protein, and some for specific acids, gums or colors. But the chemist who wishes to analyze a processed food, either to duplicate it or determine its legal status is adrift on an endless sea. With the number of options in ice cream, for instance, the number of combinations approaches infinity, and since these options are included in the standard, he may be sure somebody has used them either experimentally or, more likely, commercially. The Official Methods for ice cream contain procedures for weight per unit volume, total solids, nitrogen, fat, lactic acid, gum, alginates, gelatin, coloring matter and phosphatase.

One company advertises 28 flavors. What is each flavor composed of? Coloring is relatively simple because there are only a few legal ones. The identification of the milk fat is routine, but estimating the amount of mono-

and diglycerides may be a problem and several of the other optional ingredients have quantitative limits.

The varieties of cheese are also a problem. The Official Methods contain procedures for moisture and fat, which are the usual quantitative determinations. Salt content is simple, but how does one distinguish one kind of cheese from another? Of the common varieties in the American market, Roquefort is the only one made from ewe's milk and may be identified by the higher Polenske number of the fat and the lack of carotene. After a routine determination of moisture, fat, etc., the microbiologist may be able to identify the ripening organism, but the chemist must depend largely on taste and appearance, either of which will prove to be controversial in case legal action is involved. Objective distinction between varieties of cheese is a challenging field for research.

BIBLIOGRAPHY

ANON. 1972. Cheeses of the World. Dover Publications, New York.
ANON. 1975. Code of Federal Regulations 21: Food and Drugs. Parts 10–199. U.S. Govt. Printing Office, Washington, D.C.
ANON. 1965. Federal and State Standards of the Composition of Milk Products, Agricultural Handbook No. 51, U.S. Dept. Agr., Washington, D.C.
ARBUCKLE, W. S. 1972. Ice Cream, 2nd Edition. Avi Publishing Co., Westport, Conn.
ASSOC. OFFIC. ANAL. CHEMISTS. 1975. Official Methods of Analysis, 12th Edition, Assoc. Offic. Anal. Chemists, Washington, D.C.
DAVIS, J. G. 1941. Enzymes in Cheese. Chemistry and Industry. 259–265.
EITMILLER, R. R., VAKIL, J. R., and SHAHANI, K. M. 1970. Production and properties of Penicillium roqueforti lipase. J. Food Sci. 35, 130–133.
GARARD, I. D. 1974. The Story of Food. Avi Publishing Co., Westport, Conn.
GILLIES, M. T. 1974. Whey processing and utilization. Noyes Data Corp., Park Ridge, N.J.
GILLIES, M. T. 1974. Dehydration of Natural and Simulated Dairy Products. Noyes Data Corp., Park Ridge, N.J.
GOULD, I. A., and TROUT, G. M. 1936. The Effect of Homogenization on some of the characteristics of milk fat. J. Agr. Res. 52, 49–57.
GOVINDARAJAN, S., and MORRIN, H. A. 1973. Pink discoloration in cheddar cheese. J. Food Sci. 38, 675–678.
HALL, C. W., and HEDRICK, T. J. 1975. Drying of Milk and Milk Products, 2nd Edition. Avi Publishing Co., Westport, Conn.
HALL, C. W., and TROUT, G. M. 1968. Milk Pasteurization. Avi Publishing Co., Westport, Conn.
HENDERSON, J. L. 1971. The Fluid-milk Industry. Avi Publishing Co., Westport, Conn.
HERB, S. F., MAGIDMAN, P., LUDDY, F. E., and RIEMENSCHNEIDER, R. W. 1962. Fatty Acids of cow's milk. B. Composition by gas-liquid chromatography aided by other methods of fractionation. J. Am. Oil Chemists' Soc. 39, 142–146.
KINSELLA, J. E. 1975. Butter flavor. Food Tech. 29, No. 5, 82–98.
LEE, F. A. 1975. Basic Food Chemistry. Avi Publishing Co., Westport, Conn.
MACY, I. G., KELLY, H. J., and SLOAN, R. E. 1953. The composition of milks. Composition and properties of human, cow and goat milk, colostrum and transitional milk, Publ. 254. Natl. Acad. Sci., Washington, D.C.
MARRE, E. 1906. Le Roquefort. E. Carrere, Rodez, France.
MARTH, E. H. 1963. Microbiological and chemical aspects of cheddar cheese ripening (Review) J. Dairy Sci. 46, 869–890.
MORR, C. V. 1976. Whey protein concentrates: an update. Food Tech. 30, No. 3, 18–22.

PATTON, S. 1964. Volatile acids of Swiss cheese. J. Dairy Sci. *47*, 817–818.

SCHWARTZ, M. E. 1973. Cheese Making Technology. Noyes Data Corp., Park Ridge, N.J.

SHERMAN, H. C. 1948. Food Products, 4th Edition. Macmillan Co., New York.

WATT, B. K., and MERRILL, A. L. 1963. Agriculture Handbook No. 8, Composition of foods. ARS-USDA, Washington, D.C.

WEBB, B. H., and WHITTIER, B. O. 1971. Byproducts from Milk, 2nd Edition. Avi Publishing Co., Westport, Conn.

CHAPTER 12

Oil and Fat Industries

Food fats are commonly known as cooking fats, butter, salad oils and shortenings. Butter and shortenings are plastic fats, salad oils are liquids and cooking fats may be either. Any salad oil may also be used for frying or sautéing. The word "shortening" has two meanings: to the housewife, it is any fat used in baking; in the industry, it is a manufactured plastic fat but not lard, which has been the chief shortening agent used in this country from its beginning.

Fat contributes to the taste of food, to the texture of bakery products and to speed in cooking. In the diet, fats contribute to satiety; they supply 9 kcal per gm and the unsaturated fats are essential nutrients.

The two fats of longest use in the United States are butter and lard. Butter has been the favorite fat because of its flavor, and was used mainly to spread on bread, which needs something to improve its palatability. Some was used in cooking or used to season mashed potatoes or sauces for meats or vegetables. It was too scarce for use as shortening except in some kinds of cake. The production of butter was described in Chapter 11. The importance of the various fats in this country is indicated by Table 12.1.

LARD

Most animal tissues contain fat; the leanest of beef contains 5% fat and the leanest pork, 9%. The edible portion of a fat hog is about 40% lean and 60% fat. The separable fatty tissue averages 11% water, 3% protein and 85% fat. Lard is the pure pork fat rendered from this tissue.

There are two processes for rendering lard. Most of it is rendered in heavy steel cylinders with steam under pressure and is known as prime steam lard. Some is dry rendered by heating in a closed kettle with a steam jacket. There is some hydrolysis of the fat because of the steam or the water in the tissues. Neutral lard is made in a jacketed kettle and is rendered at a low temperature only slightly above the melting point of the fat.

Within the body cavity of a fat hog, there is a layer of fat. This, together with some other interior fat is usually rendered by the dry process and is called *leaf lard*. It is the highest grade of lard.

The hog stores the fat of its diet to a considerable extent in the body fat. Consequently, if the hog is fed a high-fat diet, such as peanut or cottonseed press cake, the fat becomes characteristic of the oil. In fact, some lard will show a test for cottonseed oil. In both cases, the lard will have a higher iodine value and a lower melting point.

TABLE 12.1

CONSUMPTION OF FATS AND OILS IN THE UNITED STATES

Fat	Lb Per Capita	
	Ave. 1957–1959	1972
Butter	8.2	5.0
Margarine	8.9	11.3
Lard	9.3	3.6
Shortening	11.4	16.9
Other fats and oils	10.8	19.5
Total	45.3	53.1

TABLE 12.2

EFFECT OF DIET ON HOG FAT

Feed	Corn and Tankage	Peanuts	Soybeans	General
Oil in feed %	4.3	33.1	17.5	?
Iodine value	126	95	128	?
Analysis of Body Fat				
Iodine value	60.8	80.6	93.2	62–65
Index of refraction	1.4580	1.4626	1.4630	1.4609–1.4620
Melting point °C	39.1	19.4	26.6	33–46
Titer	40.3	28.1	27.2	37–46.6
Saturated acids (%)	39.3	20.0	27.2	45.5
Oleic acid	52.1	58.0	32.0	43.5
Linoleic acid	8.6	21.4	32.0	10.5

Source: Data of first three columns from Ellis, and Isbell (1926).

The only refining that lard receives is a clarification and possible bleaching. Fuller's earth or activated carbon is added in small quantity, mixed well and filtered out.

From the data in Table 12.2, it is obvious that some batches of lard are too soft for use in shortening. In some cases, this is corrected by partial hydrogenation of a portion to a titer of 55 to 60 and adding it to unhydrogenated lard.

Hydrogenated Lard

In recent years, a hydrogenated lard has become popular. A soft lard is refined and then treated with sodium bicarbonate or sodium hydroxide to reduce the free acids to about 0.02%, filtered, and the entire batch hydrogenated to an iodine number of 60. It is then deodorized and packaged.

Spoilage

There are two types of spoilage that affect lard—mold and rancidity. Lard is shipped to bakers and other food processors in steel drums. The moisture in the head space condenses on the lid and drops to the surface of the lard; there is enough protein in the lard to support the growth of mold.

Far more common than mold, however, is oxidative rancidity, which occurs both in packages of pure lard and in products that contain it; the latter are especially vulnerable. Prepared mixes, frozen pie crust and many similar products need a shelf life of several months, and many such products are in cardboard containers that "breathe" with changes in temperature. This keeps them supplied with air and moisture in addition to that supplied by the original mixing.

The Kreis test reveals the rancidity of lard and, of course, the odor does also, but the problem of the processor is: how long will the lard keep before it becomes rancid? Several tests for this property have been devised; all of them accelerate the rancidity by producing optimum conditions for oxidation.

The Schaal Test heats a sample in an oven at 60°C. Another method (unpublished) exposes small samples to sunlight in individual containers so that a sample can be removed at intervals and given the Kreis test. Probably the best test for the keeping time of lard is called the Swift Stability Test, which was devised by chemists in the Swift laboratory (King, Roschen, and Irwin, 1933). In this method, air is bubbled through a sample of the fat heated to 97.7°C. A special apparatus is required, from which samples are withdrawn from time to time and given the Kreis test.

A sample of lard may keep for 10 hr or more under the conditions of the Swift test, but it remains to equate this keeping time with the stability of the product, which can only be established by shelf-life tests at room temperature, under refrigeration or under other conditions to which the product may be exposed during marketing.

The keeping time of a pure fat is not necessarily the same as that of the same fat in a composite product that may contain flour, starch, cornmeal, sugar, salt, baking powder or other ingredients. A lard may keep only 8 hr in the Swift apparatus and a hydrogenated shortening may last 100 hr, but mix them with some of the ingredients just mentioned and the keeping time may be reversed. The Swift test is not generally applicable because some vegetable shortenings develop a disagreeable odor and taste that is not the typical odor of oxidative rancidity, but which makes them unfit for use long before they show the Kreis test.

Several antioxidants have been patented or proposed. BHT, BHA, gum guaiac resin, propyl gallate and thiopropionic acid are on the GRAS list and are the most frequently used.

$$\begin{array}{c} COOC_3H_7 \\ \\ HO-\!\!\!\bigtriangleup\!\!\!-OH \\ \\ OH \end{array}$$

Propyl Gallate

$$C_2H_5-\overset{\overset{\displaystyle S}{\|}}{C}-OH$$

Thiopropionic Acid

The amount of antioxidant that may be used is limited to 0.02% of the weight of the fat except in the case of gum guaiac, which is limited to 0.01%. The use in each product must be determined separately, for an antioxidant may prolong the life of a pure fat, but the ingredients of a mixture may extend or shorten the effect of it.

The production of lard in 1972 was 1.5 billion lb.

TALLOW

In 1972, there were over 22 trillion lb of beef produced in the U.S. and 542 million lb of lamb and mutton. Tallow is the rendered fat of beef or mutton. Mutton tallow has a strong unpleasant flavor and is not much used in food processing.

Beef tallow is bland, almost flavorless, which makes it useful in the manufacture of margarine, but almost useless as a cooking fat. As a shortening, it is too firm to be easily mixed with the flour. About 200 million lb are produced annually, but the food industries consume far less and most of that is separated by warming and pressing. The liquid fraction is *oleo oil* and the solid fraction *oleo stearin*. Tallow has a saponification value of 193–199, an iodine value of 36–40 and a melting point of 48°C. Oleo oil has about the same saponification value, an iodine value around 45–48 and melts at 33°C; oleo stearin has an iodine value about 25. The constants of these fractions are variable depending on the extent of the pressing.

VEGETABLE OILS

Because of the competition of the soap industries for fats, the supply of edible fats is limited. For centuries, olive oil was the chief vegetable oil used for food. The seeds of tropical and subtropical plants are frequently rich in oil, so that residents of tropical countries have used them locally as sources of oil but only a few have reached commercial production.

Worldwide, the principal edible vegetable oils are cottonseed, olive, peanut, rapeseed, soybean, sunflower, corn, safflower and sesame.

Olive Oil

The olive tree is *Olea europaea* and, according to Jamieson (1943) there are over 300 varieties of it. Not only the yield of olives but also their oil content varies enormously, and the yield of both fruit and oil and the quality of the oil vary with the tree variety and the climate in which they are grown. Some olives are produced in California and Arizona, but most of them are pickled; a little oil is made from the small and imperfect ones. Our main supply of olive oil is imported from the countries around the shores of the Mediterranean. Spain, Italy, Tunisia and Greece are the main sources. Some of the oil from Spain and France comes from olives grown in Africa.

The oil content of olives varies from 14% to 40% and increases as the olives ripen. Olives mature from October to December according to climate and are harvested for oil at early maturity. Green olives contain a bitter substance that appears in the oil if they are harvested too soon; if they are too ripe, the oil may be rancid.

There are both large and small operations in the olive growing countries, so olive oil is subject to considerable variation. The exporters often treat or blend the oils to obtain a uniform product. The olives are crushed and pressed at low pressure and the oil thus obtained is of the highest grade and is called *virgin olive oil*. The press cake is then ground with a little water and pressed again at a higher pressure; the second and third pressings produce the bulk of the commercial olive oil. Even after three pressings, the press cake may contain 15% oil, which is extracted with carbon disulfide or other solvent. This oil is seldom refined for food use but goes to the soap industry. The press cake is fed to animals or returned to the groves as fertilizer.

When the olives are pressed, considerable juice is obtained, for the olives are about 80% water. The pressings are run into tanks where the oil collects on the surface and is drawn off or separated with a centrifuge. The oil is then put through a special filter to clarify it and remove the water, which promotes rancidity. In some cases, the oil is washed with water before filtration to obtain a milder flavor, but that and the filtration are all the refining edible olive oil receives. The constants and acid composition of olive oil are presented in Table 12.3.

Because of its unique flavor, scarcity, and costly production, olive oil is expensive and, therefore, liable to adulteration with cheaper oils. In 1974, the price of olive oil was $10 a gallon compared to $2.40 for cottonseed oil. The common food oils in the American market—cottonseed, corn, peanut and soybean—have constants far enough from those of olive oil that their determination will usually enable the chemist to estimate the extent of any adulteration. A more difficult problem is a salad oil that claims to be a definite percentage, say 20%, olive oil.

TABLE 12.3

PROPERTIES OF OLIVE AND TEASEED OILS

	Olive	Teaseed
Specific gravity, 25°	0.9100–0.9150	0.908–0.912
Saponification value	185–200	190–195
Iodine value	77–94	80–87
Index of refraction, 25°	1.4660–1.4690	1.4691
Titer	17–26	13–14.5
Acids		
Oleic	69–84%	83.3
Linoleic	4–12	7.4
Palmitic	9–15	7.6
Stearic	1.4–3.0	0.8
Unsaponifiable	0.6–1.3	0.2

Europe and Asia produce other vegetable oils that are cheaper than olive oil. The most common is probably sesame oil, but the most troublesome adulterant of olive oil is teaseed oil. Teaseed contains up to 60% oil and the constants are shown in Table 12.3. The constants vary somewhat with the variety of tea, but they are close enough to those of olive oil to make detection difficult. In 1936, Fitelson devised a test for teaseed oil, which is now one of the Official Methods. In general, the test consists of the addition of acetic anhydride, chloroform and sulfuric acid whereupon teaseed oil gives a red color. There are other tests for specific adulterants of olive oil. In the Halphen test for cottonseed oil, a 1% solution of sulfur in CS_2 is mixed with an equal volume of amyl alcohol. This reagent is then mixed with an equal volume of the oil and heated for 1 to 2 hr, when as little as 1% cottonseed oil gives a red color.

The Renard test for peanut oil is a determination of arachidic acid, which is characteristic of peanut oil, but not present in olive oil.

For sesame oil, the Villavecchia test is useful. A mixture of 10 ml of the oil and 10 ml HCl containing a trace of furfural is shaken in a test tube for 15 sec and a crimson color develops if sesame oil is present.

If court action or a dispute between seller and buyer is involved, every effort should be made to establish the presence or absence of an adulterant.

Cottonseed Oil

The first edible vegetable oil to be produced in the U.S. appeared with the establishment of a factory in 1861 to produce cottonseed oil. The oil is a by-product of the textile industry, but the cotton plant, *Gossypium* produces about 900 lb of seed to 500 lb of cotton fiber, so that there is an abundance of seed beyond the requirement for replanting. The seeds contain 18 to 20% oil. The production has averaged over a billion pounds annually in this country for the past 60 years.

The seeds are crushed and the oil obtained by pressure in a hydraulic press or an expeller or by extraction. The crude oil has a strong disagreeable odor and a dark brown color. It contains about 2% of substances other than glycerides; these are free fatty acids, carotene, sterols, chlorophyll, proteins, carbohydrates and phosphatids. One substance, gossypol, is very toxic. It is a yellow powder, a phenolic compound that melts at 200°C. The crude pressings also contain considerable water and are run into settling tanks from which the oil can be drawn off.

Refining.—Crude cottonseed oil must be refined before it can be used for food. The first step is the laboratory determination of free fatty acids. The refining kettle is a tall, cylindrical steel tank fitted with steam coils and a mechanical stirrer. It may hold 60,000 lb of oil. The oil run into it is weighed, and from this weight and that of the free acids, the amount of NaOH required to neutralize them can be calculated. In the batch process, the agitator is run at full speed as the alkaline solution is added so that the alkali will react with the free acids before it builds up a local concentration that will react with the glycerides. The oil and the aqueous solution become emulsified by the agitation and so the mixture is heated to 50° to 60°C and the agitator slowed to facilitate breaking the emulsion. As soon as the break begins, the agitator is stopped and the emulsion allowed to stand until the two phases separate. Then the oil is siphoned off. The soap fraction (or *foots*) is sold as soap stock.

A continuous method of refining is now in use in many plants. The principles are the same, but differ in engineering details. The NaOH solution is fed into the oil stream in the proper proportion and the foots are separated with a centrifuge.

Bleaching.—To remove the color, the oil is run into a tank, heated to 110°C and an agitator started to remove water because the presence of moisture promotes rancidity. The oil is then mixed with some kind of bleaching clay with or without a small amount of activated carbon. The latter is more effective than the clay but much more expensive. After the adsorbent is thoroughly mixed with the oil, it is removed by filtration. The adsorption of a material reaches an equilibrium, so that it is impossible to remove all the color from an oil by the process. The carotenoid pigments are readily removed, but chlorophyll and certain brown colors formed from heating an oil that contains carbohydrates and proteins are not. If cottonseed oil has been obtained from seeds that have been wet or otherwise damaged, the brown color is greater and difficult or impossible to remove.

Deodorizing.—Neither the refining nor the bleaching removes the odor and flavor of crude cottonseed oil. The odorous and flavoring substances are only 0.1–0.2% of the weight of the oil and are much more volatile than the glycerides, consequently, they can be removed by steam distillation. The oil is pumped into a vertical steel tank that is filled only partly full to prevent the oil splashing and going off with the vapors. The head space of the tank is evacuated, the oil is heated to 200°C or more and steam is admitted through perforated plates in the bottom of the tank. The time required is 3 hr or more according to the condition of the oil.

The oil is transferred to a cooling tank and cooled below 50°C. The vacuum is not broken between batches because the hot oil on the sides of the tank would oxidize and become rancid on exposure to air.

Winterization.—When a salad oil is kept in a refrigerator, it should remain clear. To achieve this, cottonseed oil and some others must be winterized. The reason for the cloudiness is that the oil contains saturated glycerides of high molecular weight and limited solubility in the liquid fraction. Therefore, if the oil is to remain clear at low temperatures, these glycerides must be removed. A salad oil should remain clear for 5.5 hr at 0°C.

Winterizing is simple in principle, but difficult to achieve. The oil is cooled until the insoluble glycerides crystallize, and then filtered. There are two main difficulties. At a low temperature, the liquid oil is very viscous and hard to filter. Moreover, the filtration must be done at low pressure or the filter cake will pack solid and the liquid will not penetrate it. The other difficulty is that the crystals must form slowly to be large enough to be removed by filtration. Crystals of glycerides are unlike those of a salt but are plastic and easily distorted by pressure.

According to Swern (1964), the oil at room temperature is run into tanks, cooled by brine in coils within the tank to 55°F and held for 6 to 12 hr. It is then cooled to 45°F and held for 12 to 18 hr and may be further reduced to 42°F and held for 12 to 20 hr. The exact conditions used vary with the condition of the oil and the degree of winterizing desired.

A cottonseed oil of iodine value 108 may yield 65–75% oil and 35–25% stearin. The oil increases to 110–111 in iodine value and the stearin drops to 90–98.

The stearin is used to make shortening although it is liquid at room temperature and therefore does not contribute to the firmness of the shortening. The physical and chemical constants and acid composition of the oils in this chapter are recorded in the tables of Chapter 2.

Uses.—Of the 817 million lb of edible cottonseed oil consumed in 1970, 501 were used for salad and cooking oils, 203 went into shortening, 65 into margarine and 48 found miscellaneous uses.

Corn Oil

Corn oil is also a byproduct, and the amount of it available depends on the demand for other corn products. The oil is in the germ of the kernel of the corn, *Zea mays*. The germ is 50% fat, but that obtained from the milling of corn meal has part of the endosperm attached and so contains only 20% oil. In the wet milling process, the separation of the germ is more efficient and a yield of 1.25 lb of oil is obtained from a bushel of corn against 0.5 lb in the dry milling process.

The oil is pressed from the germ by a hydraulic press or an expeller. The crude oil is dark in color and may contain 1.5% free fatty acids, and as much as 2% other nonglyceride material. It contains a higher percentage of oleic acid in its glycerides than cottonseed oil and less palmitic. However, it does contain about 1% of wax that forms a cloud when the oil is refrigerated and so it must be winterized. The composition of this wax is typical of waxes in general. It contains myricyl alcohol, $C_{14}H_{29}OH$, cetyl alcohol, $C_{16}H_{33}OH$, and myricyl esters of behenic and lignoceric acids.

Uses.—Of 391 million lb of corn oil in the food industry in 1971, 193 million were used for salad oil, 186 million to make margarine, 5 million in shortening and 7 million for all other purposes.

Peanut Oil

Peanut oil is pressed or extracted from the common peanut, *Arachis hypogaea*. In 1972, the United States produced nearly 33 billion lb of peanuts. The Virginia variety is grown largely for the roasted nuts, for they are larger than the other varieties. The smaller runner and Spanish varieties supply most of the oil. They are about 15% hulls and the shelled nuts contain nearly 50% oil. The press cake is about 45% protein and is used chiefly as cattle feed. It is a possible source of human food because of the shortage of protein.

The crude oil is refined, bleached and decolorized by the processes used for cottonseed oil. It has a cloud point around 40°F, but it is seldom used as a salad oil and so it is not winterized.

Uses.—Of the 176 million lb of peanut oil used in foods in 1970, 152 were used as salad and cooking oils, 16 million in shortening and 8 million for all other food uses.

Test.—There is no color test for peanut oil, but the amount of the three fatty acids above stearic amount to 6.6% of the total. This is characteristic of peanut oil. The Renard test is among the Official Methods. It is a determination of arachidic acid, which probably includes some acids of higher molecular weight. The amount of peanut oil in a mixture is considered to be 20 times the weight of the arachidic acid, and the method is said to detect 5–10% of peanut oil in a mixture.

The relative prices of the vegetable oils varies with economic factors and, sometimes, peanut oil is the most expensive and liable to adulteration. In this case, the chemist must depend on the oil constants, for the Renard test is incapable of determinating the amount of peanut oil in a mixture when that oil predominates.

Soybean Oil

Although the soybean has been raised in the Orient as a source of food since the beginning of history, it was not raised at all in this country until 1804, and for over a century, it was a forage crop. Not until after World War I was it grown for human food. In 1922, 1.7 million lb of oil were produced in this country. The government's attempts to aid the farmer during the 1930s launched the soybean as a food. The production increased from 7.3 million lb in 1928 to over a billion in 1942. The production of soybean oil in 1970 was 8,231 million lb.

Soybeans differ in oil content from 11% to 25% according to variety and growing conditions. Some oil is obtained by the use of presses or expellers but, because of the low oil content of the seed, most of it is extracted. The solvent is removed by steam distillation and reused. The crude oil contains 0.5% free fatty acids and 1.5% to 2.5% other nonglycerides consisting mostly of lecithin and cephalin. These lipids are washed from the oil with water and are used as emulsifiers in several food industries. The soybean meal contains 40% to 50% protein, which has been used chiefly as cattle feed, but in 1970, 17 firms were processing it for human consumption.

Uses.—The soybean is our largest source of vegetable oil and the oil has the most diverse uses. In 1970, 2,288 million lb were used as salad and cooking oils, 2,077 million as shortening and 1,387 million in margarine.

For use in shortening and for frying, soybean oil must be partly hydrogenated to produce the texture of the shortening. The oil also decomposes with heat and produces a disagreeable odor.

Soybean oil has a flavor of raw beans, which is removed in the refining process, but it reappears as the oil is in storage or in use. This is called *reversion.* The odor and taste that develop on standing may not be that of the components that caused the previous odor, but may be that of deterioration products of the linolenic acid. Whatever the cause may be, reversion is a problem in the food use of soybean oil.

Safflower Oil

In 1930, nutritionists discovered that unsaturated fatty acids are essential in the human diet. As this information became widely known, there developed a great demand for highly unsaturated food fats. The essential acid is eicosatetraenoic acid, also called arachidonic acid, unsaturated at the 5, 8, 11 and 14 positions.

The acid apparently does not occur widely in oils, but the body synthesizes it from linoleic acid, $C_{17}H_{31}COOH$. Linolenic acid, $C_{17}H_{29}COOH$, is not converted into arachidonic in the body but into docosahexaenoic acid with unsaturation at 4, 7, 10, 13, 16 and 19. It occurs in the brain. Other highly unsaturated acids are synthesized by the body from those of less unsaturation.

The importance of linoleic acid resulted in an increase in the production of safflower oil in which the acids are 65–77% linoleic. Of the common oils on the market, corn oil with 53% linoleic and soybean with 51% are the closest rivals of safflower. Nuts, especially walnuts and pecans, as well as poultry fat, supply linoleic acid to the diet.

The production of safflower oil was 135 million lb in 1965, but it declined to 40 million in 1971. In addition to its use in food, it is a valuable paint oil because it does not turn yellow with age. Its constants are: specific gravity 25° 0.9243, N_D 25° 1.4744, saponification value 188–194, Iodine Value 140–150, unsaponifiable 0.5–1.3 and titer 16–17°.

Coconut Oil

The seeds of many species of palm are rich in oil; the most important of them is the oil of the common coconut, *Cocos nucifera*. In several tropical regions, the nuts are harvested and the ½ in. thick endosperm adhering to the shell is removed and dried; the dried product is called *copra*. The copra is dried in the sun to less than 8% moisture to prevent molding and enzyme action. Some copra is dried by artificial heat, but it is not necessarily of higher grade than the sun-dried product. Fresh coconut contains 30–40% fat and 50% water.

Production and Importation.—Coconut oil has always been a culinary product in the tropical countries where the natives crushed the coconut and boiled it with water; the oil collected on the surface and the fiber settled to the bottom. The process gave a good product but did not extract all the oil.

Copra yields about 60% oil, obtained with presses or expellers. Some is extracted. It is very difficult to dry copra, especially in the sun, without the development of hydrolytic rancidity and other changes in composition. Therefore, the oil is refined, bleached and deodorized by the usual methods.

In 1971, 269 million lb of oil were obtained from copra crushed in the United States, and 629 million lb of oil were imported.

Under refrigeration, coconut oil is a plastic fat, but it is classed as an oil because in its native tropics, it is a liquid. Among the vegetable oils, coconut oil is unique in acid composition (Table 2.2). It is more closely related to milk fat than to any other because of the 15% of caproic, capric and caprylic acids. Then lauric and myristic make up over 60% of the total acids.

Uses.—The use of coconut oil has varied considerably through the years, partly because of the tariff. In 1934, a tariff was placed on the oil, which varied with the point of origin; from some countries, it was as high as 5 cents a pound and made the oil too expensive for some of its uses. Over the past 40 years, the tariff has varied and since July 4, 1974, it is uniformly 1 cent a pound.

In 1971, 56 million lb went into shortening, 7 million into margarine; other food uses consumed 393 million lb for a total of 456 million lb. It was used for frying foods, in confectionery, for coatings on cookies and for many miscellaneous uses. The soap, detergent and other nonfood industries used 417 million lb.

OTHER OILS

Sesame seed and rapeseed oils are common food oils in Europe and Asia, but they are rare in the United States. In 1971, we imported 11 million lb of rapeseed oil and 2 million lb of sesame seed oil. Linseed oil is used as a food in Russia, but none is used for food here.

There are several byproduct oils produced in small quantity and used for special purposes: among them are grapeseed oil, grapefruit seed oil and walnut oil.

HYDROGENATION

For many years the plastic fats—butter, lard and tallow—have been in such demand for food and industrial uses (e.g., soap) that the supply was inadequate. There was an abundance of vegetable fat but it was nearly all liquid. Consequently, there was a great demand for the conversion of these oils into plastic fats.

Although the hydrogenation process was discovered in 1902, commercial development was slow and the process was not used for oils in the United States until 1911. Aside from the time required for the process to become generally known, a process for the production of hydrogen had to be developed as well as special equipment for the hydrogenation. The hydrogen must be under pressure, and the oil heated and agitated.

Several types of equipment have been developed in the past 60 yr, but, in general, hydrogen under a pressure of 1 to 4 atmospheres is passed into the oil at a temperature of 120 to 210°; a nickel catalyst is suspended in the oil. When the oil has reached the desired consistency, the catalyst is filtered out and the oil deodorized.

Chemistry

Theoretically, the hydrogenation is selective, that is, the most unsaturated acid radicals are attacked first. However, there is a fantastic number of possibilities. Consider the extreme case of oleolinoleolinolenin

$$CH_2OOC(CH_2)_7 \overset{1}{CH} = CHCH_2 \overset{2}{CH} = CHCH_2 \overset{3}{CH} = CHCH_2CH_3$$
$$|$$
$$CHOOC(CH_2)_7 \overset{4}{CH} = CHCH_2 \overset{5}{CH} = CH(CH_2)_4CH_3$$
$$|$$
$$CH_2OOC(CH_2)_7 \overset{6}{CH} = CH(CH_2)_7CH_3$$

Presumably, bond 3 is the first to be hydrogenated to produce oleodilino-lein. If so, the next hydrogen goes to bond 2 or 5. Suppose the first molecule of hydrogen added to bond 2 instead of 3; it would produce an entirely different glyceride. Since a natural fat is composed of many of simple and mixed glycerides, the possibilities are enormous.

There are still other possibilities. Oleic acid is the *cis* form of 9-octade-cenoic acid; the *trans* form is *elaidic* acid.

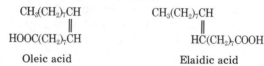

$CH_3(CH_2)_7CH$
$\|$
$HOOC(CH_2)_7CH$

Oleic acid

$CH_3(CH_2)_7CH$
$\|$
$HC(CH_2)_7COOH$

Elaidic acid

Elaidic acid does not occur in the natural fats, but under various conditions, oleic changes to elaidic and forms an equilibrium mixture containing 34% oleic and 66% elaidic. Elaidic acid melts at 44.5°C and oleic at 13°C. Olive oil has been converted into a fat of the consistency of tallow by this shift in geometric isomers. There are also geometric isomers of both linoleic and linolenic acids; linoleic has 2 *cis* and 4 *trans* forms and linolenic has 3 *cis* and 6 *trans* forms. The *trans* forms have the higher melting points.

There is still another complication that has been demonstrated with the free acids. During the hydrogenation of oleic acid, the double bond, which is at 9, shifted to both the 8 and 10 positions.

There still remains much information to be learned about the changes that occur during the hydrogenation of an oil, but the composition of an oil is so complex that progress is slow. However, it is well established that the results differ with the conditions of the hydrogenation, and considerable practical information has been obtained. The nearer a double bond is to the end of the acid radical, the more rapidly it is hydrogenated. Although the other bonds are hydrogenated more slowly, they do undergo hydroge-nation before the end bonds are saturated.

Effect of Conditions

There are four conditions to be considered in the hydrogenation of an oil: the temperature, the pressure of the hydrogen, the extent of the agita-tion and the nature and amount of the catalyst.

Temperature.—Like all other reactions, the rate of hydrogenation approximately doubles with each increase of 10°C. However, the maximum velocity of hydrogenation is reached at about 200°C and at higher tem-peratures, the velocity decreases.

Since there are several reactions proceeding simultaneously in the hy-drogenation of an oil, a variation in the various speeds will result in a dif-ferent composition of the final product. The higher temperature also pro-motes the shift of the acid radicals from *cis* to *trans* isomers. A cottonseed

TABLE 12.4

COMPOSITION OF A HYDROGENATED OIL

Temp. °F	Time Min	Catalyst 0.05% Percentage Composition of Acids			
		Saturated	Isooleic	Oleic	Linoleic
Original oil	—	26.9	—	27.1	46.0
250	72	39.6	8.5	39.3	12.6
300	31	34.1	11.2	47.5	7.2
350	19	32.5	14.3	47.6	5.6

oil was hydrogenated to an iodine number of 65.7 at three temperatures and the resulting compositions are shown in Table 12.4.

Pressure and Agitation.—According to Henry's Law, the solubility of a gas is directly proportional to the pressure of the gas at the surface of the liquid. Consequently, the greater the pressure of the hydrogen, the greater its concentration in the oil.

Agitation must be considered along with the pressure of the gas because both conditions influence the rate at which the gas reaches the surface of the catalyst as well as the uniformity of distribution of hydrogen throughout the oil. If there were no agitation, the hydrogen could only reach the lower portions of the oil by diffusion, and the concentrations would gradually decrease from the surface downward. With different concentrations of hydrogen, the rates of reaction would differ throughout the oil.

Another effect of agitation is to circulate the oil around the particles of the catalyst. The reaction occurs at the surface of the catalyst, and to get a uniform reaction, new oil and new hydrogen must reach the catalyst continuously. This effect is similar to that of stirring when dissolving a solid in water. In the hydrogenation, the amount of agitation possible varies with the size of the batch of oil being hydrogenated. It is very difficult to agitate a large batch effectively.

Catalyst.—Several metals have been tried experimentally as catalysts for the hydrogenation of oils; the most successful is nickel. It is very difficult to determine the effect of the amount of catalyst on the composition of the finished product, because the greater the amount of catalyst, the more rapid the hydrogenation. Thus the temperature effect is different because of the different length of time required to hydrogenate the oil to a given iodine number or consistency. However, Bailey studied the effect of varying the amount of catalyst in the hydrogenation of cottonseed oil at 300°F to an iodine number of 65.7 and concluded that the time required for hydrogenation had very little effect on the composition of the product.

Since the catalytic action of the metal is a surface function, the greater the surface, the greater the catalytic action of a given weight of nickel, which means that the smaller the particles of nickel, the greater the activity. However, the activity is not directly proportional to the extent of surface of the catalyst, for not all the surface is active; the activity depends on the nature of the surface.

Several theories have been advanced to explain the variation in activity of nickel catalysts. One of them assumes active spots on the surface of the metal caused by different electronic states of the nickel atoms at those points. Whatever the correct explanation may be, the fact is that the activity differs with the method by which the catalyst is prepared.

Preparation.—Although nickel wire and turnings have been used in some modifications of the hydrogenation process, the usual catalyst is finely divided nickel, which offers a much greater surface than any form of massive nickel. The nickel is generally attached to some inert material so it can be more completely removed from the hydrogenated fat.

1. Catalysts may be prepared from nickel formate, $Ni(OOCH)_2$, which is suspended in two to four times its weight of coconut oil, or some other oil, and the suspension heated to 190–200°C. At this temperature, the nickel formate decomposes rapidly.

$$Ni(OOCH)_2 \rightarrow Ni + 2CO_2 + H_2$$

A stream of hydrogen is bubbled through the suspension to remove the carbon dioxide and the oil becomes completely hydrogenated during the process. The heating takes about an hour to produce a catalyst of maximum activity.

After the reduction is complete, the mixture is cooled below 100°C and kieselguhr is added and thoroughly mixed. The catalyst is separated from the original oil and added to the oil to be hydrogenated. Catalysts prepared in this manner are called *wet-process* catalysts. They are very active, but are likely to contain some colloidal nickel, which cannot be filtered out of the finished product.

2. Another method of preparing a nickel catalyst is the *dry-reduction* process. Nickel is dissolved in nitric acid and diatomaceous earth is added in an amount about twice the weight of the nickel. A solution of sodium carbonate or sodium hydroxide is added and precipitates nickel carbonate or hydroxide on the silica. The solids are then filtered from the solution and washed well to remove the soluble nitrates. This involves several washings. The catalyst is finally removed, dried and ground to a fine powder. It is then reduced in an atmosphere of hydrogen at about 500°C and discharged under oil to avoid contact with the air, which would oxidize it instantly to nickel oxide. Most of the details of this process affect the activity of the catalyst.

3. An *electrolytic* method of preparing a catalyst has some advantages, mainly better control of conditions. In this method, kieselguhr is suspended

in a 1% NaCl solution and the solution electrolyzed with a sheet of nickel as the anode. The following reactions occur:

$$2NaCl + H_2O \rightarrow 2NaOH + H_2 \text{ (at the cathode)}$$
$$2Cl^- \rightarrow Cl_2 \text{ (at the anode)}$$
$$Cl_2 + Ni \rightarrow NiCl_2$$
$$NiCl_2 + 2NaOH \rightarrow Ni(OH)_2 + 2NaCl$$

The pH of the solution is controlled by bubbling CO_2 through it. The temperature and rate of precipitation are also controlled. The Ni-$(OH)_2$–kieselguhr precipitate is treated in the same manner as in the dry process.

4. The nickel-alloy process is unique. It was invented and first patented by Raney in 1925, and the product is, therefore, often called *Raney nickel*. In the process, nickel is alloyed with aluminum in about equal proportions. The alloy is then powdered and added to a 20% solution of NaOH. Aluminum reacts readily with strong alkalies.

$$2Al + 2NaOH + 2H_2O \rightarrow 2NaAlO_2 + 3H_2$$

The alkali dissolves the aluminum according to the above equation and leaves the nickel as fine particles, which settle to the bottom of the reaction vessel. After the reaction is complete, the alkaline solution is drawn off and the nickel washed thoroughly to remove the last traces of alkali that would otherwise form soap when the catalyst is added to oil. Soap inactivates a nickel catalyst.

Raney nickel is pyrophoric; it oxidizes in air with enough energy to produce light. Consequently, it must be handled in an inert atmosphere. It is covered with oil as soon as the last wash water is drawn off and then heated in a vacuum or in an inert atmosphere to remove the last traces of water. Kieselguhr is usually added although the catalyst is coarser than the others and can be removed from the oil by filtration without the aid of a support.

Theory of Catalysis

Catalysis, especially heterogeneous catalysis, it governed by no known laws. Nobody knows how or why it works. There are many theories, of course, that explain some feature of the phenomenon. For example, there is the active spot theory to explain why catalysts that have the same surface area do not have the same activity.

The most important question regarding heterogeneous catalysts is why or how they accelerate reactions. In the case of hydrogenation with nickel, it is well known that the nickel catalyst absorbs, or probably adsorbs, great quantities of hydrogen and a likely theory is that the H_2 molecules are broken into atoms distributed on or throughout the nickel. Atomic hydrogen is much more active than molecular hydrogen and would react more rapidly with the oil, which may also be adsorbed on the surface of the nickel.

It is known from experience that some substances promote catalysis and some retard or prevent it. Metallic aluminum or copper and probably the silicon of the kieselguhr increase the activity of a nickel catalyst, but other substances, called *poisons,* retard it or stop it completely, possibly by combining with the nickel or by preferential adsorption over the hydrogen. Among the poisons of a nickel catalyst are H_2S, SO_2, CS_2= CO, soap, lecithin, Zn, Pb and Hg.

Oils are usually refined before they are hydrogenated and many possible poisons removed, so that not much is known about the activities of the natural impurities in the crude oils. Metals may be picked up by the oil from the equipment in which it is transported, stored or processed.

Hydrogen

There are several methods by which hydrogen can be produced in large quantity.

1. The simplest process is the electrolysis of an aqueous solution of NaOH or other electrolyte.

$$2H_2O \xrightarrow[\text{Elec.}]{\text{NaOH}} 2H_2 + O_2$$

The process produces pure hydrogen, but it is expensive unless very cheap electricity is available. Oxygen is a byproduct and, if there is a market for that, the cost may be tolerable.

2. The *steam-iron* process is cheaper than the electrolytic. Superheated steam is blown over hot iron and the overall reaction is

$$3Fe + 4H_2O \rightarrow Fe_3O_4 + 4H_2$$

Most iron contains carbon and sulfur and so the hydrogen may contain H_2S, CO, CO_2 and possibly other impurities that must be removed before the hydrogen can be used with a catalyst.

3. The *water-gas* process starts with the action of steam on hot coke.

$$H_2O + C \rightarrow CO + H_2$$

The mixture of the two gases, which is called *water gas,* is then mixed with more steam and passed through a bed of iron oxide or other catalyst and the CO is oxidized to CO_2 with the production of more hydrogen.

$$CO + H_2O \rightarrow CO_2 + H_2$$

Most of the CO_2 can be removed by washing with water and the remainder by washing with a solution of primary amines or other alkali.

4. In the *hydrocarbon* process, natural gas, which is mostly methane, CH_4, a little ethane, C_2H_6 and perhaps some higher homologues, is mixed

with steam and passed over a catalyst at a high temperature. The reaction produces CO_2 and H_2.

$$CH_4 + 2H_2O \rightarrow CO_2 + 4H_2$$

The CO_2 may be washed from the mixture as in the previous method.

Very pure hydrogen is required for the hydrogenation of oil and so the hydrogen from any of the methods of production must be further purified; even water vapor interferes.

The food chemist is not usually concerned with the selection of a method for the preparation of either the catalyst or hydrogen unless he is in a small plant with no engineers. If he must become an engineer, his engineering will probably consist of searching advertising or consulting colleagues at a convention for the names of engineering firms that sell catalysts or install systems for the production of hydrogen. Salesmen will give him an idea of cost and convenience.

The chemist is concerned with the hydrogenated product, which may be used for shortening, margarine or the manufacture of soap. He must also be familiar with all the conditions that may affect the hydrogenated product even though his main concern is the use to which the fat is to be put.

MARGARINE

Butter has always been a relatively expensive food, especially in Europe where pasture land is scarce. This led the French government to offer a prize for the invention of a cheap substitute for it. Mege-Mouries won the prize in 1869 and named the product *oleomargarine*. The process spread rapidly throughout Europe and the first American patent was issued December 30, 1873. The name at different times and places has been oleomargarine, margarine and simply oleo; *margarine* now predominates.

The principle of margarine manufacture is simple: mix high- and low-melting fats to obtain the consistency of butter fat, then sour some milk to churn with the fat until it is emulsified in the fat to produce the flavor. Butter and margarine are water-in-oil emulsions.

Since margarine was to be the poor man's butter, the fat had to be cheaper than butter fat. For the first 30 years or so, the fat used was a high-grade beef tallow, which had the desirable property of very little taste. The tallow was rendered at 90°F, held until it developed large crystals and then pressed at low pressure. The liquid fraction is called *oleo oil* and varies in properties with the extent of the pressure. The solid fraction is called *oleostearin* and also varies in composition with the pressure used.

The two fractions of tallow could be recombined in any proportion and thus approximate the consistency of butter fat. After refined cottonseed oil became available, it began to replace oleo oil and through the years, the fat used has changed, partly for economic reasons and partly because hy-

TABLE 12.5

FATS USED IN MARGARINE

Fats	Million Lb		
	1917	1950	1970
Beef fat	90.2	9	90
Lard	42.4	4	9
Butter	3.5	—	—
Coconut oil	19.0	—	—
Cottonseed oil	63.7	418	68
Peanut oil	10.6	7	—
Soybean oil	—	312	1,410
Corn oil	—	1	185
Safflower oil	—	—	22
Other	—	20	11[1]
Total	239.2	764	1,795

[1] Mostly peanut and coconut.

drogenated oils became more satisfactory than oleostearin. This shift in the use of fats is indicated by Table 12.5.

Over 2 billion lb of margarine were made in the United States in 1970, over 6 times the amount produced in 1917. The growth in popularity had been gradual through the years in spite of restrictive legislation by Congress and several of the states. In fact, it remains the most restricted of all foods (Garard, 1974; Riepma, 1970).

Definition and Standard

The federal standard defines margarine as follows:

"Oleomargarine, margarine is the plastic food prepared with one or more of the optional fat ingredients named in subparagraph 1, . . ."

1. (i) The rendered fat or oil or stearin therefrom of cattle, sheep, swine or goats or any combination of two or more of such articles. These articles may also be hydrogenated. (ii) Any vegetable food fat or oil or stearin derived therefrom, any or all of which may be hydrogenated, or any combination of two or more such articles.

2. Cream, milk, skim milk, liquid sweet cream, buttermilk and various combinations.

3. Artificial coloring, sodium benzoate or benzoic acid up to 0.1%, vitamin A, vitamin D, artificial flavoring, lecithin, mono- and diglycerides, butter, salt, citric acid, isopropyl citrate, potassium sorbate, calcium disodium ethylene diamine-etra acetate, EDTA, BHA and BHT. In addition to the names of the permitted ingredients, there are detailed specifications and label requirements for many of them. If vitamin A is added, the content must be at least 15,000 IU per pound.

There is also a standard for *liquid oleomargarine, liquid margarine,* which is the same as for margarine except that the product is a liquid and not plastic.

Anyone who plans to manufacture margarine should study the details of the standard and the unique requirements of the law.

Manufacture

Margarine is a water-in-oil emulsion that requires considerable skill to achieve the three essentials: a uniform emulsion, a proper flavor and the right consistency. It must be firm but must melt in the mouth.

The selection of the fat is the first problem because both the properties of the fat and economic considerations enter into the selection. Oils can be hydrogenated to the proper consistency and the product used in making a single-fat margarine, such as a corn oil or a safflower oil margarine.

More commonly, the fat mixture is made by adding a solid fat to a liquid oil as Mouries did a century ago. Now, however, the solid fat may be oleostearin but is more likely to be a fully hydrogenated vegetable oil. Frequently, the consistency is obtained by mixing a hydrogenated product with the cheapest vegetable oil on the market, which is usually soybean oil. Obtaining the right proportions of the solid and liquid fats is a difficult problem, and each manufacturer has one or more formulas that has been carefully devised in the laboratory. When margarine is sold under a brand name, every pound must be the same in texture, color and flavor.

The fats are warmed and well mixed if two or more are used. Color, vitamins and emulsifiers are added to the fat. The colors are annatto, which has been used to color butter for decades, and carotene of more recent application. Although carotene is a natural color, it is now synthesized commercially. Carotene is also provitamin A, and, in case vitamin A is added, the carotene is included in the amount expressed on the label.

The most common emulsifier is lecithin, which also affects the properties of the finished product; the mono- and diglycerides are used to a less extent. Formerly, the butter flavor all came from the soured milk, but now artificial flavor, which is chiefly diacetyl, or acetylmethylcarbinol, is added to the fat.

Although most margarine is handled under refrigeration, and consumed within a few weeks of manufacture, a preservative is usually added to the fat. Sodium benzoate to 0.1% of the weight of the finished product is probably the most common, but the other preservatives named in the standard may be used and declared on the label. Citric acid may also be added to the fat.

The aqueous phase consists of milk, cream, buttermilk or dried milk and water. The milk is pasteurized and may be soured to a pH of 4.5 or lower. Salt is added to the aqueous layer to the extent of 1.5–3% of the weight of the finished product.

The two phases are warmed to 100°F and passed through a machine that agitates them violently and thus produces the emulsion. In the batch process, the two phases are weighed and in the continuous process, they are fed automatically to the emulsifier in the proper proportion. The finished margarine must contain 80% fat. Following emulsification, the margarine is cooled to permit crystallization of some of the glycerides to form the required plastic texture.

SHORTENING

To a food processor, shortening is a mixture of animal or vegetable fats or both with properties similar to lard, which is the traditional shortening; he distinguishes between lard and shortening. In fact, food processors sometimes apply the term to all 100% fats and oils (Weiss 1970).

Most baked cereal products require the presence of a fat to make them soft enough to be acceptable. A pie crust, for example, contains from 30% to 40% shortening; in some cases even more.

Since baked cereal products constitute a large part of our food, the demand for shortening has been high for years. Lard was the common shortening and about the only one in the United States; butter was used somewhat in cakes. But lard is also used as a cooking fat and for the manufacture of soap; consequently, there was not enough lard to supply the demand and substitutes were developed.

The first shortening compounds were made from mixtures of oleo oil and oleo stearin, or tallow, and the manufacturer simply tried to duplicate the texture of lard. But as early as 1888, cottonseed oil began to replace the oleo oil; today, there are several types of shortening compound simply called *shortening* and, like margarine, it is no longer considered a substitute but simply another food product.

The shortenings now are of two types: blended shortenings and all hydrogenated shortenings. The latter are generally made from cottonseed or soybean oil by hydrogenating to the most desirable properties for the purpose for which it is intended, which is difficult. The processor tries to reduce the iodine number to about 60 and the isooleic acid content to around 10%, but under some hydrogenating conditions, the iodine value may be as high as 75 and the isooleic acid as high as 20%. In this case, the proper texture is reached with a high degree of unsaturation and poor keeping qualities.

The blended shortenings may be of animal, vegetable or mixed fats. An all vegetable shortening is made by hydrogenating some oil completely and then mixing about 20% of the product with 80% oil. This practice produces the proper consistency but poor keeping qualities because of the high degree of unsaturation.

Some shortenings are made by blending oleo oil and oleostearin with or without lard. Or oleostearin may be blended with vegetable oils. Shortenings containing animal fat come under the control of the meat inspection service of the USDA, and are generally made by the meat packers.

TABLE 12.6

FATS USED IN SHORTENING

Fat	Million Lb	
	1950	1970
Soybean	841	2,182
Cottonseed	549	276
Palm	—	85
Coconut	—	45
All other vegetable	40	35
Lard and beef fat	208	976
Vegetable stearin	89	—
Total	1,727	3,599

The shortening sold in the grocery is usually of the all hydrogenated type. The others are used by restaurants, bakers and other food processors who select the shortening most suited to their process and the most economical. Selection of a shortening for a cereal product that must keep for several months is more critical than the selection for a restaurant that bakes its own pies or for a local baker who sells his product daily.

The keeping time of a shortening in its package may be very different from that of the same product after it is mixed with flour and the other baking ingredients. Prepared cake, pie crust and muffin mixes are one problem. Frozen biscuits or pie crust are another. Cookies and crackers are still another, for the shortening has been heated. Manufacturers make special shortenings for the cookie and cracker baker that are hydrogenated to a high degree to reduce the linoleic acid content below 3%. These shortenings are too hard for the housewife to use at room temperature, but the baker has better facilities and control of conditions. Crackers and cookies have a rather short shelf life at best.

Shortenings are also used as cooking fats for frying doughnuts, potatoes, potato chips and various snacks. For these purposes, some processors use a hydrogenated shortening because it is more stable and produces less free fatty acids, acrolein and smoke.

Table 12.6 lists the fats used in making shortening.

BIBLIOGRAPHY

ANDERSON, A. J. C., and WILLIAMS, P. M. 1965. Margarine, 2nd Edition. Pergamon Press, New York.

ASSOC. OFFIC. ANAL. CHEMISTS. 1975. Official Methods of Analysis, 12th Edition. Assoc. Offic. Anal. Chemists, Washington, D.C.

CHANG, H. C. and BRANEN, A. L. 1975. Antimicrobial effects of butylatedhydroxyanisole (BHA). J. Food Sci. *40*, 349–351.

ELLIS, N. R., and ISBELL, H. S. 1926. The influence of the character of the ration upon the composition of the body fat of hogs. J. Biol. Chem. *69*, 219–248.

GARARD, I. D. 1974. The Story of Food. Avi Publishing Co., Westport, Conn.

GILLIES, M. T. 1974. Shortenings, Margarines and Food Oils. Noyes Data Corp., Park Ridge, N.J.

HARRIS, R. S., and KARMAS, E. 1975. Nutritional Evaluation of Food Processing. Avi Publishing Co., Westport, Conn.

JAMIESON, G. S. 1943. Vegetable Fats and Oils. Reinhold Publishing Corp., New York.

KING, A. E., RASCHEN, H. L., and IRWIN, W. H. 1933. An accelerated stability test with the peroxide value as an index. Oil and Soap *10*, 105–109.

KINSELLA, J. E. 1974. Grapeseed oil. Food Tech. *28*, No. 5, 58–60.

KINSELLA, J. E. 1975. Lipids and fatty acids in foods: quantitative data needed. Food Tech. *29*, No. 2, 22–24.

KRISHNAMURTHY, H. G. *et al.* 1967. Identification of 2-pentyl furane in fats and oils and its relationship to reversion flavor of soybean oil. J. Food Sci. *32*, 372–374.

KUMEROW, F. A. 1975. Lipids in atherosclerosis. J. Food Sci. *40*, 12–17.

MARKLEY, K. S. 1950, 1951. Soybeans and Soybean Products: Their Chemistry and Technology, Vol. 1 and 2. John Wiley & Sons, New York.

MATTIL, K. F. 1959. The visible fats in our diet. Food Tech. *13*, 46–49.

RIEPMA, S. F. 1970. The Story of Margarine. Public Affairs Press, Washington, D.C.

SANDERS, J. B. 1959. Processing of food fats. (Review) Food Tech. *13*, 41–45.

STANSBY, M. E. 1967. Fish Oils. Avi Publishing Co., Westport, Conn.

SWERN, D. 1964. Bailey's Industrial Oil and Fat Products. John Wiley & Sons, New York.

WEISS, T. J. 1970. Food Oils and Their Uses. Avi Publishing Co., Westport, Conn.

WOERFEL, J. B. and BATES, R. W. 1958. Blending and the measurement of the properties of shortening. Food Tech. *12*, 674–676.

CHAPTER 13

Cereal Products

The cereals have been the staple food for mankind for generations—rice in the East, wheat, rye, barley and corn in the West. They are sources of energy and protein—after water, the two main requirements of the human diet. Also of major importance is the low water content, which enables them to keep from season to season; in fact, apparently indefinitely when the moisture content is less than 14%.

The relative importance of the common cereals in the United States is indicated by the production and consumption, which are indicated in Table 13.1.

Table 13.2 shows the approximate composition of the common cereals.

The values in Table 13.2 are average values, but there is considerable variation in each of them.

WHEAT

There are many varieties of wheat and great variation in composition. The varieties grown in the United States vary somewhat as follows: protein 8.6–17.2%, fiber 1.7–3.7%, ash 1.4–2.4% and carbohydrate other than fiber 66.7–76.1%. These are gross analyses and include gums, phosphatids, vitamins and small amounts of other constituents.

Composition and Standards

The USDA classifies wheat as: (1) hard red, spring wheat; (2) hard red, winter wheat; (3) soft red, winter wheat; (4) white wheat and (5) Durum

TABLE 13.1

PRODUCTION AND CONSUMPTION OF CEREALS (1971)

Grain	Production (Million Metric Tons)		Per Capita Consumption (Lb)
	U.S.	World	U.S.
Corn	143.3	291.0	7.4
Wheat	44.0	322.6	110
Oats[1]	12.8	54.9	1.0
Barley[2]	10.1	130.4	2.0
Rice	3.9	299.4	7.7

[1] Only about 10% of the crop is used as human food.
[2] Used mainly as malt in the baking and the alcoholic beverage industries.

TABLE 13.2

COMPOSITION OF CEREALS

Grain	Water %	Protein %	Fat %	Carbo-hydrate %	Fiber %	Ash %
Corn	13.8	8.9	3.9	72.2	2.2	1.2
Buckwheat	11.0	11.7	2.4	72.9	9.9	2.0
Rice	12.0	7.5	0.6	77.4	0.9	1.2
Rye	11.0	12.1	1.7	73.4	2.0	1.8
Wheat, hard	13.6	13.0	2.2	69.1	2.3	1.1
Wheat, soft	14.0	10.2	2.0	75.4	1.9	1.7
Wheat, Durum	13.0	12.7	2.5	70.1	1.8	1.7
Barley	6.5	11.5	2.7	72.7	3.8	2.9

Source: Watt and Merrill (1963).

wheat. In general, spring wheats are grown in the northern tier of states and winter wheat in a wide area from New Jersey to Colorado and as far south as Texas and northern Florida.

The wheat kernel consists of an outer layer of flat, brown cells that constitute the *bran*. Next to the bran is a layer of coarse cells called the *aleurone* layer. The main portion of the wheat kernel is the *endosperm* which consists of small particles of starch and protein. At one end of the kernel is the filamentous *germ*. The bran is about 5% of the weight of the kernel and the aleurone layer 8%. The endosperm constitutes about 82% of the kernel and the germ, 5% or less.

Each of the parts of the wheat kernel has a different chemical composition. Most of the fiber is in the bran, the fat in the germ, the minerals and vitamins mainly in the aleurone layer, which is also rich in protein. The endosperm is mostly starch with some protein.

Flour.—From colonial days to the 1870s, wheat was ground in stone mills and screened into three fractions: bran, flour and middlings. The bran and middlings were used mainly as animal feed.

With the introduction of the roller mills and their gradual reduction process, it became possible to make several kinds of flour from the same wheat.

Flour is commonly of 72% extraction, that is, 72% of the weight of the kernel; in war time, the extraction was increased to 80%.

In the gradual reduction process, the wheat passes through several sets of rollers; the number differs from mill to mill. After each set, the product is screened and the fine portion emerges as flour, which means that there are many streams that can be blended to produce flours of different properties. This procedure, together with the different varieties of wheat, which

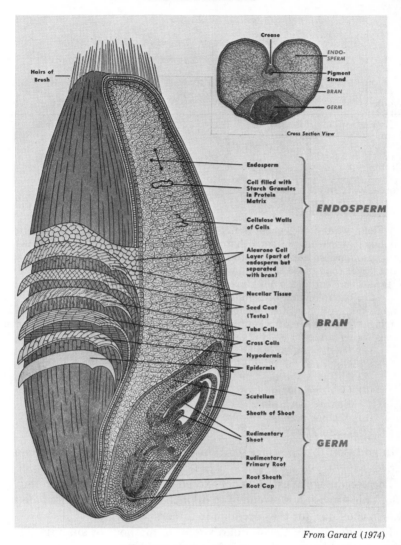

From Garard (1974)

FIG. 13.1. WHEAT KERNEL

may also differ with the season or the geographic source, results in an enormous number of flour possibilities, so that the chemist for a baker or other food processor who buys flour, may select the flour with the properties most suitable for his purpose.

There is a federal standard for flour, but because of the great number of varieties, its provisions are very general as the following excerpts will indicate:

TABLE 13.3

COMPOSITION OF WHEAT FLOURS

Flour	Water	Protein	Fat	Carbo-hydrate	Ash	B_1	B_2	Niacin
Whole	12.0	13.3	2.0	71.0	1.7	0.55	0.12	4.3
80% Extraction	12.0	12.0	1.3	74.1	0.65	0.26	0.07	2.0
Family	12.0	10.5	1.0	76.1	0.43	0.06	0.05	0.9
Bread	12.0	11.8	1.1	74.7	0.44	0.08	0.16	1.0
Cake	12.0	7.5	0.8	79.4	0.31	0.03	0.03	0.7
Gluten	8.5	41.4	1.9	47.2	1.0	—	—	—

Source: Watt and Merrill (1963).

Flour, white flour, wheat flour, plain flour, is the food prepared by grinding and bolting cleaned wheat, other than durum wheat and red durum wheat.

... not less than 98% of the flour passes through a cloth having openings not larger than those of woven wire cloth designation "210 micron (no 70) .... Its moisture content is not more than 15%.

... the following bleaching ingredients may be added in a quantity not more than sufficient for bleaching or, in case such ingredient has an artificial aging effect, in a quantity not more than sufficient for bleaching and such artificial aging effect:

(1) Oxides or nitrogen
(2) Chlorine
(3) Nitrosylchloride
(4) Chlorine dioxide
(5) Benzoyl peroxide
(6) Acetone peroxides
(7) Azodicarbonamide

In addition to the above, certain enzymes and ascorbic acid may be added.

The percentage of ash may not exceed $\frac{1}{20}$ the weight of the protein plus 0.35. This provision is intended to limit the amount of bran and aleurone cells in the flour, for they are high in mineral content.

The remainder of the standards refer to methods of analysis. The fat content is not limited although the presence of fat invites rancidity.

The composition of some common flours is reported in Table 13.3.

Flours do not contain any vitamin A or C and, from the ash content, it is obvious that the content of calcium, phosphorus and iron cannot be very high except in the whole wheat flour, which contains 1.7% ash.

Protein.—The protein of wheat is called *gluten* because of its characteristic texture when it is hydrated. Put some white flour into a muslin bag and knead it under water. The starch passes through and the gluten remains in the bag as a glue-like or gummy mass. It is this characteristic that makes wheat flour useful for making yeast bread. During the handling, it holds the dough together and retains the gas to make the bread porous. A cake or pastry flour has much less gluten and therefore, it yields a much more tender texture in the product.

Family flour and bread flours are generally bleached, not that the color is important but the bleaching agents are also aging agents. A freshly milled flour does not make very light bread because the gluten is not very elastic and does not stretch well during the rising of the dough. If the flour is stored several months, the gluten develops elasticity. The bleaching agents accelerate the process and therefore, make a long storage period unnecessary. However, the bleaching agents permitted by the federal standard are oxidizing agents and destroy the natural flavor of the flour, which consists mostly of aldehydes and is therefore easily oxidized, but there is so little fat that the rancid odor is slight and volatile and therefore is removed during baking.

Gluten was first separated from flour in 1728 and almost a century later, in 1820, it was found to be a mixture of two proteins, which came to be called *glutenin* and *gliadin*. The percentage of protein is important, not only because it determines the use of the flour, but also the price of it. The protein content of flour and grains is determined by some modification of the Kjeldahl method for nitrogen and multiplying the nitrogen content by 5.7, a factor obtained from the average percentage of nitrogen in the wheat proteins.

The analytical values in Table 13.3 are average values and are intended to show the differences between the types of flour. But flours differ in protein content with the variety of wheat and from season to season in a growing area so that a consumer of flour must devise his own specifications. Family flour, sold by grocers, is made by blending and is reasonably uniform.

The quality as well as the quantity of the gluten in a flour varies. Just why one gluten is better than another is not known, but it is known that gluten is a mixture of four proteins: gliadin, edestin, glutenin and leucosin. The nomenclature, however, is not totally agreed upon.

Gliadin is soluble in 70–80% alcohol but not in water. At one time, it was thought that the ratio of gliadin to glutenin was the reason for the difference in baking properties of gluten and, after 1900, several chemists undertook to determine this ratio. Concentrations of alcohol from 20% to 90% were tried at several temperatures and each combination gave a different result. Even with the same concentration and temperature, the length of time the alcohol was in contact with the protein made a difference.

Osborne published a monograph, *The Proteins of the Wheat Kernel*, in 1907, which reported the results of an enormous amount of research on gluten. He reported four principal proteins in the mixture; albumin, globulin, gliadin and glutenin. The albumins were water soluble, the globulins soluble in salt solutions, the gliadin in alcohol and the glutenins in alkali but insoluble in all neutral solvents.

Solubility was the only method for distinguishing between proteins, but solubility changes with temperature and with the concentration of the reagent. In the case of salt solutions, it also varied with the salt used.

Osborne determined the amino acid content of his four proteins and showed them to be different, but subsequent work by many research chemists showed that these four fractions were not single chemical substances. Even the gliadin-glutenin ratio that was popular for so many years was never established beyond doubt. The Official Methods at times have had procedures for the determination of the different fractions, but later abandoned them. In the 12th Edition (1975), there are only methods for total nitrogen and for water soluble protein precipitable by 40% alcohol.

About all that can be said at present, with confidence, is that gluten is a mixture of simple proteins that vary in molecular weight, amino acid composition and solubilities. Some think there may be hundreds of them.

Nutritionists are concerned only with the nutritive value of the wheat protein and the amount of it. It is known to be deficient in lysine.

The baker is interested in the colloidal properties of the protein. He knows that the gluten from different sources behaves differently in the baking process, and technologists have invented instruments that measure the force required to mix the dough. By making a dough for a standard, the others can be evaluated. The instruments draw a curve on paper, which shows the character of the gluten and how long it will hold its consistency during mixing. The value of such instruments is indicated by a statement of the chief chemist for a large bread-baking company who said that he would rather have this instrument than all the rest of his laboratory equipment combined.

Starch.—The endosperm of the wheat kernel, which constitutes the bulk of the flour, is about 90% starch. It occurs in grains of different sizes, mainly three: large, 15–35 microns in diameter, medium, 7–15 microns and small, less than 7.5 microns; over 90% of the starch by weight consists of the large grains.

It is difficult to describe the appearance of the starch grains even with diagrams or photomicrographs. The chemist who has occasion to work with cereal products should obtain authentic samples of the different starches to be used for comparison when the identification of a starch becomes important.

The starch of wheat is insoluble in water although it does absorb a little water when it is immersed for some time. At 50°C, it begins to swell, and at 65°, it is dispersed to a viscous paste.

Wheats differ somewhat in starch content, but average about 50% to 55%.

Flours differ with the milling practice, and commercial flours contain from 65% to 75%. There is a procedure in the Official Methods for the determination of starch in the various cereal products.

Untold hours of research have been spent on the nature of wheat starch and about the only point of agreement is that it occurs in granules of different sizes. All agree that it is a mixture of amylose and amylopectin, but the relative proportions have been reported with such great differences that no one ratio has enough support to justify recording here.

The starch contains phosphate of the order of 0.05% to 0.07%. Whether the phosphate is chemically bound to the polysaccharide or adsorbed on the grains is controversial. Also, if it is chemically attached, whether it is combined with the amylose or the amylopectin is also in controversy.

Lipids have been reported as 0.58% of the weight of wheat starch and from them have been obtained oleic, linoleic and linolenic acids; here again, the nature of the lipid and whether it is combined or adsorbed is unknown. Fortunately for the daily duties of the food chemist, the precise composition of wheat starch seems to be of little importance, but the effect of the composition on the properties of the starch may be more important than now appears.

Sugars.—Soluble carbohydrates constitute about 1% of wheat flour. One chemist reported about 0.15% reducing sugars, 0.30% sucrose and 0.55% of a fructoside called *levosine* for which the empirical formula is $C_{48}H_{40}O_{40}$. Raffinose has also been reported. The different values that have been reported for the soluble carbohydrates may be due to poor methods, different varieties of wheat or the stage of maturity of the wheat kernel. Early in the development of the kernel, levosine is three times as abundant as starch. Also an examination of wheat on June 22 showed over 6% levosine; the same crop on July 25 contained only 0.85%. Analysis of sprouting wheat disclosed the increase of soluble sugars from 0.9% in the original wheat to 2.83% when the sprouts were $2\frac{1}{2}$ times the length of the kernel.

Hemicelluloses and Gums.—Research chemists do not agree on the amount of these substances in wheat and the nature of them. The integument of the wheat kernel contains cellulose and pentosans that are mainly arabans although xylans and other hemicelluloses have been reported. One chemist reported four kinds of hemicellulose in bran for a total of 8.5%. Hydrolysis of these products gave arabinose, xylose, glucose and a uronic acid. The gums separated from wheat flour appear to consist of pentosans only.

Lipids.—The lipids from the wheat grain are very complex. The percentage differs with the method of extraction and with the variety of wheat. Watt and Merrill reported the fat content of five wheats from 1.8% to 2.5%, crude bran 4.6% and the germ 10.0%. The white flours contain about 1% fat.

Several investigators have studied wheat lipids. The method of extraction of the lipid makes a difference both in the amount of it obtained and the composition of the extract. In one case, petroleum ether extracted 1.16% from a flour, ethyl ether 1.2%, and after a previous extraction with alcohol, 1.99%; with the same methods, bran gave 4.18%, 4.36% and 4.99% and the germ gave 7.82%, 8.20% and 10.31%.

In another investigation, ether extracted 5.6% lipids from bran and 12.05% from germ, while an alcohol-ether extraction gave 7.36% and 15.24% respectively. The phosphorus content was about five times as high in the alcohol-ether extract, which indicates that the increased yield consists mostly of phospholipids.

Completely extracted lipids consist of glycerides, phosphatids (mostly lecithin) and sterols. The latter are chiefly sitosterol and dihydrositosterol with a small amount of ergosterol.

Commercial wheat germ oil has the following constants: specific gravity 25° 0.9248–0.9268; N_D 20° 1.4762–1.4851; saponification value 180–189; iodine value 115–131; unsaponfiable 3.5–5.7; saturated acids 12% to 16%. One oil contained: oleic acid 26.6%, linoleic 39.1%, linolenic acid 9.6%, palmitic 12.15%, stearic 0.84% and lignoceric 0.3%. The oil is used mostly in pharmacy because of its vitamin content.

Minerals.—The ash content of wheat is of the order of 1.7%; that of flour 0.4%. Bran contains 6.75% ash and the germ about 5%. The ash of the wheat kernel is mostly potassium and magnesium phosphates with trace amounts of other metals.

The ash content of flours increases with the percentage extraction from as low as 0.3% in a 62 extraction flour to 0.45% in a 75% extraction and to 1.7 in a whole wheat flour.

The percentage of ash and its composition differs with the variety of wheat, the stage of maturity, the fertilizer and the amount of rainfall. If a chemist needs to know the ash content or composition of a given wheat product, he must analyze it.

Phytic Acid.—The phosphorus in wheat is distributed among lecithin 4%, nuclein 15%, phytin 71% and inorganic salts 11%. The salts of phytic acid are important in nutrition because they are insoluble and thus diminish the available calcium and iron. It is an inosite hexaphosphate.

Phytic Acid

The phytic acid and phytates are mostly in the bran and germ of the wheat kernel. One sample of 70% extraction flour contained 0.051% and an 85% extraction flour contained 0.127%. Another sample of such flours contained 0.023% and 0.11% respectively.

The enzymes of yeast hydrolyze phytic acid, but that does not improve the calcium content of bread because calcium phosphate is also insoluble.

Pigment.—The pigments in flour constitute only 2 ppm; they are carotenoids, mostly xanthophylls. The carotene content is very slight. The pigments in the bran are a complex mixture of carotenoids and flavones.

Enzymes.—There are several enzymes in wheat, some in the embryo and others in the endosperm. Quantitatively, they differ with the variety of the wheat and the extent to which it may have germinated.

There are two starch-splitting enzymes, alpha and beta amylase. Some of these enzymes are combined with other proteins in the wheat and can be extracted only after treatment with a proteolytic enzyme.

Alpha-amylase is a liquefying enzyme. Added to a starch paste, it reduces the viscosity rapidly and hydrolyzes bonds in the starch to produce dextrins of low molecular weight, which give no color with iodine. It produces very little sugar.

Beta-amylase does not attack whole starch grains but reduces molecules of gelatinized starch by splitting off maltose units from the ends of the chains. However, only about 60% of the starch is hydrolyzed to maltose; the other 40% consists of a dextrin that gives a red color with iodine. To achieve 60% reduction to sugar requires the action of both amylases. There is a very small amount of protease in wheat and wheat flour. It is of the papain type, that is, its activity is increased by reducing agents. The most effective reducing agents that have been tried are hydrogen sulfide, cysteine, cyanides and sulfites.

Flours are bleached, or conditioned, with NO_2, Cl_2, Cl_2O, $KBrO_3$ or other oxidizing agent, which inactivates the protease. The effect of this inactivation on the breadmaking quality of the flour is controversial. Some investigators have thought that the protease injures the gluten, and that the inactivation of the enzyme is the reason for the improvement; others think that the inactivation of the protease is incidental and that the improvers act directly on the protein. The latter opinion seems to be the current one and that the small amount of protease is insignificant in the properties of the flour.

The fat-splitting enzymes of wheat have not been thoroughly investigated. Wheat contains about 3% fat and it appears in all the mill products with the germ containing some 15%, bran 6.5% and flour from 1% to 3% according to the variety of wheat and the milling process. Since the wheat kernel is the plant seed, these lipids require a lipase, and wheat does contain a slight amount. It is most active at pH 7–8 and increases with the water content above 14%. There appears to be no activity below that water content. Strangely, in attempts to locate the enzyme in the wheat kernel, the least activity was found in the germ and the most in the least refined flours. Apparently, it is present in greatest amount in the aleurone layer. If the lipase affects the properties of flour in baking or in other uses, its effect has not been discovered.

There are several oxidative enzymes in the wheat kernel that are concerned with the germination of the seed, but no role in the quality of the flour has been demonstrated.

Vitamins.—The important vitamins in the wheat kernel are thiamin, riboflavin and niacin. Vitamins A, C and D are absent. The actual amount of the B complex vitamins differs with the variety of wheat and with the growing conditions. The average content of the hard wheats is: thiamin 23 mg, riboflavin 5.4 mg and niacin 19.5 mg per lb. Table 13.4 shows the distribution of the vitamins in the wheat kernel.

There are several tocopherols of different vitamin E activity, therefore, the tocopherol content shown in the table is greater than the vitamin E content. Some investigators have found wheat to contain as much as 17 mg of tocopherol per lb with a vitamin E activity of 10 mg.

Enrichment

Nutritional surveys disclosed that bread made from white flour constituted a large portion of the food of Americans and that there were many dietary deficiencies among us. Consequently, the government decided in 1941 to enrich the flour with the minerals and vitamins that were eliminated in the milling. From 1943 to 1946, enrichment was mandatory. Some states still require it, but in most of them, enrichment is now optional.

In addition to the requirements and options in Table 13.5, enriched flour must conform to the general definition of flour, and optional ingredients must be stated on the label.

TABLE 13.4

VITAMIN DISTRIBUTION IN WHEAT

	Mg per Lb				
Cereal	Thiamin	Ribo-flavin	Niacin	Vitamin E	Toco-pherol
Hard wheat	23.00	5.40	19.3	10.0	4.13
Flour	2.70	0.23	4.1	2.7	0.14
Bran	3.27	1.59	95.4	—	1.36
Germ	9.10	3.09	19.1	—	71.91

TABLE 13.5

ENRICHED FLOUR REQUIREMENTS

	Mg per Lb Required	
	Minimum	Maximum
Thiamin	2.0	2.5
Riboflavin	1.2	1.5
Niacin	16.0	20.0
Iron	13.0	16.5
	Optional	
Vitamin D	250 units	1000 units
Calcium	500 mg	625 mg
Wheat germ	—	5%

BREAD

Durum wheat is used to make macaroni, spaghetti and other pasta, but the vast majority of wheat in the United States is used to make bread, cake, pastry or other baked products.

There are three kinds of bread: unleavened, chemically leavened and yeast leavened. The unleavened bread was undoubtedly the first to be made, but its present importance is much less than that of either of the other two.

Yeast Bread

From the earliest days in this country, the housewife made a dough of flour, water and distiller's yeast, *Saccharomyces cerevisiae*. The flour was sifted to eliminate lumps and foreign matter and to aerate it, for the yeast is a plant and requires air to grow. The dough was mixed in a large pan and allowed to remain there overnight. The enzymes of the yeast generated carbon dioxide, which produced a sponge-like structure in the dough and caused it to rise in the pan and frequently to overflow the edges. The texture of the sponge was not uniform, for the gas bubbles were of different sizes including some very big ones.

The dough was transferred to a breadboard and worked by hand to reduce the gas bubbles to a small and uniform size. The dough was then divided into portions to make loaves of a convenient size. These loaves were placed in individual pans and allowed to remain there until they rose to about twice their volume and then they were baked.

Obviously, making bread in the home was an art. The housewife judged from experience how much of each ingredient to use, how long the dough was permitted to rise, baked it at whatever temperature a wood, coal or gas stove would provide and guessed when it was done.

The commercial bakers simply standardized and mechanized the housewife's procedure and have tried to improve the quality of the bread. The federal standards permit the addition of ingredients in 17 categories.

Standards.—The definition is for bread or white bread and rolls or buns made by baking a dough made by mixing flour with water, yeast and salt. Flour has been defined and for making bread, it should contain at least 10% gluten although the standard does not require that. During the fermentation process, changes occur besides the generation of carbon dioxide; a characteristic flavor is produced and the nature of the gluten changes (develops); it becomes more elastic and retains the gas better, but the chemical nature of the changes is unknown.

Optional Ingredients

1. Shortening, which may contain lecithin or mono- and diglycerides, but not more than 20% of the weight of the shortening. The shortening produces a more tender bread and greater loaf volume; 4 gm of shortening per 100 gm of flour increased the loaf volume from 800 cc to 955 cc and also improved the grain of the bread in one experiment conducted to establish its use. The emulsifiers improve the mixing qualities of the dough.

2. Milk in any form may be added and improves the flavor and nutritional quality of the bread.

3. Any kind of buttermilk may be added.

4. Eggs in any form are permitted.

5. Several kinds of sugar, including honey, dextrose, corn syrup and molasses, but not the black strap variety.

6. Malt, enzymes from *Aspergillus oryzae,* amylases from *Bacillus subtalis.* These enzymes are for the hydrolysis of starch to provide sugar for the yeast when no sugar has been added.

7. Three kinds of yeast may be used.

8. Lactic acid bacteria.

9. Corn flour or one of several starches. Two or more of them can be used in combination, but the total may not exceed 3 parts to 100 of flour.

10. Ground soybeans, with or without the fat, but with their enzymes. The amount may not exceed 0.5 parts per 100 parts of flour.

11. Calcium sulfate, lactate, carbonate, phosphate or chloride, alone or in combination to the extent of 0.25 parts per 100 of flour. These salts are added as yeast food. Yeasts are plants and require fertilizer as well as a source of energy.

12. Potassium and calcium bromate or iodate or calcium peroxide alone or together with the others, but not to exceed 0.0075 parts per 100 of flour.

13. Monocalcium phosphate, vinegar, calcium or sodium propionate, sodium diacetate or lactic acid. The propionates are mold inhibitors; the acids for the adjustment of pH for the proper enzyme activity. Each of these additives is limited to a specific, small amount.

14. Spice and spice extracts.

15. Polysorbate 60 and several other complex emulsifiers, not to exceed 0.5 part per 100 of flour.

16. L-cysteine not to exceed 0.009 part to 100 of flour.

17. Ascorbic acid to 0.02 part per 100 of flour.

(b) 1. Bread is baked in units of ½ lb or more. Buns and rolls are the product weighing less than ½ lb.

2. Wheat gluten not to exceed 2 parts per 100 in loaves of bread or 4 parts per 100 in buns.

There are also standards for macaroni and noodles but not for bread raised by chemical agents.

Staling

A day or so after bread has been baked, it changes in character; the flavor changes and the texture becomes harder, tougher, more crumbly and more opaque. Originally, these changes, commonly called *staling,* were thought to be simply a drying out, but as long ago as 1852, Bouissingault, the great French agricultural chemist, showed that stale bread contains just as much water as fresh bread. The change in odor and flavor is probably caused by the loss of volatile compounds by evaporation or oxidation but the cause of the hardening of the crumb is harder to discover. Dozens of chemists have worked on the problem and there are many opinions as a result of all this work. In fact, in a system consisting of the two fractions of starch, several proteins, water, various lipids, emulsifiers and an assortment of salts and sugars, it would not be surprising if staleness were caused by several different reactions.

The chief agreement is that the changes are in the starch. If a dilute starch paste is allowed to stand for several hours, the paste will begin to cloud and then a precipitate forms. This is called *retrogradation.* The cause is thought to be the combination of amylose molecules with each other by hydrogen bonding to form large insoluble aggregates. Presumably, this occurs with the amylose in bread even though the molecules are less free to move about and collide than they are in a liquid suspension. Investigators agree that wheat starch contains amylose but the percentage of it is estimated from 10% to 75% according to the method used to determine it.

The possibility that staling may be caused by the shifting of water or interaction between starch and protein or other components of the system remains to be established.

Baking-powder Bread

The chief advantages of a chemically leavened bread are a characteristic flavor and less time required to make it. Yeast bread requires 12 hr or more to rise, but chemically leavened bread rises in minutes. Since the bread is eaten shortly after it is baked, usually while it is still warm, it is a product of the home kitchen or the restaurant.

Originally, the housewife used baking soda, $NaHCO_3$, and sour milk. This required great skill because she had no way of knowing how much acid there was in the milk, and if she used too much acid, the bread was sour; if too much soda, the bread was yellow and tasted like soap. The invention of baking powder was a great improvement in the making of chemically leavened bread.

Baking Powder.—There are four kinds of baking powder and also powders that contain two or more of the basic types. All of them contain sodium bicarbonate, corn starch, dried to less than 4% water, and they all yield 12% of their weight of carbon dioxide. The oldest of the baking powders are the *tartrate* powders in which the acid ingredient is powdered cream of tartar, $KHC_4H_4O_6$. The starch, soda and tartrate are intimately mixed and, since the reaction is ionic, there is none as long as water is absent. The starch serves two purposes: it separates the particles of soda and acid and it absorbs moisture, even so, there will be a slow loss of gas when the moisture exceeds 4%. When water is added to a tartrate powder the reaction is

$$KHC_4H_4O_6 + NaHCO_3 \rightarrow KNaC_4H_4O_6 + CO_2 + H_2O$$

Some of the tartrate powders have tartaric acid added to increase the acidity. Both tartaric acid and the potassium acid salt are fairly strong acids and all the gas is generated very quickly.

The *phosphate* powders contain dihydrogen phosphate or other acid phosphate as the acid component. When water is introduced, the reaction is

$$3Ca\,H_4(PO_4)_2 + 8\,NaHCO_3 \rightarrow Ca_3(PO_4)_2 + 4NaHPO_4 + 8CO_2 + 8H_2O$$

Here the $H_2PO_4^-$ is weaker than tartaric acid and dissociates as the H^+ is removed to form water, therefore, the generation of gas is slower than that of the tartrate powders.

The *alum* powders contain sodium aluminum sulfate (sodium alum), $NaAl(SO_4)_2$, which is acid by hydrolysis. The overall reaction is a combination of several that occur in succession.

$$2NaAl(SO_4)_2 + 6NaHCO_3 \rightarrow 4Na_2SO_4 + 2Al(OH)_3 + 6CO_2$$

The alum powders are the slowest of the three and many of them are *alum-phosphate* powders, that is, they contain both sodium alum and calcium acid phosphate.

While these four powders are sold for home use, a fifth type is used commercially in prepared mixes and in self-rising flour. The acid ingredient is sodium acid pyrophosphate. Its reaction is

$$Na_2H_2P_2O_7 + 2NaHCO_3 \rightarrow 2Na_2HPO_4 + H_2O + 2CO_2$$

Baking powders are used in the home to leaven cakes as well as bread. For them, the flour used need not have as much protein as that required for yeast leavened bread because the gas has less time to escape and the breads and cakes have a more tender texture.

Industrially, baking powders are used in prepared dry mixes for biscuits or cakes and both the acid ingredient and soda are added to the flour to make it self-rising. The standard for self-rising flour permits the use of monocalcium phosphate, sodium acid pyrophosphate and sodium aluminum phosphate.

The baking powders differ mainly in cost and the speed with which they deliver the CO_2. Although they all yield 12% CO_2, the amount required for a given purpose differs because of the speed with which the gas is evolved. Their relative merits for a given purpose must be determined by experiment; the best powder for a biscuit mix may not be the best for a cake mix. In one case, the author found that one powder was cheaper and required less than another so that the cost of it was less than one-third that of the more expensive powder.

From nutritive considerations, the calcium phosphate powders are favored because both calcium and phosphate are dietary requirements. The starch present supplies about 1 kcal per gm of baking powder.

CORN

Corn (*Zea mays*) is our biggest cereal crop. In 1972, the United States produced over 5.4 billion bushels compared to some 1.5 billion bushels of wheat. Human consumption of corn meal and corn flour, however, in 1972 was only 7.4 lb per capita while that of wheat flour was 110 lb per capita. In addition to the 7.4 lb of ground corn, we consumed 21.4 lb as corn syrup and corn sugar—out of a production of 147 lb per capita, we consume 28.8 lb. The remainder is exported or fed to domestic animals. The nutritive value of corn as compared to wheat and rice is shown in Table 13.6.

The values in the table are based on 100 gm of the cereal. Corn meal differs somewhat from corn flour because the meal may contain part of the germ. Wheat flours also vary, but the figures are for the commercial all purpose flour.

The chief difference in corn and wheat flours is the lower protein and the higher content of iron and vitamin A of the corn. Recently, the plant breeders have developed a variety of corn of higher protein content, but the protein is not gluten and is not so adhesive. A simple corn bread (pone) crumbles so easily that it cannot be spread with butter. Corn muffins, which

TABLE 13.6

NUTRITIVE VALUE OF THREE CEREALS

	Bolted Corn Meal	Wheat Flour	White Rice
Moisture (%)	12.0	12.0	12.0
Kcal/100 gm	362	364	363
Protein (%)	9.0	10.5	6.7
Fat (%)	3.4	0.5	0.4
Carbohydrate (%)	74.5	76.1	80.4
Ash (%)	1.1	0.43	0.3
Ca (Mg/100 gm)	17	16	24
Phosphorus	223	87	94
Iron	1.8	0.8	0.8
Thiamin	0.30	0.06	0.07
Riboflavin	0.08	0.05	0.03
Niacin	1.9	0.9	1.6
Vitamin A (IU)	480	0	0

have largely succeeded it, are made from about equal parts of corn meal and wheat flour. The consumption of corn meal breads has declined enormously in the past century. In 1889, the per capita consumption of corn meal was 115 lb compared to 7.4 lb in 1972. The early settlers raised corn because it was easier to grow than wheat among stumps or dead trees so that it naturally became the main cereal in America. Europeans are unfamiliar with it and, in the food scarcity of World War I, we shipped corn meal to Europe, but the French and British would not eat it and so our government urged us to eat more corn and thus leave more wheat for export. The increased prosperity, the move to cities and the influx of European immigrants contributed to the decline in the use of corn for food in this country.

As for the nutritive value of the protein, the amino acid content of corn protein as compared to that of wheat is shown in Table 13.7.

TABLE 13.7

ESSENTIAL AMINO ACID CONTENT OF THE CEREAL PROTEINS

	Percentage		
	Wheat	Corn	White Rice
Arginine	14.06	9.36	8.4
Histidine	7.89	0.75	2.6
Isoleucine	0	0.8	4.5
Leucine	33.91	31.22	8.4
Lysine	5.88	6.71	3.5
Methionine	4.85	2.35	2.5
Phenylalanine	8.40	9.34	4.5
Threonine	2.0	2.90	3.9
Tryptophane	8.14	2.07	1.2
Valine	3.82	1.88	6.5

The determination of the amino acid content of a protein is a difficult assay and so there is considerable variation in the reported results. The values in Table 13.7 are average values of several reported analysis.

Compared with wheat proteins, those of corn are low in histidine and tryptophane. The latter is of special importance because tryptophane is a precursor of niacin.

Corn Products

In 1972, we consumed 16.2 lb of corn as corn syrup, 5.2 as corn sugar, a small amount as breakfast cereals and a considerable amount in the form of Bourbon whiskey.

The syrup and the sugar are from the wet milling process. The corn is cleaned and soaked in water for two days. The kernel contains approximately: 15% moisture, 60% starch, 9% protein, 5% oil and 5% hull. Sulfur dioxide is added to the water to supress germination and the growth of organisms and to aid in loosening the hull. The grains swell considerably during the steeping and the H_2SO_3 bleaches them.

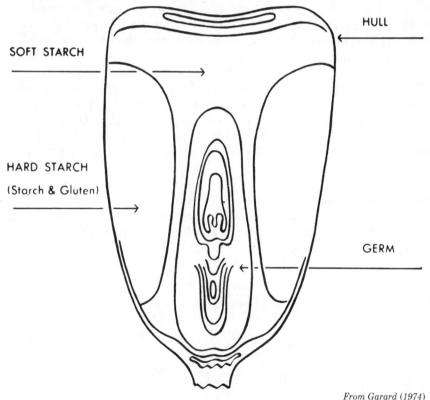

From Garard (1974)

FIG. 13.2. DIAGRAM OF THE CORN KERNEL

The swollen and softened kernels go to a machine that loosens the hull and the germ. The germs are separated by flotation, dried and go to an oil extractor. The corn is then ground on a stone mill, suspended in water and passed over a bolting cloth to remove the hulls. The suspension of starch and protein is separated by a centrifuge. The steep water, solids, protein, hulls and press cake from the germ are combined for cattle feed.

The starch is washed with water or further processed and used for a variety of purposes. A large part of it is hydrolyzed to produce syrup or sugar. The process is carried out in bronze converters. HCl is added to approximately 0.1% and steam passed into the aqueous suspension at 30 psi (120°C) or more and the hydrolysis occurs in a few minutes. As we noted in Chapter 3, starch is hydrolyzed by stages through dextrins and oligosaccharides to dextrose. The reaction can be stopped at any stage and the mixture cooled and neutralized with Na_2CO_3 to pH 4–5. Alkalinity must be avoided because sugars decompose in alkaline solution.

A great variety of products is produced. Formerly a first product, mostly dextrin, and called "glucose" was made and sold to confectioners to improve the texture of fondants and other candies. It was about the consistency of taffy ready to pull and, therefore, difficult to handle. The use of the commercial term "glucose" encouraged chemists to use the name *dextrose* for the hexose sugar, although glucose and dextrose are still used interchangeably for the name of the sugar.

The main product today of starch hydrolysis is *corn syrup,* which is made in so many different compositions that the term does not indicate a specific product. Sucrose is added to some corn syrup to increase the sweetness. Recently a corn syrup has been made by hydrolyzing the starch with an enzyme mixture, one of which converts glucose to fructose and thereby produces a sweeter syrup. Table 13.8 shows the composition of some corn syrups. Besides the retail package of table syrups, large quantities are sold to confectioners, bakers, ice cream manufacturers and other food processors, which is obvious from its inclusion in so many food standards. The syrups are also dehydrated and the processor may buy his corn syrup under a great variety of specifications.

TABLE 13.8

COMPOSITION OF SOME CORN SYRUPS

	Water %	Dextrins %	Maltose %	Dextrose %	Sucrose %	Fructose %
1.	25.3	35.4	22.2	7.5	7.1	—
2.	25.2	9.1	25.5	28.0	—	—
3.	25.4	36.0	22.0	7.4	—	—
4.	29.0	—	6.0	50.0	—	42.0

The reducing sugars in a corn syrup may be determined by one of the copper reduction methods and reported as anhydrous dextrose. The value is known in the industry as the *dextrose equivalent* (DE) and increases with the degree of hydrolysis. The change in oligosaccharide content with the degree of hydrolysis is indicated by the following examples:

DE	Mono	Di	Tri	Tetra	Penta	Hexa	Hepta	Higher
10	2.3	2.8	2.9	3.0	3.0	2.2	2.1	81.7
45	21.0	14.9	12.2	10.1	8.4	6.5	5.6	21.3
65	42.5	20.9	12.7	7.5	5.1	3.6	2.2	5.5

Corn Sugar

If the hydrolysis of starch is carried to completion, the product is dextrose. During the neutralization, any local excess of alkali may cause coloration in addition to that produced by the heat. The color is removed with boneblack, the resulting syrup evaporated in vacuum pans and the sugar crystallized. The sugar is mainly hydrated dextrose, $C_6H_{12}O_6 \cdot H_2O$ and may be mixed with a little maltose. Its sweetness is about 60% that of sucrose.

Dextrose is a more highly refined sugar and mostly anhydrous. During World War II, dextrose was packaged for retail sale in the groceries because of the shortage of sucrose. But it was an unfamiliar product and did not sell, so it was withdrawn after the war. It is sold to many other food industries as the food standards indicate.

RICE

The consumption of rice in the United States is 7 lb per capita or about the same as that of corn meal, but the production is much less than that of corn or the other common grains. The world production of rice (299.4 metric tons) in 1972 was second to wheat (322.6 metric tons).

In the United States, rice is hardly a competitor of corn and wheat in the diet. It is mostly boiled and eaten as a vegetable just as it is in the Orient. It is more of a competitor of potatoes although some of it does appear as a dessert.

Tables 13.6 and 13.7 indicate the nutritive value of rice. The protein is poorer than that of wheat and corn in five of the essential amino acids. There is some compensation for the tryptophane deficiency because of the greater niacin content.

Because of the difference in thiamin content in brown and white rice, those foods played the leading role in the discovery of that vitamin and the cause of beriberi (Garard, 1974).

Very little rice is ground into flour. The brown outer layer (bran) is ground off mechanically, hence the term "polished rice."

Nutritionists favor brown rice just as they do whole wheat and for the same reason. Brown rice contains more calcium, phosphorus and iron and each of the three B vitamins. The processors and grocers are not so enthusiastic because of insect infestation and the consumers do not like the brown rice as well as the white variety.

BARLEY

The growing of barley differs from that of the other grains in several respects. It grows well in semiarid regions, but not with heavy rainfall. It withstands both heat and cold and can be raised much farther north than either wheat or corn. It yields more to the acre than wheat. The production of barley in the United States in 1972 was 423,461,000 bushels compared with 1,544,775,000 bushels of wheat. North Dakota grew about one-fourth the total followed by Montana, California, Idaho and 34 other states.

Although we produce a large amount of barley, it is not a major cereal food; we consume a small amount as pearled barley and barley flour.

Pearled Barley

The barley kernel resembles a grain of wheat. The outer coats are polished off much as rice is polished and the remainder, which is about half the seed, is sold mainly for use in soups.

Barley Flour

A little barley is ground into flour, which is used mainly to make infant foods and breakfast cereals; barley bread is rare in this country.

The flour contains water 10.3%, protein 8.0%, fat 1.7%, carbohydrate 80%, ash 0.65% and is practically devoid of nutrient minerals and vitamins. The protein is not glutinous, hence the scarcity of barley bread in the U. S. Some European and Asiatic countries do have barley bread because of the agricultural advantages of barley over wheat, but it is a very different product from wheat bread.

Malt

Probably the greatest consumption of barley in the U.S. is in the form of malt and its products.

When the seed sprouts, barley produces far more amylase than it needs to hydrolyze its own starch, and chemists have taken advantage of that fact to provide enzymes for several processes. The barley is steeped in cold water for 2 or 3 days or until the kernels have fully swelled. It is then aerated for 4 or 5 days at a constant temperature to permit sprouting. The sprouting is stopped before it is complete by heating in a kiln hot enough to stop the growth but not enough to inactivate the enzyme. The extent of the sprouting differs according to the intended use of the malt.

There are four malt products: malt, malt extract, malt syrup and malt sugar. Malt amylase converts starch to maltose and, if the process is allowed to go to completion, extraction and evaporation produces the malt syrup, which is used as a sweetener in baked goods and a few other products. If the syrup is refined and the sugar crystallized, the product is malt sugar, which consists of maltose and some dextrin. It is used to make malt vinegar and as a general sweetener. Malt is simply the sprouted grain. Malt extract is made by grinding the malt and extracting it with water and then adjusting the enzyme content to the desired level, or the extract may be evaporated to dryness for shipment and is still called malt extract. The following figures are for comparison of dried malt and dried malt extract:

	Water	Protein	Fat	Carbo-hydrate	Ash
Malt	5.2	13.1	1.9	72.4	2.4
Malt extract	3.2	6.0	trace	89.2	1.6

Malt extract is used in the manufacture of malt liquors: beer, ale, porter and stout. For this purpose, the malt is ground and mixed with other ground grains, which may be corn, rice, rye or wheat according to the availability and price of the grain, or some brewers feel that certain grains produce a better flavored product. The ground grains are mixed with water and heated to gelatinize the starch and then kept warm until the starch is hydrolyzed to sugars. The sugars are then fermented with yeast. The grains used and the engineering details differ with brewers. At one stage, hops are added to produce the characteristic bitter taste of the malt liquors.

The use of malt in the manufacture of whiskey is essentially the same as that of the manufacture of beer. In the manufacture of Scotch whiskey, the malt is dried, with peat as part of the fuel, and the smoke flavors the malt. The flavor is carried through the distillation. Some malts get more smoke than others and, therefore, some batches of whiskey have a much stronger flavor of smoke than others, and since the whiskey is made in small batches, dealers buy several batches and blend them to get a uniform taste to all the samples of their brand. Both Scotch and Irish whiskey are made from barley. The grain for Bourbon whiskey must be at least 51% corn. Blended whiskies may contain alcohol from fermented molasses (Garard, 1974).

Malted Milk

This product is made by treating wheat flour with malt, then extracting the mixture with water. The extract, containing sugars, dextrins and soluble proteins is added to milk and the mixture is dehydrated. An analysis of malted milk is: water 2.6%, protein 14.7%, fat 8.3%, carbohydrate 70.8% and ash 3.6%; Ca 288, P 380, Fe 2.1, thiamine 0.33, riboflavin 0.54% and niacin 0.3 mg in 100 gm of the milk; vitamin A 1020 IU per 100 gm.

OATS

The 1972 oats crop was over 694 million bushels, which was down more than 206 million bushels from the average of the previous five years. It is grown in all the states except Alaska, Arizona, Hawaii and the New England states. It is a hardy crop that matures in about five months. About a fifth of the oat crop is used for food, the remainder is cattle feed. The chief food use is as a cooked breakfast cereal with less in a prepared cereal and as flour used to make cookies. The protein of oats is not suitable for making yeast bread.

The oat crop is a large potential source of food if other methods of using it can be devised and accepted. The composition of rolled oats is much the same as that of the whole kernel and contains: water 7.7, protein 16.7, fat 7.3, fiber 1.3, other carbohydrates 66.2 and ash 2.1%. The 100 gm portion contains Ca 53, P 405, Fe 4.5, thiamin 0.6, riboflavin 0.14 and niacin 1.4 mg. These data indicate the nutritional superiority of oats over the other cereal grains.

RYE

Like oats and barley, rye is a great potential food supply. In 1972, the United States produced nearly 30 million bushels. Since rye weighs 56 lb to the bushel and oats 32, the rye crop is less than one-tenth of the oat crop.

Most of the rye crop goes into cattle feed and the manufacture of whiskey; the per capita consumption is less than 3 lb. The proteins are similar to those of wheat and the flour is used to make both leavened bread and a hard bread of the Swedish type.

Rye is hardier than wheat and exceeds wheat in the diets of northern European people. There are three grades of rye flour: light, medium and dark. The composition of the light flour is: water 11, protein 9.4, fat 1.0, carbohydrate 77.9 and ash 0.7 percent; it is inferior to oats in all the other nutrients. The dark flour, however has: water 11, protein 16.3, fat 2.6, carbohydrate 68.1 and ash 2%. The 100 gm portion contains: Ca 54 mg, P 536 mg, Fe 4.5 mg, thiamin 0.66 mg, riboflavin 0.22 mg and niacin 2.7 mg. The medium flour has about the average composition of the other two.

USE OF CEREALS

With the pending shortage of food, thought must be given to more efficient use of the present and potential foods. Table 13.9 shows the per capita consumption of the various foods.

The chief trends in recent years have been a 20% increase in the consumption of beef and pork and an 11% drop in poultry, eggs and dairy products, a 23% drop in fresh fruit, a 6% increase in sugar and a 12% increase in potatoes and sweet potatoes.

TABLE 13.9

PER CAPITA CONSUMPTION OF FOODS IN THE U.S.

	1957—1959 Average (lb)	1972 (lb)
Meats	156.6	188.8
Fish	10.5	11.5
Poultry and eggs	356.0	315.0
Dairy products	378.0	295.0
Fats and Oils	45.3	53.1
Fresh fruits	95.5	77.7
Processed fruits	47.8	48.1
Fresh vegetables	104.1	98.2
Processed vegetables	49.9	61.2
Corn, wheat and rice	148.6	148.7
Cacao beans	3.5	4.4
Peanuts	4.6	6.1
Dry beans	7.7	5.9
Sugar	96.1	102.4

The changes in the high protein foods are in the wrong direction both nutritionally and economically; eggs and dairy products are superior to beef and pork nutritionally, and the decline in fresh fruit is also a negative trend. Economically, beef and pork are expensive. Cattle convert only 4% of their feed into food and hogs only a little more, yet most of all the grains except wheat and rice are used for feed. The cow converts 36% of her feed into food, and poultry is a much better converter than cattle; but poultry, eggs and milk are all declining in use.

Normal diets require about 12% protein, and beans, peas, lentils and peanuts are the chief vegetable sources of this nutrient. The cereals are next in order as vegetable sources of protein and their consumption has not changed, but cornmeal has declined and been replaced by the purely carbohydrate products, starch and its hydrolytic products. The protein from the corn goes into animal feed.

A promising trend is the increase in the consumption of sea food, which freezing and improved transportation has made possible; this is protein that does not draw on the cereals.

The food chemist must recover more of the protein of the cereals and oil press cakes that now goes into feed, and make it acceptable to the consumer.

BIBLIOGRAPHY

ANDERSON, J. A. 1946. Enzymes and Their Role in Wheat Technology. Interscience Publishers, New York.

ANON. 1973. New standards for flour enrichment. Food Tech. 27, No. 12, 72–73.

ASSOC. OFFIC. ANAL. CHEMISTS. 1975. Official Methods of Analysis, 12th Edition. Assoc. Offic. Anal. Chemists, Washington, D.C.

BAILEY, C. H. 1944. The Constituents of Wheat and Wheat Products. Reinhold Publishing Corp., New York.

BANKS, W., and GREENWOOD, C. T. 1975. Starch and its Components. John Wiley & Sons, New York.

BOHN, R. M. 1957. Biscuit and Cracker Production. American Trade Publishing Co., New York.

BOOKWALTER, G. N., WARNER, K., BREEKE, O. L., and GRIGGIN, E. L. JR. 1974. High-lysine corn fractions and their characteristics. J. Food Sci. *39*, 166–170.

CLAUSI, A. S. 1971. Cereal grains as dietary protein sources. Food Tech. *25*, No. 8, 63–67.

COOK, A. H. 1962. Barley and Malt. Academic Press, New York.

DANIELS, R. 1970. Rice and Bulgur Quick-cooking Processes. Noyes Data Corp. Park Ridge, N.J.

FANCE, W. J. 1969. Bread Making and Flour Confectionery. Avi Publishing Co., Westport, Conn.

GARARD, I. D. 1974. The Story of Food. Avi Publishing Co., Westport, Conn.

HARRIS, R. S., and KARMAS, E. 1975. Nutritional Evaluation of Food Processing. Avi Publishing Co., Westport, Conn.

HERZ, K. O. 1965. Staling of bread—a review. Food Tech. *19*, No. 12, 90–103.

HULLINGER, C. H., VANPATTEN, E., and FRECK, J. A. 1973. Food applications of high amylose starches. Food Tech. *27*, No. 3, 22–24.

INGLETT, G. S. 1974. Wheat Production and Utilization. Avi Publishing Co., Westport, Conn.

LEE, F. A. 1975. Basic Food Chemistry. Avi Publishing Co., Westport, Conn.

LOWRIE, R. A. 1970. Proteins as Human Food. Avi Publishing Co., Westport, Conn.

LORENZ, K. 1972. Food uses of triticale. Food Tech. *26*, No. 11, 66–74.

MATZ, S. A. 1972. Bakery Technology and Engineering. Avi Publishing Co., Westport, Conn.

MATZ, S. A. 1969. Cereal Science. Avi Publishing Co., Westport, Conn.

MATZ, S. A. 1970. Cereal Technology. Avi Publishing Co., Westport, Conn.

MCWILLIAMS, N. M. 1969. Wheat flour components. J. Food Sci. *34*, 493–496.

OSBORNE, T. M. 1907. The Proteins of the Wheat Kernel. Carnegie Institution of Washington, Washington, D.C.

OSMAN, E. M. 1975. Interaction of starch with other components of food systems. Food Tech. *29*, No. 4, 30–35.

PETERSON, R. F. 1965. Wheat. John Wiley & Sons, New York.

POMERANZ, Y., and SHELLENBERGER, J. A. 1971. Bread Science and Technology. Avi Publishing Co., Westport, Conn.

SENTI, F. R., and DIMLER, R. J. 1959. High-amylose corn—properties and prospects. Food Tech. *13*, 663–667.

SHOUP, F. K., POMERANZ, Y., and DEYOE, C. W. 1966. Amino acid composition of wheat varieties and flours varying widely in bread-making potentialities. J. Food Sci. *31*, 94–101.

SINHA, R. N., and MUIR, W. E. 1973. Grain Storage. Avi Publishing Co., Westport, Conn.

SULTAN, W. J. 1969. Practical Baking, 2nd Edition. Avi Publishing Co., Westport, Conn.

WARDRIP, E. K. 1971. High fructose corn syrup. Food Tech. *25*, No. 5, 47–50.

WATT, B. K., and MERRILL, A. L. 1963. Agriculture Handbook No. 8, Composition of foods. ARS-USDA, Washington, D.C.

WEISS, T. J. 1970. Food Oils and Their Uses. Avi Publishing Co., Westport, Conn.

CHAPTER 14

Sugars and Related Industries

Next to salt, if not ahead of it, sugar is our favorite flavor. In the years 1957–1959, the per capita consumption of sugar in the United States was 96 lb and by 1972, it had increased to 104 lb. This was the consumption of refined sugar and in addition to corn sugar, honey, sorghum syrup, maple syrup and the sugars in fresh fruits and vegetables. This sugar was eaten primarily for its flavor rather than for the 4 kcal/gm of energy that it supplies.

CANE SUGAR

Most of our refined sugar comes from the sugar cane, *Saccharum officinarum,* which is one of the largest grasses; it grows to a height of 20 ft and a diameter of approximately 2 in. The Spanish explorers brought it to tropical America in the 17th century and by 1794, it was growing in Louisiana. Today, Florida and Louisiana produce about a million tons of sugar annually and the beet growers in some 20 states produce another 3 million tons; we import over 75 million tons from Hawaii, Puerto Rico, the Philippines and the islands and coasts of the Caribbean countries. Table 14.1 shows the composition of these two sources of sucrose.

Originally, small cane mills produced the sugar. There is said to have been 1000 of these mills in Louisiana. Some are still in use in the less industrialized countries, but most of the cane sugar of commerce is produced in large steam powered mills.

Manufacture

With the modern mill, the fields are cut clean and the blades stripped off and left in the field for fertilizer. In some places, the fields are burned to remove the blades and save labor.

TABLE 14.1

SUGAR CONTENT OF CANE AND BEETS

	Cane %	Beet %
Water	67–75	75–85
Dry substances		
Fiber	10–15	4–6
Sucrose	11–16	12–16
Invert sugar	0.5–1.5	0–0.3
Ash	0.5–1.0	0.8–1.5
Protein	0.4–0.6	1.5–2.5
Gums	0.2–0.5	0.3–0.6
Wax and fat	0.4	—

The stripped cane is hauled to the mill by train or truck and dumped into a pit where elevators carry it to two large rolls that cut it into short pieces and crush it. From these, the cane goes through three or more mills, each consisting of three rollers. After the first and each succeeding set of rolls up to the last, the cane is sprayed with water before it passes through the next set. This is a familiar extractive principle, which the chemist uses to wash a precipitate and the housewife uses to wash soap out of a sponge; wet the sponge, squeeze it, wet it, squeeze it again, and repeat as often as necessary. The *bagasse* from a modern mill has no sweet taste whatever. However, the chemist must decide how much water to add. The more the better for the extraction, but the water must be evaporated later in the process and the problem becomes whether the additional sugar is worth the energy it takes to evaporate the extra water.

The bagasse from some mills is used to make wall board; others use it for fuel for the boilers that supply heat and energy for the process.

The several streams of juice are combined and the total goes to a clarifier because it contains acids, proteins, waxes, gums, phosphatids and bits of cane. The removal of colloidal matter is extremely important, because it interferes with the crystallization of the sugar. Lime is added and the clarifier removes the solid matter, which is returned to the field for fertilizer.

The clear cane juice is evaporated in a series of vacuum evaporators until a sample drawn from the last evaporator crystallizes satisfactorily. It is then run into crystallizers. When it has cooled, the crystals are separated from the mother liquor, or molasses, by centrifuge. The molasses may be refined for use in food, but the crude "black strap" molasses goes into cattle feed or to the distilleries for the production of alcohol. The molasses contains about 30% sucrose, 30% invert sugar, 1.5% amino acids and considerable calcium, phosphorus and iron, but the flavor is not a favorite with the American public and so only a small amount of it appears in the food markets.

The raw sugar varies around 95% sucrose and was formerly shipped to other companies for refining. In the United States, this meant that the refineries were in the coastal cities from Boston to Seattle. Even the sugar from the Everglades was shipped to Savannah to be refined. In recent years, there is a tendency to build the refineries closer to the source of the sugar, sometimes as part of the factory.

The evaporation of cane juice involves more than the removal of water. Vacuum pans are used so the water can be evaporated at a lower temperature, because heating sucrose with water hydrolyzes it to invert sugar. The raw cane juice contains about 1.5% invert sugar and the first molasses contains anywhere from 10% to 35% even at the reduced temperature. Invert sugar does not crystallize and is mostly an economic loss because of the low value of the molasses.

In addition to inversion, there is a loss of sugar by caramelization. The pH of the juice when it enters the evaporators is extremely important; if it is too low, it increases the inversion of the sucrose and if it is too high, the sugars decompose. Furthermore, the juice contains amino acids that react with the monosaccharides to form brown substances (Maillard reaction).

The engineering details of a sugar factory are an excellent example of the application of engineering to food technology, for the economical production of pure sucrose from sugar cane is a difficult operation.

Refining

The product of the sugar factory, at best, is a light brown crystalline product that still contains a little molasses on the surfaces of the crystals. It is hygroscopic and, exposed to the atmosphere, it will soon form a solid cake. Both industry and the housewife demand a free-flowing sugar and so the raw sugar must be refined.

The refining is a continuous process and to save energy, the refinery is a tall building. The sugar is elevated to the top and thereafter flows from one process to the next by gravity.

The sugar is first washed with a little water and separated by a centrifuge. It is then dissolved in water at 150° to 170°F to form about a 50% solution. The solution is neutralized to prevent inversion and remove any gums, acids or other colloidal material. The precipitate is removed but the clear solution, which is still brown, is filtered through a large cylindrical tank of boneblack, which removes the color by adsorption. The boneblack soon becomes covered with the adsorbed impurities and must be removed and recharred, but boneblack is porous and much of the surface is within the pores and these become filled up, so that after several recharrings, little adsorption occurs and new boneblack must be supplied.

Just after World War I, the activated carbon that had been developed for gas masks was used in some refineries, but apparently did not prove as economical as the boneblack.

After the syrup has been decolorized, it goes to vacuum pans similar to those used in the factories. From the evaporators, the syrup goes to crystallizers and centrifuges where the refined crystals are separated from the *refiner's syrup*. The latter contains less impurity than molasses and finds use in some food industries.

The crystals are dried and screened to get uniform size in the several grades of granulated sugar; the very large and very small crystals are ground to powdered or confectioner's sugar.

Many industries that use sugar dissolve it and incorporate it into their product as a syrup. Consequently, considerable sugar is now marketed as syrup, which shortens the refining and saves one step for the industry that uses it.

BEET SUGAR

In 1972, the United States produced 28,523,000 tons of sugar beets. The sugar beet, *Beta vulgaris,* grows well in temperate climates, but does not produce as much sugar to the acre as sugar cane does. It has been grown in Europe since 1769 at least, but not in this country until 1870.

The sugar beet is shaped somewhat like a parsnip, is yellow in color and weighs a pound or more. It contains about 15% sugar. The other components of the beet are different from those of the cane and so the process of manufacture is very different.

Extraction

The beet tops are removed in the field and are an excellent cattle feed. The beets are brought to the factory and washed. They are then sliced into thin ribbons and the sugar extracted by dialysis; the cell walls serving as the semipermeable membrane. The beets are treated with water at about 85°C, as warm as possible without cooking the beets, which would destroy the cell walls. The higher temperature accelerates the rate of diffusion.

To illustrate the principle involved in the extraction and keep the mathematics simple we have an equal weight of water and beet slices in contact; the beets contain 20% sugar. The sugar and other low molecular weight solubles pass through the cell walls into the water until equilibrium is reached when the water and the beets each contain 10% sugar. The water passes on to an equal weight of fresh beets and at equilibrium, those beets and the water each contain 15% sugar. Meanwhile, fresh water reduces the first beets to 5% sugar. After several portions of water have passed through the dialyzers, the first beets are reduced to a small fraction of 1% sugar and the water is increased to close to 20%. Some factories have as many as 12 dialyzers, each of which holds several tons of beets. The engineering is usually arranged so that ten are dialyzing while one is being emptied and another being filled. Unfortunately, the theoretical extraction described above is not achieved, for less than an equal volume of water is used and the water does not make perfect contact with all the slices. In practice, the syrup contains 12–15% sugar and the residual beets about 2%.

Refining

The dialysis is the first step in the refining of the sugar, for the high molecular weight gums, proteins and other compounds are left in the cells.

Many cells are damaged in the slicing and so the syrup does contain some of these colloidal substances.

The syrup is filtered to remove particles of beets, and lime is added to neutralize free acids and coagulate the organic substances. The excess lime is then precipitated by passing CO_2 into the solution; the precipitate is removed and the carbonation repeated. The syrup may then be bleached with SO_2 and concentrated in vacuum pans in the same manner as cane juice is concentrated. Beet sugar is refined by the same process used for cane sugar. The refinery and factory are usually one complex.

There is no difference between well-refined beet and cane sugar. Official standards that call for sugar call it "sucrose" and do not distinguish between the sources of it.

MAPLE SUGAR

The sugar maple tree, *Acer saccharum,* is a native of our northern states and Canada; it thrives as far west as Wisconsin and as far south as West Virginia and the Ohio river. In February and March when the sap begins to rise, it contains 2% to 4% sugar. The trees in colonial days supplied the sugar needs of most families of New York, New England and Pennsylvania with all their sugar needs. Today, it is a luxury and marketed chiefly in the form of syrup; the sugar is a rare confection.

Before the first spring thaw, holes were bored in the trees 1 to 4 ft above the ground. A *spile,* or tube was made from a straight piece of elder or sumac and driven into the hole. The spiles were 8–10 in. long. A pail was set under the end of the spile. Sometimes, two or even three spiles were placed close together to drain into the same pail, and a tree might be tapped on opposite sides.

When the first thaw came, the sap began to rise in the tree and on a warm day, a spile might run a constant stream of "sugar water." When the nights no longer dropped below freezing and the leaves began to appear, the sap became milky and the season was over—2 to 6 weeks according to the weather. The sap was collected and boiled down in open kettles over a wood fire.

Making syrup was an art, for the farmer had no way of telling when it was finished except to dip out a small sample, cool it and examine its viscosity. A commercial syrup had to weigh 11 lb to the gallon; at 12, some sucrose would crystallize and at 10, organisms would spoil the syrup. Because of the difficulty of determining the proper concentration of the syrup and the scarcity of jars to contain it, many farmers evaporated it further and stirred the syrup until it solidified to a crystalline mass, which soon set up to a solid cake. There is no refining of maple sugar except to remove a scum that collects on the surface during the boiling and to decant it from the "sand" that settles as the boiling continues. The sand is mainly salts of malic and other acids.

The solid sugar was easy to store and, since it was used mostly as syrup, a supply could be made as needed by dissolving the sugar in hot water.

Few improvements have been made in the maple sugar industry through the years. Spiles have become metal tubes and steel hooks are sometimes driven into the trees above the spiles so the pails can be hung high enough above the ground to keep the small animals from drinking the sap. Tanks on sleds for collecting the sap succeeded men and boys with a pail in each hand and, when a "bush" or grove, was large enough to justify it, an evaporator consisting of a flat pan some 3 by 8 ft and 6 in. deep or larger, set on a stone furnace, succeeded the kettles. These pans have partitions part way across so the sap moves slowly from one end to the other as the water evaporates. A good operator has the evaporation finished when the sap reaches the end of the pan, so that evaporation has become a continuous process.

The United States Department of Agriculture has done considerable research (Underwood and Willits 1963) on the technology of maple syrup production. Among other improvements, they have substituted a system of plastic tubes to collect the sap instead of the spiles, buckets and tank sleds. They have also substituted oil for wood as fuel and have installed instruments to determine the final concentration of the syrup. At 11 lb to the gallon, the syrup contains 65.5% sugar.

The color and flavor of maple syrup develop during the boiling. The sap contains some invert sugar and is slightly alkaline so the sugar is decomposed to products that produce color and flavor by the Maillard and probably other reactions.

Maple sugar is seldom marketed because of its scarcity and tendency to form very hard cakes. In tourist sections of New England, the syrup is stored and sugar is made from it daily and supplied to roadside stands and other tourist outlets. The fresh sugar is an excellent confection but hardens in a few days.

Because of the short season, labor, heat requirements and unique flavor, maple syrup is expensive and, therefore, vulnerable to adulteration with the cheaper sucrose or corn syrup; consequently, the Official Methods contain several analytical determinations. The methods of identity depend mostly on the fact that maple syrup contains 0.7% ash and malic acid or its salts, whereas the cheaper syrups contain neither. It may also contain up to 4% reducing sugars, so that adulteration can readily be detected unless it is expertly done by a well informed chemist, which is unlikely.

SORGHUM SYRUP

A fourth source of sugar within the United States is sorghum. Sorghum is a member of the grass family and there are many varieties of it; some are grown for cattle feed and some for the sugar in the sap. It is raised mostly in the southern states but grows well as far north as Pennsylvania. It is grown from seeds and matures in about 5 months.

The canes are cut, stripped of their leaves and topped in the field. Many of the mills are of the primitive type such as those formerly used for sugar cane, but some are more like a modern cane mill, if somewhat simpler.

The cane juice contains 10–12% sugar, but the earlier it is cut, the greater the amount of invert sugar, which may reach half the total; the remainder is sucrose. Because of the high content of invert sugar, crystalline sucrose is not made from sorghum. The juice is evaporated to 11.25 lb/gal. and the finished syrup contains about 68% sugars and 2.4% ash. The ash is especially rich in iron.

Most of the sorghum syrup is made in open pans and is very dark because of the Maillard reaction, decomposition of hexoses and iron content. In some of the more modern factories, the juice is evaporated in vacuum pans and some of the color removed with boneblack. The syrup, especially that from the primitive mills, has a strong flavor that is not generally liked.

HONEY

There are as many kinds of honey as there are flowers that produce nectar. There are said to be 50 crops in the United States that depend on bees for fertilization of the blossoms. In addition to the crops, there are many wild and domestic flowers.

Much of the honey is of mixed origin. Each species of plant has a short blooming season. Apple trees, peach trees and citrus trees all have a short blooming season even when all varieties are included, but in the fruit country, there may be millions of trees. Consequently, in a fruit area a distinctive type of honey may be produced. Clover, alfalfa, buckwheat and orange blossom honey are the most common distinctive varieties. The first two of these furnish about half the commercial honey. In some parts of the country, there are large areas of these forage crops and they have a longer blooming season than fruit trees. The different honeys differ in flavor, for the bees collect some of the scent of the flower along with the nectar.

The plant nectar contains sucrose and other sugars and also formic, acetic, citric and malic acids, so that by the time the bee collects the nectar, stores it and fans it to a syrup heavy enough to remain there, most of the sucrose has been inverted. Sherman (1948) reported the average composition of 92 samples of honey as:

	%		%
Water	17.70	Dextrin	1.51
Sucrose	1.90	Ash	0.18
Fructose	40.50	Undetermined	3.73
Glucose	34.48		

From these data, it is obvious that as a nutrient, honey supplies mainly energy. The usual 100 gm sample supplies 304 kcal and 0.3% protein. Its chief value lies in its flavor. In the early days of the country, it was an important source of sugar; now it is a luxury.

Officially, honey is levorotatory and contains not more than 25% water, not more than 0.25% ash and not more than 0.8% sucrose. It is liable to adulteration but very little sucrose could be added because it is easily detected and if dextrose is added, it crystallizes out. In fact, it sometimes crystallizes out of pure honey causing a granulation. Official Methods cover a complete analysis of honey.

Production

The bee stores honey for food during the off season, and to preserve it, concentrates the sugars to over 70% and seals it in a wax container we call a honeycomb. The comb consists of small hexagonal cells fitted together back to back with common walls. It is a model of engineering construction and needs to be, for it requires 5–6 pounds of honey to make 1 lb of wax.

Also, bees must eat while they work and it is estimated that they consume 8 lb of honey for every pound of surplus available to the beekeeper. If the bees are robbed, they starve.

Formerly, honey was marketed in the comb, that is, the hives were fitted with square wooden frames that held about a pound of honey, but since there is no great demand for beeswax, the honey is now pressed from the comb and the wax returned to the hive. Currently, the tops of the cells are shaved off and the honey thrown out of the comb by a centrifugal machine that leaves the comb nearly intact. The bees soon repair it and proceed to fill it again.

CORN SUGARS

Corn sugar and corn syrup have been described in Chapter 13, and that completes the list of sugar sources in the United States, but with a per capita consumption of 108 lb of refined sugar, plus the other sugars and syrups described here, the food processors must give some thought to increasing the supply of saccharine products. Sugar cane is limited to the extreme southern states, Puerto Rico and Hawaii. Sorghum and corn sugar can be increased but they compete with grains, fruits and vegetables.

Sugar from wood is a possibility. There is also the 5% lactose in milk, but the different sugars have different degrees of sweetness and lactose is low on the list and, furthermore, its solubility is only 1.7 gm per 100 ml of water at room temperature. The relative sweetness of sugars is subjective and compiled by taste panels, consequently, there is some difference in the reports. Taking sucrose as a standard, the sweetness of the common sugars is:

Sucrose	100
D-Fructose	103–150
Invert sugar	78–95
Maltose	60
D-Glucose	50–60
Lactose	27–28

Nonsugar Sweeteners

Since sugar is used mostly as a flavor, its 4 kcal per gm has become objectionable to the obese and diabetics; therefore, there is a demand for a noncaloric sweetener. There is no relation between chemical structure and taste except that the H^+ ion is apparently the source of sourness, but not necessarily the only one. As a result of a lack of correlation between chemical and physiological properties, research chemists are unable to predict sweetness. Several sweet substances have been discovered accidentally, which are discussed in Chapter 17.

CACAO PRODUCTS

The familiar chocolate products are made from the seeds of the tree *Theobroma cacao,* which is native to tropical America, but is now grown

From Garard (1974)

FIG. 14.1. CACAO TREES UNDER SHADE

FIG. 14.2. CACAO WITH SEEDS EXPOSED

also along the west coast of Africa and in Indonesia. Many varieties are supplied by many countries.

The cacao tree is unique in one particular; the seed pods do not grow on or near the end of the branches, but from "cushions" on the trunk and larger branches of the tree. These production spots are in continuous production, containing blossoms and pods of several stages of maturity. The mature pod is ovoid and 5 in. or more in length. The hull of the pod is a hard, brittle shell. The seeds are imbedded in a white gelatinous substance somewhat like the seeds of a watermelon, but with more and bigger seeds and less pulp—20 to 40 seeds to the pod. The seeds are called "beans" and, when raw, they taste much like any other raw bean. They are bigger than a kidney bean.

The cacao tree, which is about the size of a peach tree, requires a rich, moist soil and a temperature between 80° and 90°F. For these reasons, it is grown under the shade of taller trees in most areas. The shade is sometimes provided by bananas, but more commonly by tall hardwood trees native to the locality.

In some countries, cacao is grown on small farms with a few trees each. In such cases, the mature pods are broken open and the beans removed. They retain some of the adhesive pulp and are spread out in the sun to dry. These beans produce a poor grade of chocolate and during the rainy season, they are likely to mold and become useless.

In the larger plantations, the mature pods are cut from the tree by a man with a machete and a burlap bag. He carries the pods to the edge of the grove where there is a highway or a railway. There the hulls are removed and the slimy beans are hauled to a processing plant where they are fermented.

Fermentation

The chief problem during the fermentation is temperature control. The beans are put into a bin and as they ferment, the temperature rises and may get high enough to injure the beans, so there must be some arrangement to stir the beans or the fermentation must be carried out in smaller units.

During the fermentation, the adhesive pulp, which is composed of polysaccharides, is hydrolyzed and the sugars fermented, first to alcohol and then to acetic acid. The beans can then be washed clean.

The changes that occur within the bean during the fermentation are more complicated than those of the external coating. Yeasts and some 30 varieties of bacteria have been isolated from the fermenting beans, which account for the alcohol and acetic acid from the surface material. Weisberger *et al.* (1971) reported the presence of phosphoric, citric, gluconic, lactic, malic, oxalic, succinic and tartaric acids in cacao beans. The greatest acid concentration was 1.32 gm per 100 gm of beans. They also found that the fermentation did not change the acid concentration appreciably. It is generally agreed that the fermentation process affects the flavor of the chocolate, but the nature of the changes has not been fully determined. Although unfermented beans are used to make chocolate, the roasted beans have a different flavor if they have been fermented.

The fermented beans are washed, dried, bagged and shipped to countries throughout the world.

Every large chocolate manufacturer buys cacao beans from several sources in order to insure a constant supply and to blend them for a uniform flavor for his products, for they differ appreciably in flavor with variety, climate, fermentation and handling.

Roasting

The beans are roasted in rotating cylinders to obtain uniformity and produce the characteristic flavor; a stronger roast produces a more bitter flavor. There is considerable variation in cacao beans but an approximate average is water 5%, protein 12%, fat 50%, carbohydrate 30% and ash 2.5%.

The hull of the cacao bean is similar to that of the chestnut and when the bean is roasted, the hull becomes very brittle.

Grinding

After they are roasted, the beans are passed through rollers that crack them and loosen the hull. However, some of the hull adheres to the endosperm and is not removed. The standards for the several chocolate products allow for this difficulty by allowing for a maximum hull content.

The cracked beans are called *cacao nibs*. The U.S. standard defines them simply as cracked, roasted cacao beans with the shell removed, but permits a shell content of 1.75% of the weight of the nibs.

The cacao nibs are ground on a stone mill, and the heat of the grinding melts the fat so the product runs over the edge of the bottom stone as a viscous liquid that is called *chocolate liquor* with the additional names of *chocolate, baking chocolate, bitter chocolate, cooking chocolate, chocolate coating* and *bitter chocolate coating*. When the liquid cools, it sets to a very hard, firm solid, but it is still called chocolate liquor.

Standards

The names included in the standard indicate some of the uses of the product. The standard defines chocolate liquor as: "The solid or semiplastic food prepared by finely grinding cacao nibs" and then specifies several conditions. It must contain not less than 50% and not more than 58% cacao fat, and such fat may be added if necessary to bring the liquor up to the standard. The product may contain: (1) ground spice, (2) ground vanilla beans; any natural food flavor, oil, oleoresin or extract, (3) vanillin, ethyl vanillin or other artificial food flavoring, (4) butter, milk fat, dried malted cereal extract, ground coffee, ground nut meats and (5) salt. The liquor may also be treated with alkali to the extent of 3% potassium carbonate or its alkali equivalent.

The standard also includes label specifications to cover any of the optional ingredients.

Other Chocolate Products

There are definitions and standards for twelve products in all. Most of them are familiar on the grocer's shelves.

Sweet Chocolate.—This product must contain at least 15% chocolate liquor, but if it is called *bittersweet chocolate,* it must contain 35%. It may be sweetened by "sugar, or partly refined cane sugar, or both"; dextrose and corn sugar, corn syrup in several forms or combinations, and honey, molasses, brown sugar and maple sugar.

It may contain the spices and flavors permitted in the chocolate liquor.

A serious problem with candy bars and several other chocolate products is the separation of the fat. If the product gets warm enough to melt the fat (about 85°F), the liquid fat collects on the surface and then when the temperature falls, the fat solidifies to a white layer on the surface of the product so that consumers think it is spoiled. Chemists solved the problem by including an emulsifier in the mixture, and the standard permits the use of: (1) lecithin, (2) mono- and diglycerides, (3) sorbitan monostearate and (4) polysorbate 60. These substances may be used in combination provided

(2) and (4) do not exceed 1% and (1) and (2) do not exceed 0.5%. Several specified dairy items may be included up to 12%.

The standard also includes labeling directions and four alternate names for the product: *semisweet chocolate, bittersweet chocolate, semisweet chocolate coating* and *bittersweet chocolate coating.*

Milk Chocolate.—The standard for milk chocolate is essentially the same as that of sweet chocolate in permissable sweeteners, emulsifiers and flavorings, but it also requires 3.66% milk fat, not less than 12% milk solids and not less than 10% chocolate liquor.

Cocoa.—The federal standard recognizes three kinds of cocoa: *breakfast cocoa, medium fat cocoa* and *low fat cocoa.*

Cocoa is made by warming the chocolate liquor and pressing it to remove some of the fat and the three grades are based on the amount of fat retained.

Breakfast Cocoa, Highfat Cocoa.—The richest cocoa must contain 22% cacao fat and may be mixed with ground spice, ground vanilla beans, vanilla extract, vanillin, ethyl vanillin, or other artificial flavoring other than chocolate flavor.

Cocoa, Medium Fat Cocoa.—has the same requirements as the high fat cocoa except that the cacao fat must be at least 10% but not more than 22%.

Low Fat Cocoa.—has the requirements of the other two but contains less than 10% cacao fat.

The federal standards include the following additional products:

Skim milk chocolate
Buttermilk chocolate
Mixed dairy products chocolate
Sweet chocolate and vegetable fat
Sweet cocoa and vegetable fat
Milk chocolate and vegetable fat
Cocoa with dioctyl sodiumsuccinate for manufacturing

Dutch Process Products.—The standards for cacao products do not mention the Dutch process, but they permit the addition of 3% potassium carbonate or its alkali equivalent in both chocolate and cacao products.

This treatment with alkali originated in Holland and is commonly known as the Dutch Process. The product is treated with a mild alkali, such as ammonium, sodium or potassium carbonate or sodium or potassium bicarbonate in about a 2% solution in water. Strong or excess alkali is avoided to prevent the saponification of the fat and consequent production of the flavor of soap.

The treatment turns the color a darker brown and reduces the bitter taste. The pH is increased from the 5 or 6 of untreated cocoa to about 7.

Cocoa Butter.—In the manufacture of cocoa, from 30% to 40% of the fat is removed from the chocolate liquor. This fat is called *cocoa butter*. It is a hard fat of pale yellow color and a slight taste of chocolate. For some purposes, it is bleached and deodorized. The physical and chemical constants of cocoa butter are: specific gravity at 50°C 0.8823–0.8830, at 25° 0.8921; N_D 40° 1.4525–1.4570; saponification value 192–198; iodine value 30–40; Reichert-Meissl value 0.3–1.0; Polenske value 0.5; unsaponfiable 0.3–0.8 and titer 48°–50°. The acid content is shown in Table 2.2.

Cocoa butter is used in the manufacture of confectionery, in some cosmetics, and in the pharmaceutical industry. There is a limited supply of it because it is a byproduct of the manufacture of cocoa. The scarcity increases the price and invites adulteration. Adulterants that have been found in it include coconut oil, palm oil, hydrogenated oils and paraffin. The available fats melt much lower than the cocoa butter but they are sometimes pressed to produce a higher melting "stearin."

ANALYSIS OF CACAO PRODUCTS

The Official Methods include 59 procedures in Chapter 13 of the 12th Edition under "Analysis of Cacao Bean and its Products." There is the usual moisture, protein ($N \times 6.25$), fat and ash; then, more specifically: alkalinity of ash, shell, pectic acid, crude fiber and chocolate liquor, a thorough examination of the fat, a special test for coconut and palm kernel oils, lecithin, milk protein, lactose and other sugars, starch and theobromine. Sampling procedures are also included.

The analysis of a chocolate product may present a problem partly because of lack of fixed composition. For example, the average theobromine content of cacao nibs is about 0.8% and over 1% in cocoa, but with the various kinds of cocoa and chocolate, what the theobromine content should be requires considerable calculation after the analysis is made. Then, cacao nibs contain about 10% starch, which will be in a different percentage in each product. If a product is adulterated by the addition of starch a microscope may detect it, but if it has been heated, only the amount present will be significant. The analysis of a cacao product for nutritive value is easy, but for legal control, specifications and duplication, it may be a major project.

BIBLIOGRAPHY

ALLEN, W. G., and DAWSON, H. G. 1975. Technology and uses of debranching enzymes. Food Tech. *29*, No. 5, 70–80.
ANON. 1975. Code of Federal Regulations *21*, Food and Drugs, Parts 10–199. U.S. Govt. Printing Office, Washington, D.C.
ASSOC. OFFIC. ANAL. CHEMISTS. 1975. Official Methods of Analysis, 12th Edition. Assoc. Offic. Anal. Chemists, Washington, D.C.

BAINBRIDGE, J. C., and DAVIES, S. H. 1912. The essential oil of cacao. J. Chem. Soc. 2209–2221.

COLE, S. J. 1967. The Maillard reaction in food products. Carbon dioxide production. J. Food Sci. 33, 245–250.

GARARD, I. D. 1974. The Story of Food. Avi Publishing Co., Westport, Conn.

INGLETT, G. E., 1974. Symposium: Sweeteners. Avi Publishing Co., Westport, Conn.

JUNK, W. R., and PANCOAST, H. M. 1973. Handbook of Sugars for Processors, Chemists and Technologists. Avi Publishing Co., Westport, Conn.

LAWRENCE, A. A. 1973. Edible Gums and Related Substances. Noyes Data Corp., Park Ridge, N.J.

MEADE, G. P. 1963. Cane Sugar Handbook: A Manual For Cane Sugar Manufacturers and Their Chemists. John Wiley & Sons, New York.

MINIFIE, B. W. 1970. Chocolate, Cocoa and Confectionery Science and Technology. Avi Publishing Co., Westport, Conn.

OSTOVAR, K., and KENNEY, P. G. 1973. Isolation and characterization of micro-organisms involved in the fermentation of Trinidad cacao beans. J. Food Sci. 38, 611–617.

ROHAN, T. A., and STEWART, T. 1967. The precursors of chocolate aroma: studies in the degradation of amino acids during the roasting of Accra cocoa. J. Food Sci. 32, 625–629.

SCHWARTZ, M. E. 1974. Confections and candy technology. Noyes Data Corp., Park Ridge, N.J.

SHALLENBERGER, R. S., and BIRCH, C. G. 1974. Sugar Chemistry. Avi Publishing Co., Westport, Conn.

SHERMAN, H. C. 1948. Food Products. MacMillan Co., New York.

UNDERWOOD, J. C., and WILLITS, C. O. 1963. Research modernizes the maple syrup industry. Food Tech. 17, No. 11, 44–47.

VAN DER WAL, B. et al. 1971. New volatile compounds of roasted cocoa. J. Ag. Food Chem. 19, 276–280.

WASSERMAN, A. E., and WILLITS, C. O. Maple syrup. XX. Conversion of buddy maple syrup into normal maple syrup. Food Tech. 15, 438–439.

WEISBERGER, W., KAVANAGH, T. E., and KEENEY, P. G. 1971. Identification and quantitative determination of several nonvolatile organic acids of cacao beans. J. Food Sci. 36, 877–879.

WIELAND, H. 1972. Cocoa and Chocolate Processing. Noyes Data Corp., Park Ridge, N.J.

WILLITS, C. O., FRANK, H. A., and BELL, R. A. 1961. Maple syrup. XIX. Flavor and color through controlled fermentation of maple sap. Food Tech. 15, 473–474.

YOKOTA, M., and FAGERSON, I. S. 1971. The major volatile components of cane molasses. J. Food Sci. 36, 1091–1094.

CHAPTER 15

Convenience Foods

We hear the expression "convenience foods" frequently and it generally means full or partial meals ready to heat and serve, but the term belongs to foods that are ready to eat hot or cold with little or no preparation. Since there is no official definition of the term, I shall use it for all foods that require little or no preparation.

There is nothing more convenient than sliced bread, milk, cheese or sugar, but they have been described in other chapters.

FRUITS

Many fruits keep long enough to be marketed in the fresh condition and require no processing except washing. Oranges and grapefruit do require a little work, but both are available as juices.

Apples

There are said to be 700 varieties of apples, but of the two dozen or so that are available in the American market, several may be eaten raw or cut up and used in a fruit salad. Apples are grown in many of the northern states from Washington to Maine, and the consumption of fresh apples in 1972 was 174 lb per capita.

The composition of apples differs with the variety but, in general, the composition is shown in Table 15.1.

In addition to the components of apples shown in Table 15.1, there are acids, which are included in the carbohydrates, and enzymes in the protein fraction, as well as color and flavor.

TABLE 15.1

COMPOSITION OF APPLES

Component	Based on 100 Gm Weight	Component	Weight
Water	84.4 gm	Iron	0.3 mg
Protein	0.2	Potassium	110.0
Fat	0.6	Thiamin	0.03
Carbohydrate	14.5	Riboflavin	0.02
Ash	0.3	Niacin	0.1
Kcal	58.0	Ascorbic acid	4.0
Calcium	7.0 mg	Vitamin A	90.0 I.U.
Phosphorus	10.0		

The acid in apples is mostly malic

$$\begin{array}{l} \text{HOCHCOOH} \\ | \\ \text{CH}_2\text{COOH} \end{array}$$

In the green apple, the carbohydrate is mostly starch, but as the fruit ripens, the starch disappears and sucrose and invert sugar appear.

Apples contain no appreciable amount of protein and the ether extract is probably mostly wax rather than fat.

There are several enzymes in the apple; the most obvious one is an oxidase that turns a cut surface brown in a few minutes, a change that can be prevented by applying a dilute solution of ascorbic acid to the surface.

The vitamin C content and the depth of color of the skin of a given variety depend on the amount of sunshine the apple has been exposed to during its development.

The pectin content of apples is 1.5% to 2.5%.

Bananas

The banana was known in India as early as 327 B.C., but it was not imported regularly into the United States until 1870. The banana plant is tropical and is not grown commercially in the continental United States; we are supplied from the islands of the Caribbean and the countries that border that sea. Some come from the Philippines.

The banana plant *Musa sapientum* is hardly a tree. It grows to a height of about 30 ft and a diameter of 6 in. Each plant produces only one stem of bananas, which may contain as many as 13 hands; each hand contains about a dozen bananas (fingers). The variety imported into the United States is the Gros Michel although there are a few finder bananas and a few plantains in the market.

Formerly, banana stems were hung in the holds of ships for transportation, but, more recently, the hands have been removed and packed in boxes. The merchant then cuts the hands into groups of two or more for the convenience of the customer.

The banana is a big source of food in the areas where it is grown, but it is never allowed to ripen on the tree, because on ripening, it bursts open and is infested with insects. Bananas are unique in that they are the only fruit that must be harvested green.

The green banana contains about 75% water, 0.9% ash, 0.9% protein and 0.2% ether extract; these constituents do not change appreciably as the fruit ripens, but the starch and sugar content change enormously as ripening proceeds.

In addition to the change in carbohydrate, the banana develops a flavor during ripening that consists of a mixture of a large number of organic

TABLE 15.2

CHANGES IN COMPOSITION IN RIPENING BANANAS

	Green	After 3 Days	After 5 Days	After 11 Days
Starch	20.65	12.85	6.00	1.20
Total sugars	0.86	7.66	13.76	17.91
Pectin	0.0	0.27	0.36	0.40
Reducing sugar	0.24	2.81	7.64	15.31

compounds, many of them in traces only, so that their contribution to the flavor is questionable. Ethyl alcohol predominates but isoamyl acetate is the most characteristic component. Also present are 11 other alcohols, one with 7 carbon atoms and 5 unsaturated. There is one other ester and there are 2 ketones.

Bananas should not be refrigerated because chilling seriously affects the flavor. Also, they should be ventilated during storage because one or more of the volatile compounds catalyzes ripening.

Oranges

The orange is a subtropical tree. It grows in all the states of our southern border and in countries throughout the world that have a similar climate. Our biggest production is in Florida and California. There are many varieties. California grows mainly navels and valencias. In Florida, the early season oranges are Hamlin, navel and Parson Brown. Midseason varieties are pineapple, temple, tangerines and tangelos and the late season from February to July are valencias. A few other varieties are grown on a small scale. All these varieties are marketed as fresh fruit, but the bulk of the crop is converted into juice.

Since 1930, the quality of concentrated or canned orange juice has been greatly improved so that the emphasis in the industry has shifted from fresh fruit to juice. This has caused some shift in varieties as old groves are abandoned and new ones planted, for the valencia is preferred for juice production.

Both varieties from California are easily eaten as fresh fruit; the skins are rather loose and the sections come apart easily. Of the Florida varieties, the tangerines, tangelos, mercots, temples and navels come apart easily, but the other varieties must be cut and the juice reamed out.

Oranges destined for the juice factory are just washed and juiced. The orange oil is in cells in the outer rind and the juicing aims to include as little of it in the juice as possible because it deteriorates rapidly, mostly by oxidation and the juice soon goes off flavor, which was one of the most serious problems in the early attempts to concentrate or can the juice.

The oranges sold in the market get a more thorough treatment at the

TABLE 15.3

COMPOSITION OF ORANGES

	California		Florida	
	Navel	Valencia	Oranges	Juice
Water (%)	85.4	85.6	86.4	88.3
Kcal/100 gm	51.0	51.0	48.0	45.0
Protein (%)	1.2	1.2	0.7	0.7
Ether extract	0.1	0.3	0.2	0.2
Carbohydrate	12.7	12.4	12.0	10.4
Ash	0.5	0.5	0.7	0.4
Ca (Mg/100 gm)	40.0	40.0	43.0	11.0
P	22.0	22.0	17.0	17.0
Fe	0.4	0.8	0.2	0.2
K	194.0	200.0	206.0	200.0
Thiamin	0.10	0.10	0.10	0.09
Ascorbic acid	61.0	49.0	45.0	50.0
Riboflavin	0.04	0.04	0.04	0.03
Niacin	0.4	0.4	0.4	0.4
Vitamin A (IU)	200	200	200	200

packing house. Oranges mold readily and they are washed and usually dipped in a solution of borax or other moldicide. They are sometimes polished with paraffin or a mixture of higher fatty acids.

Oranges mature before the chlorophyll disappears completely so that one side may be green and the other orange. Then too, a valencia orange after it is fully ripe and fully colored turns completely green again.

In California, the "green" problem is solved by using slatted field boxes and stacking them in a room that is maintained at about 80°F, 90% humidity and an ethylene content of 1 part in 5000. The green color disappears overnight. The chlorophyll is destroyed and the carotenoids remain.

In addition to the greening problem in Florida, many of the varieties are a pale yellow color and so the oranges are colored with the orange dye, Citrus Red No. 2.

The composition of oranges and orange juice are shown in Table 15.3.

When we consider variation in composition with season, variety, and degree of maturity, it is obvious from Table 15.3 that there is no appreciable nutritional difference in oranges from different locations or different varieties. An investigation, however, showed that Florida Valencias increased in sugar content from 6% in December to 10% in June. But Valencias are marketed from late February to April or May when the variation was 7.5% to 8.5% sugar; about half of it is sucrose and half reducing sugar.

The orange juice industry is relatively new; 1% of the crop was juiced in 1945 and by the 1970s, 80% of the crop was marketed as juice, either single strength or concentrate. The extracted juice contains only about half the total weight of the whole orange. The rind, seeds and "rag," or section walls, are used for cattle feed. However, the orange oil is in the outer cells of the rind and is generally recovered. It is used in compounding perfumes and flavors. It has some 25 components of which the principal one is D-limonene.

$$CH_3$$

$$H_2C \overset{C}{\diagup} \diagdown C-H$$
$$H_2C \diagdown \underset{C}{\diagup} CH_2$$
$$\overset{|}{\underset{H}{}} $$
$$H_3C-C=CH_2$$

D-Limonene

Limonene oxidizes readily and the oil changes flavor.

If orange peel is to be candied at home, it should be well scrubbed to remove any coating or added color, for Citrus Red No. 2 is not an edible dye. Commercial confectioners may be able to secure peel that has not been treated.

Oranges do not keep well in storage because of their tendency to mold, but the method of marketing does not call for long storage such as apples undergo. Oranges can be harvested in nearly every month of the year and are shipped or juiced as they are obtained.

Other Fruits

Many fruits have short seasons, although some seasons are lengthened by transportation from outside the country and sometimes domestically. Strawberries, for example, are available in Florida in February and the seasons ends in New York in July.

Grapes, peaches, pears, cherries, plums, blackberries, several kinds of melons and grapefruit are available in season and avocados and tomatoes meet them at the salad bar.

NUTS

There are many kinds of nuts already prepared for the table. Although they are highly prized by vegetarians, the general public treats them as confections, a cocktail snack or perhaps a salad component. Peanuts are the exception; they are boiled and eaten in the southern states and peanut butter is widely used. Reference to Table 15.4 will disclose the nutritive value of the common nuts. Peanuts are not true nuts but belong with them in composition and use.

Chestnuts, almonds, walnuts, Brazil nuts, pecans and several other nuts are shelled before they are marketed and are totally convenient.

TABLE 15.4

COMPOSITION OF SOME COMMON NUTS

	Almonds	Cashews	Peanuts	Pecans	Walnuts
Water (%)	4.7	5.2	5.4	3.4	3.5
Protein	18.6	17.2	26.3	9.2	14.8
Fat	54.2	45.7	48.4	71.2	64.6
Carbohydrate	19.5	29.3	17.6	14.6	15.8
Ash	3.0	2.6	2.4	1.6	1.9
Calcium (Mg/100 gm)	234.0	38.0	50.0	73.0	99.0
Phosphorus	504.0	373.0	409.0	289.0	380.0
Potassium	773.0	464.0	674.0	608.0	450.0
Iron	4.7	3.8	2.0	2.4	5.1
Niacin	3.5	1.8	15.8	0.9	0.9
Riboflavin	0.2	0.25	0.13	0.13	0.13
Thiamin	0.24	0.43	0.99	0.86	0.33
Ascorbic acid	—	—	—	2.0	2.0
Vitamin A (IU)	—	100	—	130.0	30.0
Kcal	508.0	561.0	568.0	687.0	651.0

Source: Watt and Merill (1963).

Peanuts

In 1970, the U.S. production of peanuts was 2,979 million lb of which 1,580 million lb were used for food, and the rest were exported, planted or crushed for oil.

Peanut Butter.—In the manufacture of peanut butter, the nuts are roasted at 320°F for 40 min, shelled, and screened to remove the fines. The roasting must be closely controlled; the characteristic flavor develops during the roasting and so all nuts must be roasted the same in order to have a uniform product. The color should also be uniform, and overroasting produces a darker color. If a few kernels are scorched, they contribute black specks to the butter, which the customer is likely to think is dirt.

After the peanuts are roasted, they are cooled and put through a machine that removes the paper thin brown seed coat. It also splits the kernels and removes the germ. The kernels are then screened to remove the skins and the germs. Salt, 1% to 4%, along with any other additives are added before the nuts go to the grinder. The heat generated by the grinding softens the butter so that is is filled directly into the containers in which it is marketed.

The oil in the germ is different from that of the endosperm and increases the liability of the butter to rancidity.

The federal standards list two kinds of peanut butter. One includes the skins and germs, but if it is so made, the label must say so. The other type is made as described above and may not exceed 55% fat. However, additives are permitted up to 10% of the total weight. The additive provision reads:

"Seasoning and stabilizing ingredients that perform a useful function are regarded as suitable except that artificial flavorings, artificial sweeteners, chemical preservatives, added vitamins and color additives are not suitable ingredients of peanut butter. Oil products used as optional stabilizers shall be hydrogenated vegetable oils."

Probably the most serious chemical problem with peanut butter is the tendency of the oil to separate, hence the high percentage of permitted stabilizers. Peanut oil is a liquid, and attempts to hold it in the butter involve the addition of a hard fat to increase the viscosity and melting point of the fat fraction. Other remedies constitute the addition of stabilizers.

Another problem with peanut butter is the presence of extraneous matter (filth). The Official Methods of analysis include a procedure for its determination. The filth consists mostly of insect fragments and rodent hairs and excreta. Other such matter may be ground glass or rocks that have escaped detection on the picking tables.

The most likely adulteration is the substitution of a cheaper oil than peanut oil or the addition of starch or wheat flour. There is no minimum fat requirement, but the butter must be made from peanuts that contain 40% to 50% fat. Of course, it may also be adulterated with color, preservative or other prescribed additive.

Table 15.5 shows the composition of peanut butter to be expected.

Although the seasonal variation in the composition of peanuts and addition of additives will result in peanut butters that vary from the values in Table 15.5, the variations should not be large and if they are, the product may be adulterated.

The estimated production of 33 million lb of peanut butter sandwiches in 1970 would indicate that other nut butters would be a commercial possibility. However, any nuts except peanuts require considerable land and several years to get them into production; therefore, any expansion of nuts as a major source of food is likely to be limited to peanuts, which are an annual legume crop that produces considerable forage from the vines in addition to the nuts.

TABLE 15.5

COMPOSITION OF PEANUT BUTTER

Water (%)	1.8	Calcium (Mg/100 gm)	63.0
Protein	27.8	Phosphorus	407.0
Fat	49.4	Iron	2.0
Carbohydrate	17.2	Potassium	670.0
Ash	3.8	Niacin	15.7
Salt	1.5	Riboflavin	0.13
Kcal/100 gm	581.0	Thiamin	0.13

Source: Watt and Merill (1963).

Soybean Butter.—Another legume that makes a satisfactory butter is the soybean. The beans contain only half as much oil as the peanut, but partly hydrogenated soybean oil can be added to obtain the proper texture. Such a butter has been marketed but failed to succeed largely because of its flavor. The product needs more research.

NONALCOHOLIC BEVERAGES

We need to consume 2.5 liters of water daily; in a warm climate, probably more. Many foods contain from 70% to 90% water and the average diet may contain 70% to 75%, but that falls far short of 2.5 liters. The two best sources of additional water are water itself and milk, both of which have been described in earlier chapters.

Fruit Juices

Table 15.6 shows the composition of some fruit juices that are available in the retail market. An inspection of Table 15.6 reveals that fruit juices are 80% to 90% water and the caloric value is only 19–60 kcal per 100 gm. They are good sources of calcium and iron and tomato and orange juice are rich sources of vitamin C.

There are federal standards for orange, pineapple, prune and tomato juices.

Fruit Drinks

In addition to the fruit juices, the federal standards contain definitions and standards for five fruit drinks, of which the main provisions follow.

Lemonade.—This familiar drink when commercialized must contain

TABLE 15.6

COMPOSITION OF FRUIT JUICES

	Apple	Grape	Grapefruit	Orange	Pineapple	Tomato
Water (%)	87.8	82.9	90.0	88.3	85.6	93.6
Protein	0.1	1.2	0.5	0.7	0.4	0.9
Ether extract	tr	tr	0.1	0.2	0.1	0.1
Carbohydrate	11.9	16.6	9.2	10.4	13.5	4.3
Ash	0.2	0.2	0.3	0.4	0.4	1.1
Kcal/100 gm	47.0	60.0	39.0	45.0	55.0	19.0
Calcium (Mg/100 gm)	6.0	11.0	9.0	11.0	15.0	7.0
Phosphorus	9.0	12.0	15.0	17.0	9.0	18.0
Iron	0.6	0.3	0.2	0.2	0.3	0.9
Potassium	101.0	116.0	162.0	200.0	149.0	227.0
Thiamine	0.01	0.04	0.04	0.09	0.05	0.05
Riboflavin	0.02	0.02	0.02	0.03	0.02	0.03
Niacin	0.2	0.2	0.2	0.4	0.2	0.8
Ascorbic acid	2.0	tr	38.0	50.0	9.0	16.0
Vitamin A (IU)	0.0	0.0	80.0	200.0	50.0	800.0

enough lemon juice to supply 0.70 gm anhydrous citric acid in 100 ml of the finished product. The sweeteners permitted are sugar, invert sugar and the usual sugars and syrups made from corn.

Other additives are lemon oil, sodium benzoate, sorbic acid, buffering salts and emulsifying agents when oil is added.

Limeade.—The definition and standards for limeade are the same as those for lemonade except that the juice and oil must be from limes.

Orange Juice Drink.—The definition of this product is similar to that of lemonade but has many more stipulations. "Orange juice drink is the beverage prepared by adding water to one or more of the unfermented orange juice ingredients which are specified in paragraph (b) . . ." The ingredients in paragraph (b) are 21 varieties of orange juice, or orange pulp, each one described in detail. They include fresh orange juice, frozen concentrate, canned orange juice and several less familiar products. Whichever product is used, the product must contain the equivalent of not less than 35% nor more than 70% single strength orange juice, or orange juice solids not less than 4.13% nor more than 8.26%. Total solids must be at least 12%.

There are 12 additives; many more, in fact, because several of the items are classes of substances. The additives are:

1. Nutritive sweeteners	7. Buffers
2. Organic acids	8. Orange pulp
3. Thickeners	9. Orange peel
4. Stabilizers	10. Natural and artificial flavor
5. Clouding agents	11. Natural and artificial color
6. Emulsifiers	12. Preservatives

Orange Drink.—This product is essentially a diluted orange juice drink. It must contain not less than 10% acid and not more than 35% equivalent of single strength orange juice, or not less than 1.18% nor more than 4.13% orange juice soluble solids; total soluble solids must be at least 10%. The permitted orange juice ingredients do not include fresh, frozen and dehydrated orange juice. Other ingredients are the same as those in orange juice drink.

Pineapple-Grapefruit Juice Drink.—"Canned pineapple-grapefruit juice drink is the beverage food prepared from one or both of the pineapple juice ingredients and one or both of the grapefruit juice ingredients specified in paragraph (b) . . . ."

This product has a viscosity requirement. The time of flow must be at least 30 sec when measured by a specified method.

The ingredients of paragraph (b) are pineapple juice, concentrated pineapple juice, grapefruit juice and concentrated grapefruit juice. The combined juices must constitute 50% of the final weight. The usual sugar products are permitted as sweeteners. Other additives are: orange, lemon or grapefruit oil, not less than 30 mg nor more than 60 mg of ascorbic acid in 4 fluid oz, and sodium citrate.

NONALCOHOLIC BEVERAGES

Under this head, the federal definitions and standards include all the so-called "soft drinks." They are all soda water with numerous additions.

Soda Water

"Soda water is the class of beverage made by absorbing carbon dioxide in potable water. The amount of carbon dioxide used is not less than that which would be absorbed by the beverage at a pressure of one atmosphere and at a temperature of 60°F." It may contain alcohol that is added in a flavoring extract, but it shall not exceed 0.5%. Cola beverages are limited to 0.02% caffeine.

The nutritive sweeteners are sugar in dry or liquid form, invert sugar, dextrose, fructose, corn syrup, glucose syrup, sorbitol or any combination of them.

Flavorings may be added in ethyl alcohol, glycerine, propylene glycol or edible vegetable oil. Natural and artificial color may be added.

The acid ingredient may be acetic, adipic, citric, fumaric, gluconic, lactic, malic, phosphoric or tartaric. Although these acids differ in flavor, the flavoring added to the beverage will cover the difference and so the selection of an acid is made largely on an economic basis.

Buffers to regulate the sourness of the beverage may be the acetate, bicarbonate, carbonate, chloride, citrate, gluconate, lactate, orthophosphate or sulfate of calcium, magnesium, potassium or sodium. The selection of a buffer from among these 36 salts may be a problem. For example, calcium and magnesium may form an insoluble compound with the acid or a component of the flavoring and cause cloudiness or even a precipitate. The formulation of a soft drink requires some experimentation.

These are 17 emulsifiers, stabilizers or viscosity increasing agents to choose from. These include the usual food gums, pectins, lecithin mono- and diglycerides and several synthetic compounds.

Edible vegetable oils may be used to produce cloudiness.

Foaming agents are unusual additives; they are ammoniacal glycyrrhizin, gum, ghatti, licorice, or glycyrrhiza, yucca, quillaia and modified soy protein in propylene glycol.

They may also contain caffeine to 0.02% and quinine to 83 ppm.

Preservatives, 21 in number include ascorbic, sorbic and benzoic acids, BHA, BTH, several salts and one enzyme.

ALCOHOL

Alcohol, like sugar, is a source of energy; it supplies 7 kcal/gm, which is nearly twice that of sugar. The standards for the various alcoholic iiquors are set by the Internal Revenue Service and are administered by the Alcohol and Tobacco Tax Division.

Alcohol is prohibited in confectionery and in the fruit beverages and sodas except up to 0.5% that may be added as the solvent for a flavoring. The alcoholic content of vanilla extract, for example, is not less than 35%. When such flavoring is added to food to be heated, however, the alcohol is mostly or completely evaporated.

There are four classes of alcoholic beverages: malt liquors (beer, ale, porter and stout), wines, distilled liquors (brandy, rum, whiskey and vodka) and cordials (curacao, benedictine, chartreuse and pernod). The first two classes will be described briefly; the other two are seldom the concern of the food chemist.

Malt Beverages

These beverages are made from barley malt mixed with corn, rice or other grain. The mixture of grains is ground and cooked with water to gelatinize the starch. As the mixture stands after cooking, the starch is partly hydrolyzed to a mixture of dextrins and sugars. Hops and yeast are added; the yeast converts the sugars to alcohol and the hops contribute a bitter resin. The finished beer contains 92% water, 4–5% alcohol, 3–4% carbohydrate, mostly dextrin, and 0.3% protein. Its caloric value is 40 kcal/100 gm. Its food value is limited to its energy content. Federal regulations do not require the alcoholic content to be stated on the label although some states may do so.

Wines

There are hundreds of varieties of wine, but they all are included in two classes: table wines and dessert wines.

Apple, orange and other fruit wines have been marketed, but the word "wine," both in the trade and to the general public, means fermented grape juice.

Table Wines.—Several varieties of grape are used to make wine. The chief characteristic is a high sugar content. Most of the wine grapes have blue skins, and, like litmus, the color turns red in acid solution.

The grapes are stemmed, crushed and held a few days for the violent fermentation to cease. As the sugar is converted to alcohol, CO_2 is evolved and brings the skins to the surface. The yeast is the white "bloom" on the skins of the grapes, and so the skins are submerged from time to time to insure a good inoculation. The equation for the reaction is:

$$C_6H_{12}O_6 \xrightarrow{\text{zymase}} 2C_2H_5OH + 2CO_2$$

After 3 to 5 days, the fermentation subsides and the wine is pressed to remove the skins, seeds and stems. The clear wine is still fermenting and must be left out of contact with the air. Otherwise, the alcohol is fermented to acetic acid and the product is vinegar.

$$CH_3CH_2OH \xrightarrow{O_2} CH_3CHO \xrightarrow{O_2} CH_3COOH$$

Alcohol   Acetaldehyde   Acetic Acid

In small operations, the wine is run into casks fitted with valves to permit the escape of the gas. In carbonated wines, such as champagne, the wine is bottled as soon as it is pressed so that some of the gas remains in the wine.

If white wine is made from blue grapes, the wine is pressed immediately after crushing. This not only prevents the extraction of color from the skins but of the tannin also, so that the flavor of the white wines is much milder than that of the red ones.

After the fermentation is finished and the wine bottled, it ages by the combination of alcohol with the acids from the grapes. Tartaric acid is the principal one but there are others. This esterification is a slow process and goes on for years before an equilibrium is reached. A white wine reaches maturity in 7 to 8 yr, but the red wines age for a much longer time.

The sugar in grapes is mostly dextrose and it may constitute 12 to 25% of the weight of the grapes according to variety and climate, but when the alcoholic content reaches 12–14%, it inactivates the yeast and stops the fermentation. Consequently, the wine may be dry or sweet according to the sugar content of the grapes. The table wines have an alcoholic content of 11–13%.

Dessert Wines.—Port, sherry, tokay and muscatel are the familiar dessert wines. They are made from very sweet grapes and have brandy added to them to bring the alcoholic content up to 17–20%. They are also called *fortified wines*.

Wine making differs in detail from the procedure of the man who makes a few gallons at home to that of the manufacturer who makes thousands of gallons. It also differs with the locality and with many individual wines.

OTHER FERMENTATIONS

There are many fermentations in the food field other than those that produce alcohol.

Vinegar

In the manufacture of vinegar, some source of sugar is fermented with yeast to alcohol; apple juice, grape juice and malt liquor are the principal sources of the sugars. The yeast fermentation is anaerobic and the end product is a dilute solution of alcohol. Apples and grapes carry bacteria on their skins that will convert the alcohol to acetic acid if there is oxygen available

$$CH_3CH_2OH + O_2 \rightarrow CH_3COOH + H_2O$$
$$\text{Alcohol} \qquad\qquad \text{Acetic Acid}$$

The word *vinegar* is a corruption of the French *vin aigre,* which means sour wine. Many an amateur wine maker has ended up with vin aigre. The names of the varieties of vinegar are the names of the source of it, cider vinegar, wine vinegar, malt vinegar and the like. Flavored vinegars are labeled with the name of the flavor, for example, tarragon flavored vinegar.

Cider vinegar is made from apples that are crushed and pressed. The press cake is stirred up, wet with water and pressed again to remove more of the sugar from the apples. The sugar content differs with the variety of apples but it is seldom more than 10% to 12%. The skins of the apples have yeasts, bacteria and molds on the surface and these inoculate the cider so that fermentation soon starts. In commercial production, a pure culture of *Saccharomyces ellipsoideus* is added to hasten the fermentation before other organisms can destroy the sugar with the possible production of off flavors. The farmer simply leaves the bung out of the barrel so that the acetic fermentation starts as soon as there is any alcohol to ferment.

The acid forming bacteria are some species of *Acetobacter.* The bacteria grow so well that the cider may become cloudy or form a gelatinous precipitate (mother of vinegar), which can be used to inoculate a fresh batch of cider.

The conversion of alcohol to acetic acid is an aerobic process so the bacteria must have a good supply of air. The manufacturer uses a tall cylindrical tank filled with beech wood shavings, coke or other porous material that will give a spray of cider a large surface, which produces a rapid fermentation.

The flavor of acetic acid is distinctive. A souring wine is easily detected, because the flavor of the acetic acid is very different from that of the tartaric acid that causes the tartness of the wine.

The various vinegars are made in much the same way and have an acid content of about 5%. They are usually pasteurized to stop the growth of organisms that may cause cloudiness. A more concentrated vinegar may be made for use in the food industries by freezing some of the water in the vinegar and removing the ice with a centrifuge.

In addition to the flavor of the acetic acid, the vinegars contain flavors from the apples, grapes or malt. *Distilled vinegar* is made from alcohol distilled from the fermentation of molasses and has little flavor other than that of the acid.

Lactic Fermentations

Several species of *Lactobacillus* convert sugar into lactic acid, which has a different flavor from that of acetic. *Lactobacillus* is usually associated with milk and it is used in the manufacture of buttermilk and other milk drinks. *Yogurt* is made from milk by the action of two species of lactic-acid producing bacteria. One portion of the milk is inoculated with *B. caucasius* and another with *Lactobacillus Bulgaricus*. After 10 to 12 hr, the two portions are thoroughly mixed.

Acidophilus milk is made by sterilizing milk at 212°F for 2 hr or autoclaving it a shorter time. The milk is then cooled and inoculated with *B. acidophilus* and held for 20 hr or more at 98°F. It is then cooled, bottled and marketed by milk plants. It is said to have therapeutic value because of the action of the organism on the bacterial flora of the intestines. The bacterial content of the product deteriorates rapidly in storage where it will keep only a few days.

Sauerkraut.—The word *sauerkraut* is a German word that means sour cabbage. It is made at home by shredding cabbage and packing it in jars with 2–2½% salt. Several kinds of bacteria from the surface of the cabbage cause a complex fermentation. Calculated as lactic acid the acidic concentration at the end of the 10 day fermentation is about 1.5%. However, about a third of the acid is acetic. The fermentation should be carried out under 70°F. If the temperature gets too high or if too much salt is used, yeasts will grow and cause a pink color and poor flavor.

Pickles.—To make pickles, cucumbers are put into a tank with 10% brine. The salt used must be pure, for calcium and magnesium, the common impurities, make the cucumbers tough. The bacteria that cause the fermentation are from the surface of the cucumbers and a *Lactobacillus* is among them so that lactic acid is produced.

Cucumbers are about 95% water and so the brine is diluted during the fermentation and more salt must be added from time to time. The fermentation takes from 3 to 9 months.

At the end of the fermentation, the cucumbers contain about 10% salt, which is reduced to 3.5% to 4% by soaking them in fresh water. They are then covered with vinegar and left for about a week. The time of soaking and acid content of the vinegar differ with the manufacturer. Sweet pickles are made by adding sugar and usually spices to the vinegar. Each manufacturer has his recipe. The sugar, which may be as much as 40% is added gradually over a period of two or three months.

OTHER CONVENIENCE FOODS

Most shoppers, ably assisted by promotional literature, think of convenience foods as whole dinners, ready to heat and serve, or at least, items to constitute such a dinner, such as frozen pies and French fried potatoes. They may even include rolls ready to bake or pie crust ready to be filled and baked, but none of these foods is as convenient as those mentioned above. However, there are other foods that are the ultimate in convenience.

Breakfast Foods

Early in the 20th century, cornflakes were invented. They were made by cooking corn flour with water seasoned with salt and sometimes malt, then dried and toasted. In reasonably moisture-proof packages, the product would keep for months. They became popular quickly because of the time they saved over making pancakes or frying eggs or even cooking oatmeal or farina.

Cornflakes were soon followed by a long succession of similar products that now fill rows of grocer's shelves. But at the height of their popularity, nutritionists began to refer to them as "empty calories." This was not quite fair for they do contain protein and some minerals, but they were a luxurious source of nutrients. The nutritionists' complaint brought about the addition of various nutrients to most of the products. Typical of the new dry cereal is the label content of a 1975 package of cornflakes: "Made from: milled corn, sugar, salt, malt flavoring, vitamin A, sodium ascorbate, ascorbic acid, thiamin, riboflavin, niacinamide, vitamin D, pyridoxine, folacin, iron, BHA, BHT." One ounce supplies 110 kcal, 2 gm protein, 24 gm carbohydrates, or, considered as part of a day's Recommended Allowance (RDA) 1 oz provides

Protein	2%	Calcium	under 2%
Vitamin A	25	Iron	10
Vitamin C	25	Vitamin D	10
Thiamine	25	Vitamin B_6	25
Riboflavin	25	Folacin	25
Niacin	25	Phosphorus	under 2
		Magnesium	under 2

The cost of 1 oz of this cereal at retail was 4.1 cents.

A more elaborate patented, ready to eat cereal with a proprietary name is made of rice, wheat gluten, sugar, defatted wheat germ, salt, dry milk, malt flavoring and several additives. One ounce supplies 110 kcal, protein 6 gm, carbohydrates 20 gm and the following percentages of the RDA:

Protein	10%	Vitamin D	10%
Vitamin A	25	Vitamin B_6	25
Vitamin C	25	Folacin	25
Thiamine	25	Phosphorus	4
Riboflavin	25	Zinc	4
Niacin	25	Copper	2
Calcium	under 2	Magnesium	4
Iron	25		

One ounce of this cereal costs 6.6 cents.

One thing that neither nutritionists, dietitians nor food chemists have been able to accomplish is to attach a price to the nutritive value of any food unit. Prices are determined by acceptance and convenience; if they depended on nutritive value alone, some items would be practically free.

Oatmeal, which is 100% oats and is said to cook in 1 min, contains in 1 oz 110 kcal, 18 gm carbohydrates, 2 gm fat, 4 gm protein, no cholesterol and less than 10 mg salt. One ounce supplies 6% of protein RDA, 8% of thiamin, 4% calcium, 4% iron, 10% phosphorus and less than 2% each of vitamins A, C and D, thiamine, riboflavin and niacin. Compared with the two cereals described above, an ounce of oatmeal cost slightly less than 3 cents.

All the data for these three cereals were taken from the labels of packages bought in a supermarket.

Snacks

Although snacks are seldom part of a meal, large quantities are consumed between meals, especially by young people, although their elders compete at cocktail time.

Potato Chips.—Potatoes are sliced, fried in deep fat and salted to produce the familiar potato chip. One ounce supplies: 140 kcal, 1.4 gm protein, 11 gm fat, 14 gm carbohydrate, 11 mg calcium, 0.56 gm iron, 1.4 mg niacin and traces of other vitamins. In 1975, one ounce cost 11 cents.

Popcorn.—This snack furnishes 130 kcal to the ounce and little else.

Cereal Snacks.—Various kinds of crackers with or without cheese or other additives supply from 100 to 150 kcal per ounce. Three such products bought in 1975 cost 6.9, 9.5 and 15 cents an ounce.

There are many other foods that deserve to be called convenience foods. Probably one of the most popular is the collection of baby foods. Here, convenience is the predominating factor.

An interesting exercise for a student is calculation of the cost per unit weight of the foods in the supermarket; for example, the relative cost of fresh, dried, frozen and canned fruits or vegetables.

BIBLIOGRAPHY

AMERINE, M. A., BERG, H. W., and CRUESS, W. V. 1972. Technology of Wine Making. Avi Publishing Co., Westport, Conn.

ANON. 1972. Carbonated Beverages in the United States. American Can Co., Greenwich, Conn.

ASSOC. OFFIC. ANAL. CHEMISTS. 1975. Official Methods of Analysis, 12th Edition. Assoc. Offic. Anal. Chemists, Washington, D.C.

BURG, S. P., and BURG, E. A. 1965. Ethylene action and the ripening of fruit. Science *148*, 1190–1196.

CHARLES, R. J., and TUNG, M. A. 1973. Physical, rheological and chemical properties of bananas during ripening. J. Food Sci. *38*, 456–459.

COOK, A. H. 1962. Barley and Malt. Academic Press, New York.

FIDLER, J. C., and NORTH, C. J. 1971. The effect of conditions of storage on the respiration of apples. VII. The carbon and oxygen balance. J. Hort. Sci. *46*, 245–250.

GARARD, I. D. 1974. The Story of Food. Avi Publishing Co., Westport, Conn.

GILLIES, M. T. 1973. Soft Drink Manufacture. Noyes Data Corp., Park Ridge, N.J.

GILLIES, M. T. 1974. Compressed Food Bars. Noyes Data Corp., Park Ridge, N.J.

GRIFFITHS, F. P., and LIME, B. J. 1959. Debittering grapefruit products with naringinase. Food Tech. *13*, 430–433.

GUTCHO, M. 1973. Textured Foods and Allied Products. Noyes Data Corp., Park Ridge, N.J.

GUTTERSON, M. 1971. Vegetable Processing. Noyes Data Corp., Park Ridge, N.J.

HUTTON, H. O., and PROCTOR, B. E. 1961. Changes in some volatile constituents of the banana during ripening, storage and processing. Food Tech. *15*, 440–444.

KEFFORD, J. F. 1959. The chemical constituents of citrus fruits. *In* Advances in Food Research, Vol. 9 Academic Press, New York.

KENWORTHY, A. L., and HARRIS, N. 1960. Organic acids in the apple as related to variety and source. Food Tech. *14*, 372–375.

KLIEWER, W. M. 1969. Free amino acids and other nitrogenous substances of table grape varieties. J. Food Sci. *34*, 274–278.

KNEE, M. 1971. Ripening of apples during storage. J. Sci. Food Agr. *22*, 365–367, 368–371, 371–377.

KRAMER, A. 1973. Food and the Consumer. Avi Publishing Co., Westport, Conn.

KRUMEL, K. L., and SARKAN, N. 1975. Flow properties of gums used in the food industries. Food Tech. *29*, No. 4, 36–44.

LEE, F. A. 1975. Basic Food Chemistry. Avi Publishing Co., Westport, Conn.

LEE, L. S., CUCULLU, A. E., and GOLDBLATT, L. A. 1968. Appearance and aflatoxin content of raw and dry roasted peanut kernels. Food Tech. *22*, No. 9, 81–84.

MURRAY, E. 1961. Historic and modern aspects of vinegar making. Food Tech. *17*, No. 5, 74–76.

PANGBORN, R. M., VAUGHN, R. H., and YORK, C. H. 1958. Effect of sucrose and type of spicing on the quality of dill pickles. Food Tech. *12*, 144–147.

PEDERSON, C. S. 1971. Microbiology of Food Fermentations. Avi Publishing Co., Westport, Conn.

TIMMER, R., ter HEIDE, R., WOBLEIN, H. J., and de VALOIS, P. J. 1971. Phenolic compounds in rum. J. Food Sci. *36*, 462–463.

TRESSLER, D. K., and JOSLYN, M. A. 1971. Fruit and Vegetable Juice Processing Technology. Avi Publishing Co., Westport, Conn.

WATT, B. K., and MERRILL, A. L. 1963. Agriculture Handbook No. 8, Composition of Foods. ARS-USDA, Washington, D.C.

WEBB, A. D. 1962. Present knowledge of grape and wine flavors. Food Tech. *16*, No. 11, 56–59.

WHITE, P. L., and SELVEY, N. 1974. Nutritional Qualities of Fresh Fruits and Vegetables. Avi Publishing Co., Westport, Conn.

WOBLEAN, H. J., TIMMER, H., ter HEIDE, R., and de VALOIS, P. J. 1971. Nitrogen compounds in rum and whiskey. J. Food Sci. *36*, 464–465.

WOODRUFF, J. G. 1967. Tree Nuts: Production, Processing, Products. Avi Publishing Co., Westport, Conn.
WOODRUFF, J. G. 1973. Peanuts: Production, Processing, Products. Avi Publishing Co., Westport, Conn.
WOODRUFF, J. G., and PHILLIPS, G. F. 1974. Beverages, Carbonated and Noncarbonated. Avi Publishing Co., Westport, Conn.

CHAPTER 16

Colors and Flavors

The color of food may be the natural color, such as the red color of apples; the color produced by processing, such as the brown color of toast; the addition of a natural color such as carotene or annatto to butter and cheese; or the addition of a synthetic dye to a gelatin dessert.

The contribution of color to the acceptance of food is second only to flavor. Dietitians abhor an all white meal.

NATURAL COLORS

Chlorophyll

The ubiquitous green color of leaves is that of chlorophyll, which also serves as a catalyst for the reaction of carbon dioxide and water in photosynthesis

$$nCO_2 + nH_2O \xrightarrow{\text{light}} (CH_2O)_n + O_2$$

This apparently simple reaction is the basic source of all constituents of plants and animals.

All fruits are green when they are first formed and, in most of them, the green color disappears at maturity. The general practice of calling an unripe fruit "green" is unfortunate; green apples, green grapes and green oranges and some other fruits may be fully ripe.

Leafy vegetables are harvested while the leaves are still green and it is the problem of the chemist to retain the green color during processing. Housewives sometimes add sodium bicarbonate to green vegetables while cooking them. The alkali does help preserve the color, but it destroys vitamins and should not be used in commercial processing.

Chlorophyll is produced only in the presence of light; therefore, such vegetables as lettuce and cabbage contain the pigment in the outside leaves only. Some chlorophyll is lost during cooking, but spinach and other "greens" retain most of their original content.

Originally, it was thought that the metal in chlorophyll was iron and, although it was definitely proved to be magnesium more than 40 years ago, the idea still persists. Some florists have learned better, however, and put magnesium sulfate around their plants to get a deeper green color, but the practice has not become common in agriculture.

Hemoglobin

The red color of blood and of some meats is caused by the presence of hemoglobin, a conjugated protein consisting of globin, a histone and *heme*.

$$CH_3C = CCH = CH_2$$
$$HC-C_N^{}C = CH$$
$$CH_3C-C \qquad C = CCH_3$$
$$N-Fe-N$$
$$C-C \qquad C = CCH = CH_2$$
$$HOOCCH_2CH_2 \quad HC-C^N C = CH$$
$$HOOCCH_2CH_2 - C = CCH_3$$

Heme

The structure of heme is similar to that of chlorophyll. It consists of four pyrrole groups, an atom of iron, 5 methyl, 2 vinyl and 2 propionic acid groups. The molecular weight of hemoglobin is about 66,000 of which the heme constitutes only 6%.

Aside from its color, hemoglobin performs a basic function in the organism. It carries oxygen to the tissues throughout the body; the oxygen is attached to the iron—1 gm of hemoglobin carries 1.35 ml of oxygen measured at standard conditions. The hemoglobin is within the red corpuscles and can be separated by treatment with ether and centrifuging. It was the first protein to be crystallized. The corpuscles, or erythrocytes, are extremely small—a cubic millimeter contains some 5 million of them.

The hemoglobin from different species of animals may be of different composition, for the examination of the crystals from different species shows different crystal forms. The difference is in the globin. Hemoglobin is oxidized to methemoglobin, which is brown and unable to carry oxygen; the iron is oxidized to the ferric state.

Carotenoids

These pigments take their name from the carrot, which was the source of the first one to be separated, in 1831. More than 60 carotenoids have been separated from plants since then. They are divided into two groups: *carotenes* and *xanthophylls*.

Carotenes.—There are three carotenes extracted from food; usually 84% beta carotene, 15% alpha carotene and 0.1% gamma carotene. They are all orange-red in color, but as they are diluted, the color shifts towards the yellow. The difference in the three forms is in the position of the unsaturation in the ring portion of the molecule.

The word *carotene*, like *alcohol,* is both the name of a single compound and also the name of a class of compounds. The carotene group of pigments is all hydrocarbons.

Lycopene.—Another common carotene is lycopene, the red color of the tomato. In lycopene, both the rings at the end of the chain in carotene are open so that the entire molecule is that of a straight-chain hydrocarbon. The end structure that replaces the ring in carotene is

$$-\underset{\underset{H}{|}}{C}=\underset{\underset{CH_3}{|}}{C}-\underset{\underset{H}{|}}{C}-\underset{\underset{H}{|}}{C}-\underset{\underset{H}{|}}{C}=\underset{\underset{CH_3}{|}}{C}-CH_3$$

The series of conjugated double bonds along the carbon chain is the structure responsible for the color. Changes in the ring and in the substituents result in changes in hue, for the entire color range of the carotenes is from yellow to red.

Xanthophylls.—The xanthophylls are oxygenated derivatives of the carotenes. They are mainly alcohols but there are aldehydes and ketones among them and a few acids and esters.

The carotenes, being hydrocarbons, are insoluble in water; the hydroxy xanthophylls are slightly soluble; the ketones are not.

In nature, the color of a plant product usually has a predominant pigment with a mixture of several lesser ones. Egg yolk owes its color to zeaxanthin, $C_{40}H_{56}O_2$, a dihydroxy β-carotene, which is also the yellow pigment in corn.

The pink color of the shrimp is astaxanthine, both a hydroxy and ketonic xanthophyll. The color of hair, fur, feathers, milk and eggs of animals is affected by the pigments of the diet. The colors of flamingos and canaries fade if the diet does not contain the proper pigments.

Stability.—The carotenoids, being highly unsaturated, are highly susceptible to oxidation. In the plant, they are protected by the cell walls, but in processing, much of the color may be lost. They are stable to light in the absence of air, but fade rapidly when exposed to both.

Anthocyanins

Although the carotenoids are widely distributed in nature, many of the colors of fruits and flowers are the colors of pigments of very different composition. The anthocyanins are glycosides. The sugar involved may be glucose, galactose, L-rhamnose, D-arabinose, D-xylose or any one of several disaccharides. As far as the color of the pigment is concerned, the sugar part of the molecule is incidental. The color portion of the molecule is a derivative of a phenylbenzopyran.

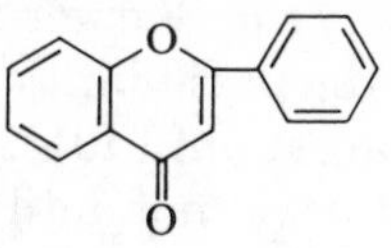

Flavone (2-phenylbenzopyrone)

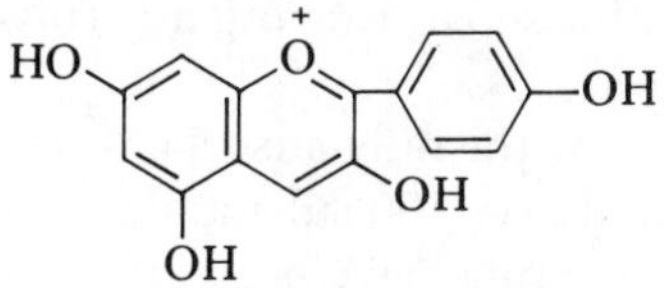

Pelargonidin

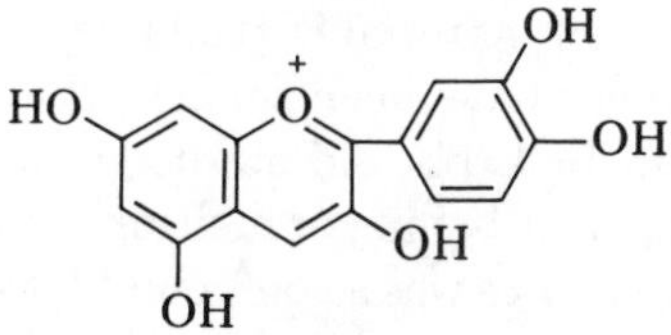

Cyanidin

The anthocyanidins are oxonium ions. The sugars are attached to the hydroxy groups in the 3, 5 positions of the benzopyran ring by a glycosidic linkage.

Pelargonidin is the chromophore in strawberries. Cyanidin is found in dark cherries and elderberries.

Many of the pigments in fruits and flowers change color with change in acidity; red at low pH and blue at high pH is common.

Pigments in the skins of Montmorency cherries were reported by Dekazos (1970) as consisting of cyanidin 1.53 mg and peonidin 0.70 mg in 50 gm of cherries. The sugars are glucose, rutinose (a disaccharide, rhamnose-glucose) and sophorose (a glucosyl glucose).

CHROMATOGRAPHY

In a paper published in 1906, Michael Tswett (1872–1919) reported that if he extracted the pigments from plants with petroleum ether and passed the solution through a glass tube packed with finely powdered calcium carbonate, the pigments were adsorbed selectively on the powder. The chlorophyll formed a green band near the top of the column and other pigments formed yellow bands farther down the column. He called the result a *chromatogram*. The pigments could be removed from the column with alcohol.

Tswett was a botanist at the University of Warsaw at the time and published in the *Proceedings of the Warsaw Society of Natural Science.* Chemists were slow to learn the possibilities of the method. Tswett continued his research on the subject until 1914, and a few chemists investigated the process mildly, but it was not until 1931 that Richard Kuhn at the Kaiser Wilhelm Institute in Berlin used it to separate the carotenoids from the xanthophylls. Other chemists developed the method and now its use has become general for separating organic substances, especially the colors and flavors of plants.

Calcium carbonate is not the only adsorbent used in chromatography. Tswett alone investigated over 100 substances and several of them have proved useful. Chromatographs have been invented for the various techniques that have been developed, and so today, chromatography is more common than distillation and crystallization, although the latter are widely used when an abundance of material is available.

Once natural substances have been separated and obtained in a pure state, they must be identified. The old method was to determine as many properties as possible, such as boiling or melting point, specific gravity and refractive index if the substance was a compound for which these properties were on record. If it were a new substance, chemical as well as physical properties had to be determined. These determinations required several grams of the substance. Chromatography provides very small quantities of the constituents of food and now the spectrograph is the common instrument of identification, for it requires very minute samples.

Colors Formed by Processing

The brown color of toast is caused by the Maillard reaction. However, other color changes occur, especially during heating. Most of them are caused by the destruction of one pigment to reveal another.

Chlorophyll and the anthocyanidins are soluble in water; the carotenoids are highly unsaturated and easily oxidized. Exposure to light will destroy many of them.

The color changes that occur during processing or storage of food depend primarily on the pigment present in the raw product and the nature of the process; each in a special case.

ADDED NATURAL COLORS

With all the natural colors there are in nature, it is surprising how few have been captured for use in coloring food. Of course, we do not know how many have been tried and rejected, but only ten are on the FDA list and they include caramel, which is artificial.

Many food products are colored simply by adding a colored food. Strawberries or cherries add a pink color to ice cream. Egg yolk, paprika, chocolate, coffee and several other colored foods serve to add color to a composite product. Such additions are not considered color additives and are not regulated legally. Natural colors, such as carotene or saffron, are additives, but do not require certification although they are officially defined and their use limited.

Caramel

Caramel is officially defined as, "The dark brown liquid or solid material resulting from the carefully controlled heat treatment of the following food-grade carbohydrates: 1 dextrose, invert sugar, lactose, malt syrup, molasses, starch hydrolyzates and fractions thereof, and sucrose. 2. Food grade acids, alkalis and salts may be used to assist caramelization. These are: acetic, citric, phosphoric, sulfuric and sulfurous acids and ammonium, calcium potassium and sodium hydroxides. The salts are ammonium, potassium and sodium carbonate, bicarbonate, phosphate, sulfate and sulfite. 3. Glycerides may be added as antifoaming agents."

From the above definition, it is obvious that caramel is a complex mixture of unknown components. Nef showed that the effect of alkali on glucose resulted in a possible 116 compounds of which he isolated several. Other sugars are permitted here, which increases the possibilities, and then these products of sugar decomposition may combine with some of the other additives. In view of the large number of alternatives, it would probably be more accurate to head this section "caramels."

Caramel may be used to color any food unless the food has a standard that eliminates it. One of the most familiar uses is in the "cola" beverages.

Annatto

Annatto extract is prepared from the seed of *Bixa arellana*. The definition specifies about a dozen liquids that may be used for the extraction. The dye may be used to color any food except where the standard for a food does not include it.

The plant from which the color is obtained is a tropical tree and the solvent used differs with the intended use of the color. It has been used for decades to color butter, cheese and margarine. For use in butter, the dye is extracted with a vegetable oil or propylene glycol. For coloring cheese, an aqueous alkaline solution is used.

The main pigment in annato is alpha bixin

$$HOOC-C=C-C=C-C=C-C=C-C=C-C=C-C=C-C=C-C=C-C=C-COOCH_3$$

Butter colors on the market usually contain from 0.25% to 0.7% carotenoid pigments of which alpha bixin predominates; cheese coloring contains a greater percentage of pigment, but it is less stable.

Turmeric

This color is defined as the ground rhizome *Curcuma longa*. The plant is a native of southern Asia and Indonesia where the young, colorless rhizomes are used for food. They develop color as they mature.

Turmeric oleoresin is an extract of turmeric made with any one or a combination of the following: acetone, methyl, ethyl and isopropyl alcohol, hexane, ethylene and methylene chloride or trichloroethylene. The solvent is removed after the extraction and commercial preparations are made by diluting the oleoresin with one of several substances including acetone, alcohol and ethyl acetate.

The pigment is the carotenoid curcumin.

β-Carotene

The pigment that is so common among the natural food colors has been synthesized on a large scale from β-ionone, which in turn is made from citral obtained from the oil of lemon grass.

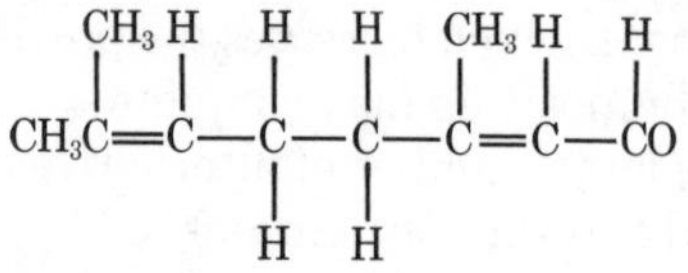

$$CH_3C\!=\!\overset{\overset{\displaystyle CH_3}{|}}{C}\!-\!\overset{\overset{\displaystyle H}{|}}{\underset{\underset{\displaystyle H}{|}}{C}}\!-\!\overset{\overset{\displaystyle H}{|}}{\underset{\underset{\displaystyle H}{|}}{C}}\!-\!\overset{\overset{\displaystyle CH_3}{|}}{C}\!=\!\overset{\overset{\displaystyle H}{|}}{C}\!-\!\overset{\overset{\displaystyle H}{|}}{CO}$$

Citral

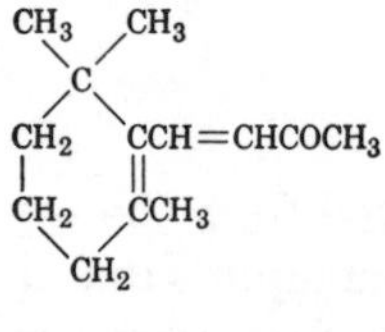

β-ionone

The molecule consists of a series of isoprene units with conjugated double bonds that supply the basic color. There are 40 carbon atoms, 11 double bonds, 2 rings and 10 methyl groups.

Carotene is now available in quantity and is permitted in foods without limitation. The pure carotene crystals are not very stable, but the color is marketed as a solution in vegetable oil in which it is much more resistant to oxidation. The solution may be added directly to fats such as shortening and water-in-oil emulsions, such as butter.

For use in aqueous foods, the carotene in oil solution is emulsified in water with gelatin or other emulsifying agent of food grade. The emulsion can then be added to foods that are mostly water.

Other Noncertified Colors

In addition to the natural colors just mentioned, several others are permitted including: paprika, saffron, grape skin extract, dehydrated beets, riboflavin and carrot oil. Some of these are flavors as well as colors.

SYNTHETIC DYES

Since Perkin discovered mauve in 1856, chemists have synthesized thousands of dyes, commonly called *coal tar dyes* because the raw materials came from coal tar. Some of them soon found their way into the food industry to color candy, soft drinks and other foods that profit by a strong color. Before 1907, dyes were used without control as long as they were not known to be extremely poisonous. Since the Food and Drugs Act of June 30, 1906 became effective, twenty-four dyes have been certified for use in foods. Upon further examination, however, over half of them have been banned. The FDA has dyes under investigation continuously and so the approved list changes as new dyes are admitted and others are banned. The approved list of 1976 is

Red No. 3

Red No. 4[1]

Red No. 40

Blue No. 1

Blue No. 2

Yellow No. 5

Yellow No. 6

Green No. 3

Violet No. 1

Citrus Red No. 2[1]

Orange B[1]

The numbers of the dyes are assigned by the FDA because the chemical names are too long and too technical for general use. The name of Blue No. 2, the simplest on the list, is 5,5-indigotindisulfonic acid. None of these dyes is fat-soluble, all are stable to light, oxidation and change of pH except Blue No. 2 and Violet No. 1. Their solubility varies from 1 gm to 20 gm in 100 ml of water, 25% in alcohol, glycerine and propylene glycol. They do not cover the whole range of color, although mixtures extend the range over that of the individual dyes.

It has been estimated that the consumption of artificially colored food in the United States amounts to 104 lb per capita. Much of it is colored with caramel or a vegetable dye.

The requirement for the certification and approval of a dye is long and detailed and includes both chemical and biological testing. A chemist who has occasion to use a food color should obtain from the FDA the latest regulations for both color certification and approved color additives.

---

[1] These three colors are restricted in their use. Red No. 4 is permitted only to color maraschino cherries; Orange B only to color sausage casings and Citrus Red No. 2 only to color the skins of oranges.

FLAVORS

The flavors of foods are even more complex than the colors. Taste is not well understood. There are the generally accepted four basic tastes: sweet, sour, salty and bitter. Whether there are more than four is not clear. The characteristic flavors of foods result largely from the volatile components present.

Other than the four tastes and numerous aromas, there are factors that affect the taste sensation. The pear and the banana are characterized by the odor of amyl acetate, but in addition to the acid in the pear, the texture causes a very different taste sensation. A friend told me that she could not eat liver but did not know why. She said she did not dislike the taste of it, but thought the taste was too strong. This effect is common in such strongly flavored foods as olives, onions, parsnips, fish and many others. Substances that irritate the tongue, such as capsicum, tannin, essential oils and even bitter chocolate produce a taste effect that is neither one of the basic tastes nor an aroma. Also the taste of some of the amino acids is not one of the usual components of flavor. However, the variations in taste may be classified; the following discussion will be limited to aroma and the four basic tastes, but any chemist who attempts to devise a new product must never forget that there is much more to acceptability than these five criteria.

Salt

Salt is a nutritional requirement as well as a flavor and all herbivorous animals (deer, cattle, horses, sheep) must ingest it, but the carnivorous and omnivorous animals get their salt from the flesh of other animals. Man belongs to the latter class. Salt, therefore, is a natural flavor of animal flesh only.

Sugars

There are several sugars of varying degrees of sweetness. Fructose is the sweetest sugar and accounts for the unusual sweetness of honey. Sucrose and glucose occur in most fruits. In some fruits, there are also fructose and mannitol and there may be other sugars. Lactose, which produces the mild

P-Methoxyphenyl
Urea
(Sweet)

O-Methoxyphenyl
Urea
(Tasteless)

sweetness of milk, is a sugar with much less flavor than either glucose or sucrose.

There is no known chemical structure that can account for a sweet taste. The natural sweet compounds are generally polyhydroxy compounds with a straight chain structure, such as the sugars and hexatomic alcohols, mannitol and sorbitol. Ethylene glycol and glycerol are sweet, but the trihydroxy benzenes are not.

Acids

The sourness of foods is invariably caused by the presence of one or more acids. Citric, tartaric and malic are the most common acids in natural foods; acetic is produced by the fermentation of alcohol and so that is common in processed foods. Vitamin C, or ascorbic acid, is present in many fruits and in some vegetables. Rhubarb, spinach and a few other products contain oxalic acid. Phosphoric acid or its acid salts are often used in food processing.

The hydrogen ion, H^+, is commonly held responsible for a sour taste and it undoubtedly is primarily responsible. However, it is not the sole factor in sourness. All the acids mentioned above, with the exception of oxalic, are comparatively weak and so there is a large portion of un-ionized acid present in foods, especially since the food also contains buffers. Sourness is not proportional to the H^+ concentration and the acids have different flavors. Anyone can detect a wine that is beginning to sour; the acetic acid formed has a very different flavor from that of the tartaric acid normal to the wine.

Bitter

The bitter taste cannot be assigned to any class of chemical substances. Bitter substances may be alkaloids, glycosides or salts as well as organic compounds of other classes.

Naringin, the bitter principle of grapefruit, is a glycoside of rutinose. Although it is a phenolic derivative and very bitter, it is not toxic.

Naringin

Amygdalin is the glycoside in bitter almonds. The sugar is gentiobiose and the molecule contains a −CN group released by hydrolysis, and so the glycoside may be toxic.

A glycoside that is responsible for the effect of mustard and horseradish is sinigrin:

$$\overset{\displaystyle OSO_3K}{\underset{\displaystyle }{C_6H_{11}O_5 - S - C = NCH_2CH = CH_2}}$$

The sugar in sinigrin is glucose.

Other glycosides are responsible for the flavor or other effects of pepper, asparagus and other natural foods of peculiar flavor.

Volatiles

The odorous compounds from natural foods are numerous. They are present in low concentrations and, before the advent of chromatography, very few of them were identified. Bananas were known to emit amyl acetate; apples, ethyl acetate; and fish, amines. Now it is discovered that the flavor of delicious apples consists of the following compounds in an amount descending in the order recorded (Schultz 1967):

1. Ethanol
2. 1-butanol
3. 2-Hexenal, trans
4. 1-Hexanol
5. 2-Methylbutan-1-ol
6. 1-Propanol
7. 2-Hexan-1-ol
8. Hexanal
9. Ethyl acetate
10. Butyl acetate
11. 2-Methylbutyl acetate
12. Isobutanol
13. Hexyl acetate
14. Propyl acetate
15. 2-Hexenyl acetate
16. Ethyl butyrate
17. Ethylpropionate
18. Acetaldehyde

Carrots have been found to contain various sapid compounds: among them are several amino acids, glucose, maltose and sucrose and the following volatiles listed in descending order:

1. Ethanol
2. Caryophyllene
3. Terpinolene
4. Gamma Bisobollene
5. Beta Bisobollene
6. Acetaldehyde
7. Limonene
8. Myscene
9. Myristicin
10. Bornyl acetate
11. Caretol
12. Sabinene

The concentration of these volatiles in raw carrots extends from 0.1 ppm of sabinene to 50 ppm of ethanol. There are several others present in less than 0.1 ppm.

Celery contains sucrose, glucose, fructose and mannitol in addition to several volatiles.

The predominating volatiles in onions are disulfides. In descending order they are

1. n-Propyl
2. n-Propyl allyl
3. n-Propyl methyl
4. Methyl allyl
5. Dimethyl
6. Diallyl

The amount of the n-propyl disulfide in fresh onions exceeds that of the other five combined.

Oil of thyme, like many other essential oils, consists largely of terpenes and their alcohols or aldehydes. A total of 24 ingredients has been identified in the oil. Among them are

> Camphene
> γ Terpinene
> p-Cyamene
> Linacool
> Carbacrol

General

Any classification of natural food flavors is of little use. As for volatiles; most fruits contain aliphatic esters, aldehydes or alcohols as the predominating aroma. Essential oils and strongly aromatic fruits and vegetables are characterized by terpene hydrocarbons and their oxygen derivatives. Very pungent foods such as onions, garlic, peppers and mustard are marked by organic compounds containing sulfur. It should also be realized that many of the compounds in a given flavor do not contribute to the aroma in proportion to the order of their presence. Different substances have different threshold values and 1% of A may contribute far more to the aroma than 10% of B. Much research remains to be done on threshold values of the many components of food flavors and the problem is complicated by the fact that the values are subjective and differ with the investigators.

Developed Flavors

Flavor components that develop during food processing may be either solids or volatiles and their origins are either of two kinds: those resulting from the fractionation of volatiles and those caused to the decomposition or other reaction of food components.

Fractionation.—Fractionation is the bane of the existence of the perfumer and the manufacturer of artificial flavors. Each substance used in the mixture has a different vapor pressure; and a delightful perfume, carelessly made, may develop a disgusting odor as the more volatile compounds evaporate. Volatile compounds may escape from foods during storage, peeling, cutting or heating. Heatherbell *et al.* (1971) found the following differences and some smaller ones between fresh and canned carrots:

	Fresh	Canned
Caryophyllene	10 ppm	5 ppm
Terpinolene	6.05	2.50
γ-Bisobolene	5.55	5.50
β-Bisobolene	1.18	0.87

Bernhard (1968) in his study of the flavor of onions found that several of the components decreased drastically when the onions were dried.

Process Flavors.—Heating changes the flavor of many foods profoundly, for example, cacao beans, coffee beans, peanuts and meat. The change may be produced by decomposition of components of the food or combination of some of the components present. An investigation of the aroma of boiled beef disclosed: 7 hydrocarbons, 10 alcohols, 3 esters, 9 aldehydes, 5 ketones, 3 acids, 4 sulfides, 8 benzene derivatives and 5 heterocycle compounds. The predominating ones were 1-Octen-3-al, 1-n-haxanal, ethyl acetate, acetoin, benzaldehyde and 2,4,5-trimethyl-3-oxazolene.

The odor of potato chips has been found to contain: 8 nitrogen compounds, 2 sulfur compounds, 14 hydrocarbons, 13 aldehydes, 2 ketones, 2 alcohols, 3 esters, 1 ether and 8 acids.

Roasted filberts yielded a flavor containing 45 compounds.

In a few cases, the sources of the flavors have been traced. The dimethyl sulfide emitted by cooking sweetcorn appears to be formed by the decomposition of methylmethionine

$$H_2NCHCH_2CH_2SCH_3$$
$$|$$
$$COOCH_3$$

Walter and Fagerson (1968) heated anhydrous glucose for 30 min at 250°C and found about 100 compounds in the distillate. They identified 15 of them among which were furane, biacetyl, acetic acid, furfural, and phenol. The others were less familiar substances.

When we consider the mixture of carbohydrates, fats, proteins and incidental compounds that constitute our common foods, an enormous number of compounds, both solid and volatile can be formed by decomposition. Furthermore, the compounds produced differ with the manner of heating.

Cabbage boiled 10 min does not taste like cabbage boiled an hour, and neither tastes like raw cabbage. Boiled peanuts do not taste like the roasted product. Boiled beef differs from that which is grilled, and the characteristic flavor of a charcoal broiled steak is the result of the decomposition of the fat that drips on to the burning charcoal.

An unusual development of flavor caused by processing is the aging of distilled liquors in an oak barrel with the inside charred. Whiskey, rum and brandy are all aged in this manner. Schaefer and Timmer (1970) investigated cognac and found 12 alcohols, 8 aldehydes, 8 ketones, 16 acids, 31 esters and 12 miscellaneous compounds in the aroma.

Wines are aged in uncharred barrels and new ones contribute to the flavor of the wine, but the main flavor development is caused by the slow reaction between the alcohol and the acids from the grapes.

Added Flavors

Flavors added to confectionery, nonalcoholic beverages and other pre-pared foods are of two kinds: essential oils or oleoresins or other extracts from aromatic plants; and synthetic substances that may or may not occur in nature.

Essential Oils.—There are about 170 of these oils in the GRAS list but, since each of them contains several chemical substances, the actual number of flavoring substances is enormous. Oil of thyme was described above. The oil of peppermint is used in a different kind of food and Cash, Hrutfiord and McKean (1971) have found the peppermint oils produced in Washington state to contain the following components:

	%
Menthol	40–52
Menthone	2.2–3.5
Menthyl acetate	4–8
Neomenthol	3–4
Isomenthone	3–6
Menthofurane	2–12
3-Octanol	1–3
Pulegone	1–2
Piperitone	1–2
Isomenthol	1
Neoisomenthol	1

The most widely used flavor additive is vanilla, which may be either a natural extract or the synthetic compound. The plant, *Vanilla planifolia* belongs to the orchid family and is a native of southern Mexico. It is culti-vated in Mexico, Indonesia and the French islands of the Indian ocean; Madagascar grows it extensively.

The fruit of the plant is a long bean pod; the raw beans do not have the flavor of vanilla until they are sweated and dried. During this curing process, the flavor develops and the beans shrink to about one fifth their original weight.

The beans are chopped into small pieces and extracted with 50% alcohol. The concentration of the alcohol is critical; too much extracts lipids and other substances that produce a cloudy extract and too little alcohol extracts an objectionable flavor. The vanilla extract is a complex mixture, but the predominant flavor is that of vanillin

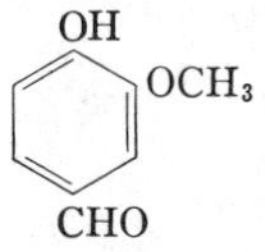

OH

OCH$_3$

CHO

Vanillin

$$H—C\!=\!O$$

$$—O—C_2H_5$$

OH

Ethyl Vanillin

Vanillin and ethyl vanillin are produced synthetically. They are in the form of small white crystals and can, therefore, be used in any amount required. They have a powerful flavor and it is easy to use too much.

Synthetic Flavors.—There are about 25 synthetic flavoring substances in the GRAS list. Many of them are known to be components of natural flavors, but some are not. The familiar flavors benzaldehyde (almonds), acetylmethylcarbinol (butter), citral (orange), eugenol (cloves), limonene (lemons) and vanillin are in the list.

Proprietary flavors for candy, soft drinks and other foods are produced by making a mixture of flavoring substances to imitate natural flavors that are difficult or impossible to isolate from the natural substances, such as strawberries and cherries. To imitate such natural flavors is a matter of art; the result is determined by taste. There is no objective method for comparing or matching flavors.

Both essential oils and synthetics must be used in accordance with "good manufacturing practice," which is defined in paragraph 121.101 of the Code of Federal regulations and includes:

1. The amount added does not exceed that reasonably required to accomplish its intended effect.

2. Substances that may become part of the food as a result of the manufacturing shall be kept as low as possible.

3. The substance is of food grade.

In addition to these requirements, a flavor may not be added to a food for which there is a standard unless the standard includes it.

Saccharin

A student at the Johns Hopkins University in 1880 accidentally noted that a compound that he had made was very sweet. It has since been estimated to be 500 times as sweet as sucrose and detectable at a dilution of 1 part in 100,000 parts of water. It has an unpleasant aftertaste and at high concentrations, it is not sweet at all.

Saccharin is o-sulfobenzoic acid imide.

It was in use in this country before 1900 and a source of controversy during the turmoil that preceded the passage of the Food and Drug Act of

June 30, 1906 (Garard, 1974). The FDA recommends that consumption be limited to 1 gm a day, or to 15 mg per kg of body weight.

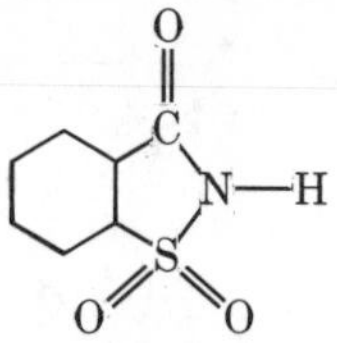

Saccharin

About 4.2 million lb were consumed in the U.S. in 1972; nearly three-fourths of it in soft drinks.

Cyclamates.—The calcium, magnesium, potassium and sodium salts of cyclohexylsulfamic acid, which are 15 to 30 times as sweet as sucrose, were used as sweeteners for several years and then banned by the FDA because someone found them to cause bladder cancer in experimental animals. The results were questioned and much more research has been done. At this writing, the subject is under review and the cyclamates may or may not be restored to use.

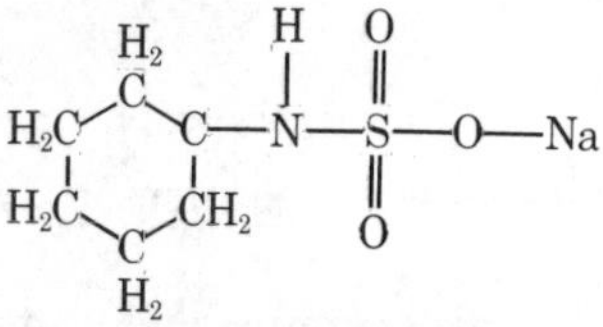

Sodium Cyclamate

It is, of course, impossible to prove a negative proposition so no substance can be proved harmless, but the Delaney amendment to the Food, Drug and Cosmetic Act provides that if a substance has been found to cause cancer in any animal species, it may not be added to food in any amount, so that the fate of saccharin, the cyclamates and all other food additives hangs on the validity of the results of the research on the substance as a carcinogen. Toxicity, of course, will also eliminate anything from the list of approved additives.

Aspartame.—Among the amino acids, glycine is sweet and so are some of the derivatives of others. In 1970 a manufacturer introduced aspartame, which is said to be 160 times as sweet as sucrose and has no unpleasant aftertaste. It has been approved by the FDA for use in foods; approved for packages in an amount not to exceed the sweetness of two teaspoonfuls of sucrose for table use. It is also in chewing gum, dry bases for beverages and in puddings and simulated dairy product toppings. It is not approved for baked or fried products because prolonged heating decomposes it. Its formula is:

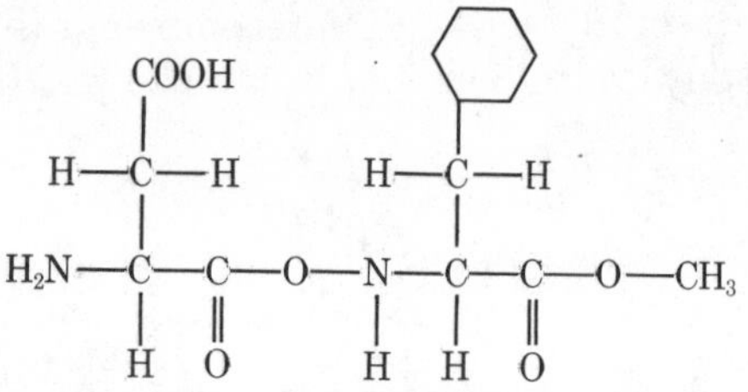

Aspartame

Other Sweeteners.—The discovery of the dipeptide aspartame has led to further search for sweeteners among the amino acid derivatives and the derived proteins.

The amino acids are nutrients but since the sweetness of some of their derivatives is so strong, the calories they add to the food are negligible.

Another sweetener among the acid group is 6-chloro-D-tryptophane

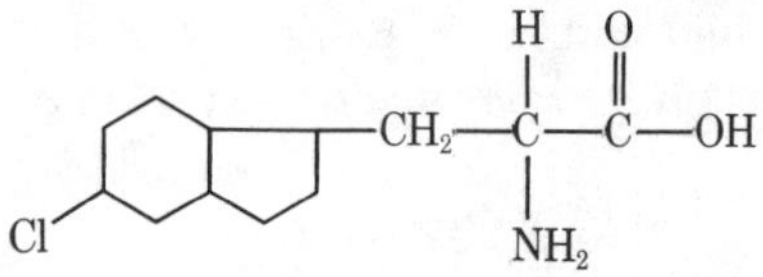

6-Chloro-D-tryptophane

It is said to be 1,000 times as sweet as sucrose and is undergoing toxicity tests.

Monellin, a protein of some 90 amino acid units is said to be 3,000 times as sweet as sucrose. It was isolated from the tropical plant *Discoreophyllum cumminsii.*

Another protein *thaumatin* has been isolated from a plant source. It has 180 amino acid units and is claimed to be 1,600 times as sweet as sucrose.

A third protein, reminiscent of sodium glutamate is *miraculin* with a molecular weight of 40,000. It has been isolated from an African plant; it is not sweet but makes an acidic food taste sweet.

Aside from their unusual sources, these proteins cannot be used in cooked products because they are destroyed by heat.

Monosodium L-Glutamate.—This salt is not a flavor, but its inclusion in many foods enhances the flavor of that food whether natural or synthetic. It is used in soups and sauces but is useless with fruit flavors. The mechanism by which it exerts its effect has never been satisfactorily explained.

The glutamic acid is obtained by the hydrolysis of wheat gluten with HCl. If it is synthesized or obtained by the hydrolysis of gluten with alkali, the DL form of the acid is obtained and the D form is ineffective.

Chapter 19 of the Official Methods (12th Edition) contains methods for the analysis of some flavors and Chapter 34 is devoted to color additives.

BIBLIOGRAPHY

AMERINE, M. A., PANGBORN, R. M., and ROESSLER, E. R. 1965. Principles of Sensory Evaluation of Food. Academic Press, New York.

AMOORE, J. E. 1970. Molecular Basis of Odor. Charles E. Thomas Publishers, Springfield, Ill.

ANON. 1974. Aspartame. Food Tech. 28, No. 10, 92–93.

AURAND, L. W., and WOODS, A. L. 1973, Food Chemistry. Avi Publishing Co., Westport, Conn.

BAINBRIDGE, J. C., and DAVIES, S. H. 1912. The essential oil of cacao. J. Chem. Soc. 2209–2221.

BAUERNFEIND, J. C., SMITH, E. G., and BUNNELL, R. H. 1958. Coloring fat-base foods with beta carotene. Food Tech. 12, 527–535.

BECKER, R. 1968. Identification of some sugars and manitol in celery. J. Food Sci. 33, 128–130.

BERNHARD, R. A. 1968. Comparative distribution of volatile aliphatic disulfides derived from fresh and dehydrated onions. J. Food Sci. 33, 298–304.

BUNNELL, R. H., and BAUERNFEIND, J. C. 1962. Chemistry, uses and properties of carotenoids in foods. Food Tech. 16, No. 7, 36–43.

BUNNELL, R. H., DRISCOLL, W., and BAUERNFEIND, J. C. 1958. Coloring water-base foods with beta carotene. Food Tech. 12, 536–541.

CARRENO-DIAZ, R., and LUM, B. S. 1969. Anthocyanin pigments in Trousseau grapes. J. Food Sci. 34, 415–419.

CASH, D. B., HRUTFIORD, B. F., and McKEAN, W. T. JR. 1971. Effect of individual components on peppermint oil flavor. Food Tech. 25, No. 11, 53–58.

CHEN, L. F., and LUH, B. S. 1967. Anthocyanines in Royalty grapes. J. Food Sci. 32, 66–74.

CO, H. J., and MARKAKIS, P. 1968. Flavonoid compounds in the strawberry fruits. J. Food Sci. 33, 281–283.

COLEMAN, R. L., LUND, E. D., and MOSHONAS, M. G. 1969. Composition of orange essence oil. J. Food Sci. 34, 610–611.

CREVELING, R. E., SILVERSTEIN, R. M., and JENNINGS, W. G. 1968. Volatile components of pineapple. J. Food Sci. 33, 284–287.

CROTEAU, R. J., and FAGERSON, I. S. 1968. Major volatile compounds of the juice of the American cranberry. J. Food Sci. 33, 386–389.

DANIELS, R. 1973. Sugar Substitutes and Enhancers. Noyes Data Corp., Park Ridge, N.J.

DECK, R. E., POKORNY, J., and CHANG, S. S. 1973. Isolation and identification of volatile compounds from potato chips. J. Food Sci. 38, 345–349.

DEKAZOS, E. D. 1970. Anthocyanin pigments in red tart cherries. J. Food Sci. 35, 237–241.

DO, J. Y., SALUNKHE, D. K., and OLSON, L. E. 1969. Isolation, identification, and comparison of the volatiles of the peach fruit as related to harvest, maturity and artificial ripening. J. Food Sci. 34, 618–621.

FULEKI, T. 1969. The anthocyanines of strawberries, rhubarb, radish and onion. J. Food Sci. 34, 365–369.

FULEKI, T. 1971. Anthocyanins in red onions. J. Food Sci. 34, 365–369.

FULEKI, T., and FRANCIS, P. J. 1968. Quantitative method for anthocyanins: Purification of cranberry anthocyanine. J. Food Sci. 33, 266–274.

HARNSTEIN, I., and TERANISHI, R. 1967. The chemistry of flavor. Chem. Eng. News 45, No. 16, 90–108.

HEATHERBELL, D. A., WROLSTAD, R. E., and LIBBEY, L. M. 1971. Carrot volatiles: 1. Characterization and effects of canning and freeze drying. J. Food Sci. 36, 219–224.

HEATHERBELL, D. A., and WROLSTAD, R. E. 1971. Carrot volatiles: 2. Influence of variety, maturity and storage. J. Food Sci. 36, 225–227.

HOLLEY, R. W., STAYLA, B., and HALLEY, A. D. 1955. Identification of some volatile constituents of concord grape juice. Food Research 20, 326.

INGLETT, G. E. 1973. Sweeteners. Avi Publishing Co., Westport, Conn.

ISLER, O., OFNER, A., and SIEMERS, G. F. 1958. Industrial synthesis of carotenoids for use as food colors. Food Tech. 12, 520–528.

KARRER, P., and JUCKER, E. 1950. Carotenoids. Elsevier Publishing Co., New York.

KAZENIAC, S. J., and HALL, R. M. 1970. Flavor chemistry of tomato volatiles. J. Food Sci. 35, 519–530.

KONIGSBACHER, K. S., HEWI, E. J., and EVANS, R. L. 1959. Application of flavor enzymes to processed foods: 1. Panel studies. Food Tech. 13, 128–131.

MACLEOD, A. J., and MACLEOD, G. 1970. Effects of variation in cooking methods on the flavor volatiles of cabbage. J. Food Sci. 35, 744–750.

MERARY, J. 1968. Food Flavorings. Avi Publishing Co., Westport, Conn.

MORGAN, R. C. 1967. The carotenoids of Queenland fruits: carotenes of the watermelon. J. Food Sci. 32, 275–278.

MOSHONAS, M. G. and LUND, E. D. 1969. Aldehydes, ketones and esters in Valencia orange peel oil. J. Food Sci. 34, 502–503.

NEF, J. U. 1914. Dissociation in the sugar group. Ann. 403, 204–383.

NELSON, P. E., and HOFF, J. E. 1969. Tomato volatiles: effect of variety, processing and storage time. J. Food Sci. 34, 53–57.

PHILIP, T., and FRANCIS, F. J. 1971. Isolation and chemical properties of capsanthin and derivatives. J. Food Sci. 36, 823–827.

RADER, C. P., TIHANYI, S. G., and ZIENTY, F. 1967. A study of the true taste of saccharine. J. Food Sci. 32, 357–360.

REITH, J. F., and GIELEN, J. W. 1971. Properties of bixin and norbixin and the composition of annatto extract. J. Food Sci. 36, 861–864.

RICHARD, H. M., and JENNINGS, W. G. 1971. Volatile composition of black pepper. J. Food Sci. 36, 584–589.

ROGER, N. F. 1961. The recovery of methylanthranalate in concord grape essence. Food Tech. 15, 309–314.

RUSSEL, G. F., and OLSON, K. U. 1972. The volatile constituents of oil of thyme. J. Food Sci. 37, 405–407.

SCHAEFER, J., and TIMMER, R. 1970. Flavor components in cognac. J. Food Sci. 35, 10–12.

SCHALLER, D. R., and VON ELBE, J. H. 1971. The carotenoids in montmorency cherries. J. Food Sci. 36, 712–713.

SCHULTZ, H. W. 1967. The Chemistry and Physiology of Flavors. Avi Publishing Co., Westport, Conn.

SHELDON, R. M., LINDSEY, R. C., and LIBBEY, L. M. 1972. Identification of volatile compounds from roasted filberts. J. Food Sci. 37, 313–316.

SHU, C. K., and WALLER, G. R. 1971. The volatile compounds of roasted peanuts. J. Food Sci. 36, 579–581.

SUN, B. H., and FRANCIS, F. J. 1967. Apple cyanins: identification of cyanidin-7-arabinoside. J. Food Sci. 32, 647–649.

SWISHER, H. E. 1963. Citrus oil flavors. Food Tech. 17, No. 7, 103–105.

VENSTROM, D. and AMOORE, J. E. 1968. Olfactory threshold in relation to age, sex or smoking. J. Food Sci. 33, 264–265.

WALTER, R. H., and FAGERSON, I. S. 1968. Volatile compounds from heated glucose. J. Food Sci. 33, 294–297.

WILSON, C. W. 1970. Relative recovery and identification of carbonyl compounds from celery essential oil. J. Food Sci. 35, 766–768.

WRIGHT, W. D. 1958. The Measurement of Color. Macmillan Co., New York.

WROLSTAD, R. E., and HEATHERBELL, R. D. 1968. Anthocyanin pigments of rhubarb. J. Food Sci. 33, 592–594.

WROLSTAD, R. E., and PUTNAM, P. B. 1968. Isolation of strawberry anthocyanin pigments by adsorption on insoluble polyvinylpyrolidone. J. Food Sci. 34, 154–155.

YANG, H. Y., and STEELE, W. F. 1958. Removal of excessive anthocyanin pigment by enzymes. Food Tech. 12, 517–519.

YOKOYAMA, H., and VANDERCOOK, C. E. 1967. Citrus carotenoids. J. Food Sci. 32, 42–48.

CHAPTER 17

Food Hazards

There are several conditions that render food repulsive or hazardous to health. In the latter case, the law often established tolerances because some nutrients, such as fluorine and copper, are highly toxic in large doses. The hazards in foods may be classed as substances that may cause mechanical damage, filth, spoilage and toxins.

NONNUTRITIONAL OBJECTS

Various nonnutritional objects are found in food occasionally and may have got there accidentally or they may have been added.

In the harvesting of lentils, small stones are often picked up from the soil. Since they may be about the same size as the lentils, they may escape detection. Consumers must examine lentils in detail. The chemist who processes them must try to eliminate stones, seed pods and other extraneous matter. A flotation brine, or some other effective device may be employed, for one stone in a can of lentils may lead to loss of sales or a lawsuit.

Food processors are plagued by fibers from cotton or jute bags, splinters from trays or crates, broken glass, nails, screws and bearings from the machinery. A flour mill uses a magnet in a conveyor to catch iron objects before they reach and damage the rolls, but many processes do not lend themselves to such a solution. Some have used an X-ray device to detect dense foreign objects in packages.

Much contamination of this sort occurs in the kitchen from broken dishes or glassware. One instance that came to my attention occurred in a cafeteria. An employee broke a thermometer in a doughnut fryer, which dumped mercury and broken glass into the fat. Both promptly sank to the bottom and the manager asked me if it were safe to use the fat. I advised her that the risk of a customer ingesting broken glass or mercury was greater than the value of a gallon or so of fat.

The only way to avoid the presence of inedible objects in food is eternal vigilance on the part of the processor.

FILTH

Filth is more common in food than is commonly supposed. It usually consists of hairs or excreta from rats, mice, birds or insects, or parts of insects. Milk may contain hairs from the cows or dust from the hay or from cleaning the barn, and a creamery or milk plant usually tests the milk it receives by filtering a sample through a white paper disk. But the milk may pick up flies or other insects during the processing.

Almost any food is susceptible to insect contamination. One of the worst problems is wheat. The crease in the seed harbors insect eggs, which hatch during storage, or the eggs may go through milling and the insect appear in the flour, especially in whole wheat flour, which includes the bran. There are machines available that destroy insect eggs with considerable efficiency in ground grains, but they increase the cost of processing considerably.

All the grains attract rats and mice and it is difficult to exclude them from storage areas. The farmer's corncrib was built about two feet off the ground, and supported by posts surrounded with sheet iron. This excluded rodents, but the method doesn't lend itself to large scale storage. Concrete silos are effective after the grain gets that far from the field.

Peanuts are even more susceptible to rodent and insect infestation than grains.

Canned sweetcorn may contain both insect fragments and corn silk. Pesticides reduce the infestation and state and federal agriculturists are hard at work trying to reduce insect and rodent infestation and destruction of food.

ADDED OBJECTS

Many years ago when a large disk of chocolate coated fondant sold for a cent, it was common to wrap a cent in waxed paper and include it in a disk—one or two to the box, so that whoever bought the disk got his money back. It was a good sales scheme, but young children often broke teeth or swallowed the cent. Also small metal toys were added to boxes of popcorn or other confection. These also caused trouble and the law now has a special clause [Section 402 (d)] designating such additions as adulteration. Before the Pure Food Law of 1906, candy containing plaster of paris or other additives was common.

Fortunately, additives of the above type are now rare and consist mostly of paper coupons. However, I recently bought a box of oatmeal that contained a glass tumbler, which is harmless, of course, unless it gets broken in the handling of the package. Its inclusion is of doubtful legality. If a processor wishes to add an inedible object to a package of food, he should read the law first.

FOOD SPOILAGE

Spoilage may occur anywhere from the field where the food is produced to the kitchen refrigerator. It is mainly, but not solely, caused by bacteria, yeasts or molds. It may also be due to staling or rancidity.

Bacteria

There are four organisms responsible for most of the bacterial spoilage: streptococcus, salmonella, clostridium and botulinus.

Bryan (1974) reported 1,703 outbreaks of food borne disease involving 97,590 cases that were reported officially in the five years, 1969–1973. This amounts to nearly 20,000 cases a year, and in addition, there are thousands of cases that are never reported officially.

From January to June, 1969, there were 66 outbreaks of food borne illness reported involving 4,078 patients; 20 outbreaks affecting 1,829 patients were caused by *Clostridium perfringens*, 20 involving 1,329 patients by *Staphlycoccus*, 14 outbreaks with 713 cases were caused by salmonella and 5 with 10 patients were caused by botulism of which 3 originated in the home and 2 were unexplained.

A survey of causes of contamination indicated that inadequate refrigeration led the list with 336 cases. Preparing foods in advance of service came next with 154, followed by infected food handlers 151 and inadequate heat processing 114. Holding food in warming devices also accounted for 114.

Most of the contamination occurs after the food is in the hands of the consumer or the feeding establishment, but not all of it. Raw food is often contaminated and the processing does not sterilize it.

Contamination.—Pathogenic organisms occur in the intestinal tracts of animals and humans. In butchering, meat is often contaminated from this source.

In a cow barn, manure dries and circulates as dust that carries organisms or spores. Sometimes the milk is infected before it is drawn. One of the most troublesome germs is *Botulinus abortus*, the organism of Bang's disease in cows and undulant fever in the human. Tuberculosis is a common disease of cows and the disease organism is transmissable to the milk. Cows with these diseases are removed from the herd, but there is always a chance of contamination before the disease is detected. In one instance, several young people died of tuberculosis in a small city in which the milk was supplied by a single herd all of which was found to have tuberculosis.

Pasteurization in which every particle of milk is heated to 145°F for 30 min will kill these pathogens and any others that may have contaminated the milk, such as *Escherichia coli, Salmonella* or *Clostridium.*

If milk is not pasteurized, the infection carries through to cheese, ice cream or other dairy products. The processor must not overlook the fact that these products are subject to contamination from sources other than the milk; frozen eggs or added gums may carry pathogens or spoilage organisms. Freezing does not sterilize any food. Pasteurization of the milk removes the tuberculosis and undulant fever hazards, but botulism along with the less serious infections may be introduced later in the process.

Contamination of foods from soils is common and hard to avoid. *Salmonella, Clostridium perfringens* and *Clostridium botulinum* are all organisms that may come along with vegetables or even with dust. Botulism is the worst problem for it is a serious disease. Mortality has been reported as high as 65% although, in more recent years, it has dropped to 30% (Wilson and Hayes 1973). Even in those who survive, recovery is a slow process.

Botulinum is the most hazardous of all the microorganisms found in food; it is spore forming and the spores are heat resistant. Boiling temperature will not inactivate the spores; the food must be heated to 115°C or more for several minutes after the heat has penetrated and raised the temperature of all the food to that level. Furthermore, the organism is anaerobic and will multiply in a sealed can if the spores are not destroyed.

Botulism has been reported from meats and other low acid foods, such as smoked and dried fish, mushrooms, beets, ripe olives, green beans and even tomatoes. The organism is unlikely to produce toxins at pH lower than 4.5, but the acidity will not destroy the toxin once it is formed. Its formation is prevented by 10% salt or 50% sucrose.

The chemistry of bacterial toxins has been slow to develop because of the extremely small amount of them; 2 μg of the botulism toxin is said to be lethal.

Bacterial toxins have been found to be proteins. That of botulism apparently has a molecular weight of 150,000. The biochemistry of proteins and toxins is in an elementary state. Why most proteins are essential foods and a few are toxins remains to be explained, which is not surprising when we consider the complexity of the former and the scarcity of the latter.

The majority of botulism attacks are caused by home processed foods. In the home, foods are generally put up in glass which is a poor conductor of heat and the temperature will not exceed 100°C unless a pressure cooker is used.

Commercial processes, however, are not immune. In a newspaper item of December 14, 1974, there was a notice of warning of canned beef stew, and the processor was trying to locate and recover all the suspected cans. Less than a year ago, a soup company recalled a large number of cans of chicken soup. Such instances are usually the result of insufficient heat treatment.

FUNGAL TOXINS

Although ergot poisoning and poisonous mushrooms have been known for centuries, intensive investigation of fungal toxins is recent—mostly in the last half of the 20th century. The molds have come in for the most investigation in the food field and the toxins are called *mycotoxins,* named from the Greek word for mushroom.

There are thousands of species of fungi, a vast number of molds and, like other plants, they differ in their chemical activity; a few, such as *P. roqueforti,* hydrolyze fat. This one is used to ripen Roquefort and blue cheese. Some molds produce antibiotics such as penicillin and streptomycin and some produce toxins. The food chemist, not being a mycologist, must avoid all molds except pure cultures of those known to be useful.

Molds grow at temperatures from 0°C to at least 30°C and seem to thrive best at 15° to 25°. They grow on any food or feed. A surprise to the author was to see seventeen drums of lard with the surface covered by a green mold. But a chemist from the company that produced the lard, said that it was a common and troublesome occurrence. Molds require moisture and the drums had been filled with hot lard and sealed. Under refrigeration, the moisture in the air of the head space had condensed and dropped onto the lard.

Although syrups usually contain too much sugar to mold, this same head space problem often results in a layer of mold on the surface where the condensed water has diluted the syrup. Mold spores travel by air and about all the chemist can do is to see that the walls and floor of the plant as well as all the products in it are free from mold.

Animals are frequently poisoned by mycotoxins. Hay that is stored before it is thoroughly cured is sometimes a source of fatal toxins and, since many products of food industries go into cattle feed, these must also be free from molds.

Prunes, raisins and other dried fruits, cheese and smoked meats are among the more susceptible foods. The molds require oxygen for growth and so are rare in canned or bottled foods, but cardboard or plastic packages "breathe" with change in temperature and are likely to bring mold spores into the package; and mold spores can be on any food that is not heat processed.

The mycotoxins are not proteins, are water soluble and relatively heat stable and so they may penetrate the food even though the mold is only on the surface. Cooking may not destroy them. Even though a mold does not produce a toxin, it generally produces a disagreeable odor and taste in the food.

Aflatoxins

In 1960, thousands of turkeys in England died and were found to have severe liver lesions. The cause was traced to the feed, which contained peanut meal. Further investigation revealed that death was caused by toxins produced by the mold *Aspergillus flavus,* which serves as the source of the name *aflatoxin.*

In the decade since this initial discovery, the mold has been found on grains, soybeans and several kinds of nuts; peanuts seem to be a favorite substrate for it. Mycologists have found that the toxins are also produced by other molds closely related to *A. flavus* and chemists have separated several toxins, determined the formulas of at least nine of them and even synthesized some.

Aflatoxin, B$_1$ R = H
Aflatoxin, M$_1$ R = OH

The other aflatoxins have formulas similar to those of B. and M. forms.

The aflatoxins have been found to affect turkeys, dogs, ducklings, rats, mice, guinea pigs, swine and trout. Sensitivity differs from species to species and also with the age of the animal. The duckling is the most sensitive of the animals tested.

Experiments with several species of animal show that in large doses aflatoxins are lethal and that small doses for a long period cause cancer of the liver.

Mold is easily observed in whole kernel nuts or other seeds but the alkali refining destroys the toxins in the oils from these sources, which leaves peanut butter the most likely source of these toxins in food. The chief industrial precaution is to see that susceptible food and feed are dry before storage and not stored in an atmosphere of high humidity.

Other Fungi

There are several species of fungi that are known to produce toxins; some are *Aspergillus,* some *Penicillium,* some *Fusarium* and there are others. All these fungal toxins have a ring structure, generally several rings and some of the toxins contain heterogeneous rings.

Ergot produces an alkaloid and the poisonous mushrooms that have been studied produce a variety of toxins and hallucinogens; only a few of them have been identified. Muscarine is a furan derivative and three other mushroom toxins are indole derivatives.

NATURAL POISONS IN FOODS

Some foods contain poisons themselves, even when they are entirely free from molds or other fungi.

Elements

One of these toxic substances is selenium. If a soil has a high selenium content, the element is absorbed and replaces sulfur in amino acids. Selenomethionine, for example, has been found in wheat gluten. When protein concentrates are made from the press cake from vegetable oil seeds, a test for selenium should be made.

Aluminum deserves mention because it has been a controversial element for over half a century and was the basis of a famous law suit over the use of aluminum salts in baking powder. Hosts of experiments have been conducted and the final decision is that the amount of aluminum in plant tissues combined with that from baking powder and from pots and pans is not toxic.

One of the virulent elements is arsenic. For many years, lead arsenate and Paris Green, $Cu(C_2H_3O_2)_2 \cdot Cu_3(AsO_3)_2$, were used as pesticides on fruits and vegetables. Organic pesticides have largely replaced them, but the organics are now under criticism and inorganics may again be encountered. Arsenic is not absorbed from the soil by plants to any great extent and so there appears to be no great danger from that source, but residues on fruits and vegetables were not uncommon when it was in use as a pesticide. A few years ago, a cargo of American apples was denied entry into England because of lead arsenate residues. The residues were generally removed by washing the fruit with dilute acid, but some shippers failed to do it.

Persons differ in their susceptibility to arsenic poisoning and the trivalent arsenic is more toxic than the pentavalent. Arsenic in the body tends to concentrate in the hair and nails. Hair analysis is used to detect arsenic poisoning. Anything over 3 ppm is considered suspicious.

In recent years, there has been much agitation about mercury in food, which has led to extensive investigation. Calomel (HgCl) was formerly used as a drug, but the dichloride ($HgCl_2$) has been known for years to be extremely toxic. The effect of a constant intake of the metal or its compounds over a period of years has not been established. The maximum concentration in a safe diet has been arbitrarily set at 0.05 ppm.

Certain bacteria convert metallic mercury into methyl mercury, CH_3Hg^+ or dimethyl mercury, $(CH_3)_2Hg$. These compounds are much more toxic than the metal or its inorganic salts. Methyl mercury is soluble in water, dimethyl mercury is not, but it does evaporate into the air.

The chief case of mercury poisoning from food occurred in Japan where mercury wastes were discharged into Mihamata Bay. Between 1953 and 1960, 43 persons in the area died, presumably from mercury poisoning. Seafood constitutes a large part of the Japanese diet, and shellfish from the bay were found to contain as much as 24 ppm of mercury on a wet basis or 27–102 ppm on a dry weight basis.

Other incidents of mercury poisoning have resulted from eating seeds intended for planting, that had been treated with a mercurial fungicide. The U.S. Environmental Protection Agency (EPA) has banned the use of mercurial fungicides on seeds.

Mercury occurs widely in nature, including the ocean, and tuna and swordfish have been found to contain over 0.5 ppm. Thus the FDA has placed 0.5 ppm as the limit for fish, which has hampered the fishing industry. Gomez and Markakis (1974) analyzed forty foods including meats, fish, cereals, fruits and vegetables and found 0.01–0.03 ppm of mercury.

The Food Technologists Expert Panel on Food Safety (1975) report that mercury in food is not a toxic hazard.

Fluorine is a sensitive element in the human diet. It is a constituent of the teeth, but if an excess is consumed over a period of years, it results in bone malformation and perhaps death. Fluorine occurs widely in nature and is absorbed by plants in limited amount and is often concentrated in seafood. Plants seldom contain more than 1 ppm.

The chief fluorine hazard is from water. The content of well water differs with the locality because of differences in fluorine content of the soils; 4–5 ppm is not uncommon in the water from deep wells and as much as 50 ppm has been reported.

The risk of too much fluorine differs with the amount of water consumed. If a food process employs water, especially if the water remains in the product, the chemist needs to consider the fluorine problem. Dehydration of foods concentrates the fluorine. However, 1 ppm is considered both safe and essential in the diet.

ADDED TOXICANTS

Several potentially hazardous substances are added to foods either accidentally or purposely and are of concern to the food chemist. The most common is salt.

Salt

Practically all food and drinking water contains sodium. In wells near the ocean, the sodium content of the water is high; at Crandal, Texas, the water is said to contain 70 mg sodium in 100 ml. If one drank two liters a day, we would ingest 3.4 gm of sodium or the equivalent of 9 gm of salt.

In addition to the sodium in foods and drinking water, salt is added to most foods as a flavor or a preservative and sodium is a fundamental part of baking powder and may constitute 10% of it. Canned vegetables contains 1.6 gm, dried fruit 8.1, green olives 2.4, bouillon cubes 24, breads 0.27–0.58, biscuits 0.63, sardines 0.76 and soups as much as 0.4 gm in 100 gm of the product.

Although sodium is a dietary essential, the actual requirement is uncertain; it may be as low as 0.5 gm of salt a day. The requirement differs with the amount of water drunk, the extent of sweating and other factors. The FNB recommends 1 gm salt for each liter of water consumed but estimates that the average American consumes 6 to 8 gm daily.

A high sodium diet is objectionable in certain heart conditions and the processing of low-salt foods deserves increasing attention.

Nitrates and Nitrites

The soil and ground waters contain nitrates and other nitrogen compounds and food grown in such soil absorbs them, for most nitrogen compounds are soluble. Any insoluble compounds are broken down by bacteria to soluble ones. The absorption of nitrate from the soil differs with the species; spinach leads the list followed by beets, radishes, green beans and lettuce in no particular order. The amount of nitrate absorbed does not increase linearly with the nitrate content of the soil although the amount is increased slightly by the use of more nitrogenous fertilizer.

Nitrates and nitrites are used in curing meats and the amount permitted is regulated by the USDA. Different amounts are allowed in different processes. A general regulation limits nitrate to 200 ppm in any finished cured meat product.

Potassium or sodium nitrate is apparently no more toxic than salt; 8 gm to 15 gm is said to be a lethal dose. This dosage, of course, could not occur from the use of either food or water. The nitrites are much more toxic. The exact lethal dose is unknown but is estimated to be about 1 gm. Furthermore, in an acid pH, nitrates react with secondary amines to form nitrosamines

$$\underset{\underset{CH_3}{|}}{\overset{\overset{CH_3}{|}}{NH}} + H{-}O{-}N{-}O \longrightarrow \underset{\underset{CH_3}{|}}{\overset{\overset{CH_3}{|}}{N}}{-}N{=}O + H_2O$$

| Dimethyl | Nitrous | Dimethylnitrosamine |
| Amine | Acid | |

The dimethylnitrosamine shown in the equation has been shown to be carcinogenic.

Pesticides

The agricultural crops all have their enemies, such as the cotton boll weevil, the Colorado potato beetle, the Mediterranean fruit fly, the corn borer and hosts of others. Then, there is ergot on rye; molds, yeasts and other fungi. Finally there are the rodents: rats, mice, chipmunks, squirrels; also raccoons, woodchucks and birds. Chemicals designed to destroy these pests have a variety of names, such as fungicides, insecticides and rodenticides, but for convenience, the agricultural chemists adopted the term *pesticide* for any or all of them. The name is more than a convenience for one chemical may destroy more than one type of pest.

Lead arsenate and copper compounds were once the chief pesticides, but chemists have been at work for years trying to find better ones and have developed several for actual use on the crops. The chlorinated hydrocarbons were widely used for several years, but DDT has been banned for most uses because it has been found to be injurious to humans, and some of the other organics are now under suspicion. Most gardeners have found that it is impossible to raise either fruits or vegetables without the use of a pesticide and so they cannot all be abandoned.

The food chemist must learn which pesticides are used on the various crops and develop a method of removing them from the food before it reaches the consumer. Since total removal may be impossible, a government agency has set a safety limit for many of the pesticides now in use. Each new one is a special problem.

Cyanogenic Glycosides

The dangerous glycosides are those that yield HCN on hydrolysis by acids or enzymes. They have been reported in over 1000 species of plants. They may occur in the roots, tubers, leaves, fruits or seeds. Farmers have known for a century or more that if cattle ate the leaves of the wild cherry, especially the wilted leaves from a tree blown down, they died.

Several human foods contain these glycosides; about 20 have been isolated from 50 species.

Amygdalin.—This glycoside has been found in almonds, and the kernels of the cherry, peach and plum. It hydrolyzes to gentiobiose $C_{12}H_{22}O_{11}$, HCN and benzaldehyde.

Linamarin.—Cassava root and lima beans have been known to cause death from the presence of linamarin. The glycosides are hydrolyzed by enzymes when the cells are broken. In the manufacture of cassava starch, the roots are crushed and left for a time for the hydrolysis to occur. Part of the HCN escapes to the air and the remainder is removed by washing. The linamarin is hydrolyzed according to the following equation

Linamarin   D-glucose

$$HO—\underset{\underset{CH_3}{|}}{\overset{\overset{CN}{|}}{C}}—CH_3 + H_2O \longrightarrow HCN + \underset{\underset{CH_3}{|}}{O{=}C}—CH_3$$

Acetone   Acetone
cyanohydrin

The amount of HCN produced by lima beans differs with the variety. An American variety produced 10 mg/100 gm of seed, but other varieties have been found that produce as much as 300 mg/100 gm.

The lethal dose of HCN lies between 50 mg and 250 mg, so that a 4-oz serving of some of these varieties could be lethal. The United States restricts the importation of lima beans to those yielding less than 20 mg HCN per 100 gm of seed.

Phenols

A huge number of phenolic compounds occur in foods and among the additives. Many of them are toxic. The anthocyanins, flavonoid pigments and the tannins are phenolic. Charred wood contributes phenols to whiskey, brandy and rum. Among the phenolic additives are vanillin and methyl salicylate (oil of wintergreen). Fortunately, most of these phenols have low toxicity. Those that have been tested have a lethal dose of 0.5 gm to 50 gm per kg of body weight. For the average adult man, this would be 35 gm to 350 gm.

One phenolic compound that deserves attention is gossypol, which occurs in cottonseed. It is a complex phenol of unknown human toxicity. The lethal dose for rats is 2.63 mg per kg of body weight; for dogs, the dose is 10 mg to 20 mg per kg of body weight. Pigs are also sensitive and chickens about the same as rats.

Cottonseed meal is coming into use as a human food because of the scarcity of protein. The subject of gossypol and its effects have been studied extensively and any chemist who has occasion to work on cottonseed meal for either food or feed should review the literature.

Oxalates

Oxalic acid, $H_2C_2O_4$, and its salts occur in many plants used as food. Spinach leads the list with 0.3% to 1.2% of the fresh weight, rhubarb is second with 0.2% to 1.3% and cocoa has 0.5% to 0.9%. The other common fruits and vegetables contain much less. Some plants that grow in pastures contain large quantities of oxalic acid and are often hazards to cattle.

Pure oxalic acid is toxic but the lethal dose is 5 gm; the salts are less toxic and calcium oxalate is insoluble. The acid in spinach is a mixture of potassium and calcium oxalates. Since the oxalate in spinach is in the form of salts, a lethal dose would be far more than anyone would care to eat. Several deaths have been reported from eating rhubarb leaves, but in each case, the actual cause of death was doubtful because of lack of an autopsy and sufficient tests for oxalates. At least 400 gm of rhubarb stems would be required to contain a lethal dose of oxalate.

Spices

According to Hall (NAS 1973), eleven potentially hazardous substances have been detected in spices: HCN, allyl isothiocyanate, umbellulone, capsaicin, glycyrrhizic acid, β-asarone, coumarin, menthol, safrol and thujone.

Of these ten substances, coumarin and safrol have been banned by the FDA although small quantities occur in natural flavors that are still permitted. HCN and allyl isothiocyanate occur in glycosides, which have been discussed. Umbellulone has been found in the leaves of the California Bay tree. It is a severe irritant and toxin. The oil, however, is seldom used. The common bay oil from *Laurus nobilis* does not contain the toxin. Asarone is present in calamus, which has been banned recently. Capsaicin is the pungent component of the red pepper; it is a powerful irritant and can be detected in water at 1 ppm.

Capsaicin

Glycyrrhizic acid is the flavor component of licorice and has been extensively used to flavor medicines and confections. It is a glucoside consisting of a complex structure of 5 fused benzene rings liberally substituted with methyl and hydroxyl groups and 2 molecules of glucuronic acid. It has been known to cause hypertensions, sodium retention and heart enlargement when large quantities of licorice candy were consumed. Licorice is a flavoring that must be used sparingly in a food that may be consumed in quantity.

Menthol is a component of oil of peppermint, which is popular in candy, cordials, chewing gum, tooth paste and cigarettes, so that an individual may consume or inhale large quantities of it. Some persons develop hives from it but it is no severe hazard.

Myristicin is present in the oil of nutmeg and mace. These spices contain from 8 to 15% oil of which some 4% is myristicin.

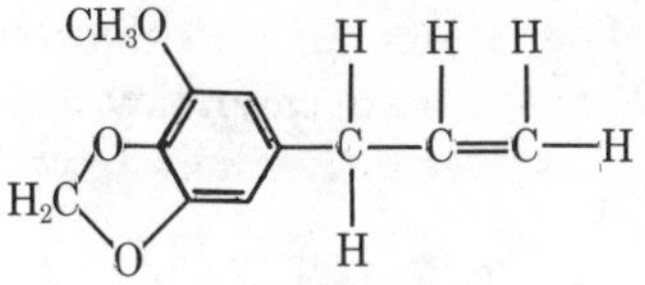

Myristicin

Myristicin also occurs in black pepper, parsley, celery, dill and carrots. Nutmeg has been used in medicine because of its narcotic effect, but this effect is followed by headache, cramps and nausea. It is probably no great hazard because a heavily flavored nutmeg food contains less than 100 mg myristicin in a 4-oz portion and the content of the common vegetables is very low.

However, myristicin is methoxysafrol, and safrol has been banned by the FDA under the Delaney amendment because 5 gm/kg in rat food caused cancerous liver tumors and 40 to 80 gm per kg of food caused extensive liver damage in dogs. The similarity in structure between myristicin and safrol does not necessarily cast reflection on the safety of the former, for it is a well known principle of chemistry that a change in chemical composition results often in a complete change in physiological properties.

Thujone, the most famous of these eleven compounds, is the major component of the oil of wormwood.

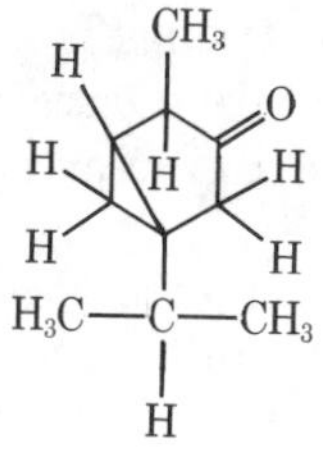

Thujone

Oil of wormwood was the main flavor of absinth until 1915 when the French banned it, and it is not on the approved list of essential oils in the United States for use in food, although slight quantities are used in vermouth. Consumption of 30 mg per kg of body weight produces convulsions and brain damage. It does not occur in any of the common foods.

Seafood Toxins

Seafood is the most hazardous of all foods. Several species of fish are poisonous. Those varieties are generally recognized and eliminated from commercial processing and are therefore only a minor hazard, but the casual fisherman may not recognize them. Shellfish are another matter. They grow near the shore, and pathogenic organisms from sewage have been known to contaminate them. The organisms and the toxins from them are destroyed by cooking. The ones that are eaten raw, even though they may have been frozen, are hazardous. Recently, there have been cases of hepatitis traced to raw clams.

The most insidious seafood poison is produced by some of the plankton on which the shellfish feed. One such organism is *Gonyaulax catenella*. When this organism reaches a concentration of 200 cells per milliliter of water, it is enough to make clams toxic. The poison has been isolated and named *saxitoxin*. It has the empirical formula, $C_{10}H_{15}N_7O_3 \cdot 2HCl$. It is a substituted tetrahydro purin of complex structure. It is estimated that the lethal dose for a man is less than a milligram and there is no known antidote for it. The organism, when it reaches a concentration of 20,000 per milliliter is the cause of the "red tide."

There are many other causes of seafood poisoning. An animal assay is the only means of detection. This may be applied to an area of production but not to processing. Furthermore, shellfish from a given area are poisonous or not according to the infestation of the water, which differs with temperature, salinity, currents or any other factor that may change the concentration of the offending organism. The frequency of poisoning has eliminated the production of shellfish in some coastal areas.

CARCINOGENS

Tumors are growths on body tissue that have little in common with the tissue on which they grow. They are classed as benign or malignant; the latter are tumors that continue to grow and to spread to adjoining tissue—cancers.

There are three main classes of carcinogens: viruses, chemicals and radiation. As food chemists, we are concerned only with the chemical agents.

Certain divalent inorganic ions and compounds have been proved to be carcinogenic, namely, Co^{++}, Ni^{++}, Cd^{++}, Be^{++}, Pb^{++}, CrO_4^{--} and some of the silicates, notably asbestos. Miller (NAS 1973) states that, ". . . perhaps as high as 80% of nonskin cancers in the human is also of environmental origin. Thus, the chemical carcinogens in our total environment, through lifetime of exposure to small amounts of these compounds may rank as major causes of human cancer."

Many hydrocarbons and other organic compounds are known to cause skin cancer, but an intensive investigation failed to discover any specific chemical structure as the cause.

In 1958, Congress passed an amendment to the Food, Drug and Cosmetic Act, which was sponsored by Representative Delaney of New York and is always referred to as the Delaney Amendment. It states: "No additive shall be deemed to be safe if it is found to induce cancer in man or animal."

Prior to the Delaney Amendment, little attention had been paid to the carcinogens among the food additives, but since then, the FDA and private laboratories have been active in tests for carcinogenicity. Rats, mice or hamsters are generally used for the tests. As a result of such tests, the cyclamate sweeteners, calamus and safrole have been banned by the FDA and several other substances are undergoing tests.

Microorganisms

In addition to toxins, several species of molds and bacteria have been found to produce carcinogens. Nitrosamines have been discussed. They appear to be involved in cancer of the colon.

Safrole

The oil of sassafras has been banned as a food additive. The oil had been used for years to flavor root beer. It may still be a hazard to those who make tea from the bark of the sassafras root.

Asarone

Asarone is the main flavor in oil of calamus and, since it has been found to be carcinogenic, it has also been banned by the FDA. The oil has been used mainly as a flavor in vermouth and bitters.

Safrole   Asarone

Carrageenan

This gum injected into rats proved to be carcinogenic although oral administration has not been found to have that effect. It is not on the present GRAS list.

Summary

The risk of cancer from the consumption of foods can only be guessed. There is no way to measure it, for carcinogenic substances in many cases must be consumed over a long time to produce tumors—in the human, this may be 25 years or more. The Delaney clause in the law is only a precaution and applies only to additives and not to the natural components of food.

Other carcinogens will undoubtedly be found among the chemicals that occur in food or are added to it, but at present, the aflatoxins and the nitrosamines seem to offer the greatest risk. Food that is moldy or otherwise spoiled should not be eaten.

SPECIAL FOODS

Lactose intolerance in people of all ages results in abdominal pain, bloating, flatulence and diarrhea. Those who are susceptible lack lactase in the intestine and the lactose is not absorbed but remains in the intestine as a substrate for gas forming bacteria. Similar sucrose intolerance has been reported.

Flatus

Production of gas in the intestinal tract has been ascribed to several foods, particularly beans of all kinds. There is always some gas in the intestines, mostly from swallowed air. This may amount to a liter or more a day. Analysis has shown it to contain: 50% nitrogen, 12% to 14% oxygen and variable amounts of CO_2, H_2 and CH_4.

Experiments in which the diet was over 50% baked beans resulted in 168 ml per hour consisting of, CO_2 49.1%, O_2 1.1%, N_2 19.4%, H_2 8% and CH_4 22.4%. The control diet containing no beans produced only 15 ml per hour and the composition was, CO_2 8.1%, O_2 3.6%, N_2 61.2%, H_2 19.8% and CH_4 7.3%. Search for the component of the beans that causes the gas production led to the conclusion that certain oligisaccharides such as raffinose and stachyose were the offending substances. The intestinal tract contains no enzyme that hydrolyzes these sugars.

Honey

Fatalities have been reported from eating honey. Bees collect honey from the nectar of flowers and from honey dew and, since there are many poisonous plants, it is inevitable that poisons will be collected along with the nectar. That this does occur has been known for more than 2,000 years. Symptoms differ with the diverse sources of the poisons and also because people differ in their reactions to the same toxin. Tingling and numbness in the extremities, loss of consciousness, lowered pulse rate and a blue face have been reported as symptoms.

Rhododendron and mountain laurel are two of the offending plants. The risk is not great in commercial honey because some poisons kill the bees, and commercial bee keepers are aware of poisonous plants in their area and the seasons in which they bloom.

Allergy

One of the greatest of all food hazards is allergy and there is nothing the chemist can do about it. Some people do not seem to be allergic to anything, others may be affected by one or a number of things and the only remedy is to avoid the allergen. Allergens are not general toxins, they are probably proteins.

It has been estimated that 20% of school children are allergic to something. The effects are mainly on respiration (asthmatic) and in the form of a rash on the skin. Food allergies may produce nausea, abdominal pains and vomiting.

Many persons are allergic to pollens of various kinds and try to avoid them. In foods, milk, eggs, butter, strawberries, nuts, grains, mangoes, chocolate and oranges have proved to be allergenic. Usually cooking will destroy the allergen, but not always. An overindulgence in a food can develop an allergy for it.

One important feature of an allergy is that a person may suffer the effects without contact with the substance. In one case, a woman who was allergic to mangoes avoided them but did not recover. On cleaning the refrigerator, she found a mango in the back of it.

Because of the heat effect, canned foods are less allergenic than the same foods fresh or frozen.

BIBLIOGRAPHY

ANON. 1972. Botulism. J. Food Sci. 37, 985–988.

ANON. 1972. Nitrites, nitrates and nitrosamines in food. J. Food Sci. 37, 989–992.

ANON. 1973. Mercury in Food. J. Food Sci. 38, 729.

ANON. 1970. Allergy. Chem. Eng. News. 48, No. 20, 84–135.

AYRES, J. C. 1972. Manioc. Food Tech. 26, No. 4, 128–138.

BRYAN, F. L. 1974. Microbiological food hazards today. Food Tech. 28, No. 9, 52–56.

CHANG, K. C. 1970. Growth of salmonella at low pH. J. Food Sci. 35, 326–328.

COON, J. M. 1969. Naturally occurring toxicants in foods. Food Tech. 23, No. 8, 55–59.

DACK, G. M. 1966. Importance of food-borne disease, outbreaks of unknown cause. Food Tech. 20, No. 10, 39–44.

DACK, G. M. 1964. Microbial food poisoning. Food Tech. 18, No. 12, 80–82.

ESCHER, F. E., KOEHLER, P. E., and AYRES, J. C. 1974. A study of aflatoxin and mold contamination in improved variety of pecans. J. Food Sci. 39, 1127–1129.

FOSTER, E. M. 1963. The problem of salmonella in foods. Food Tech. 23, No. 9, 74–78.

GOMEZ, M. J., and MARKAKIS, P. 1974. Mercury contents of some foods. J. Food Sci. 39, 673–675.

GRAHAM, H. D. 1968. The Safety of Foods. Avi Publishing Co., Westport, Conn.

HALL, R. L. 1973. Toxicants occurring naturally in spices and flavors. In Toxicants Occurring Naturally in Foods. Natl. Acad. Sci., Washington, D.C.

IFT EXPERT PANEL. 1975. Naturally occurring toxicants in foods. J. Food Sci. 40, 215–222.

LEE, L. S., CUEULLER, A. F. and GOLDBLATT, L. A. 1968. Appearance and aflatoxin content of raw and dry roasted peanut kernels. Food Tech. 22, No. 9, 81–84.

MATCHES, J. K., and LESTON, J. 1968. Low temperature growth of salmonella. J. Food Sci. 33, 641–645.

MILLER, J. A. 1973. Carcinogenic agents. In Toxicants Occurring Naturally in Foods. Natl. Acad. Sci., Washington, D.C.

MUTH, O. H. 1967. International Symposium. Selenium in Biomedicine. Avi Publishing Co., Westport, Conn.

NAKAMURA, M., and KELLY, K. D. C. 1968. Perfringens in dehydrated soups and sauces. J. Food Sci. 33, 424–426.

NATL. RESEARCH COUNCIL. 1973. Toxicants Occurring Naturally in Foods. National Academy of Sciences, Washington, D.C.

SCHINDLER, A. F. et al. 1974. Mycotoxins produced by fungi isolated from in shell pecans. J. Food Sci. 39, 213–214.

WILSON, B. J. 1969. Mycotoxins and other plant poisons. Food Tech. 23, No. 6, 70–77.

WILSON, B. J., and HAYES, A. W. 1973. Microbial toxins. In Toxicants Occurring Naturally in Foods. Natl. Acad. Sci., Washington, D.C.

Index

Acetic acid, 60, 247, 264
Acetylmethyl carbinol, 175
Acids, and taste, 279
  fatty, 9–12
  in natural fats, 12
  nomenclature, 11
  polyunsaturated, 10
  saturated, 9–10
  unsaturated, 10–11
Acree-Rosenheim reaction, 67
Acrolein, 32
Adamkiewicz reaction, 67
Additives, 134–147, 260
Adsorption, 131–133
Agar, 54–55
Alanine, 60
Albumin, 67
  egg, 131
Albuminoids, 68
Alcohol, 73, 262
  cetyl, 198
  energy from, 73
  myricyl, 198
Allergy, 304
Aluminum, 295
Amino acids, 59–61, 285–286
  essential, 64, 75–76, 228
Amygdalin, 298
Amylopectin, 52–53
Amylose, 52–53
Analysis, additives, 139–140
  colors, 143–144
  control, 113
  extraneous matter, 113
  for nutrition, 109–113
  general, 114–115
  mineral, 83–85
  official, 4–5
  official methods, 6, 13, 18, 21, 89, 93, 106, 109, 111
  sampling, 115–116
Anemia, 98
Annatto, 275
Anthocyanins, 272–273
Anticaking agents, 135
Antioxidants, 30–31
Apples, 252–253
Arabic acid, 55
Arabinose, 36, 41, 55, 272
Arachidonic acid, 10, 199
Arginine, 60, 64
Arsenic, 295
Asarone, 303
Ascorbic acid, 3, 65, 86, 102–104

Aspartame, 135, 285–286
Aspartic acid, 60
Association of Official Analytical Chemists (AOAC), 4–5
Astaxanthine, 272
Avidin, 100
Azelaic acid, 29

Babcock method, 111, 175
Bacteria, 96–97, 99–100, 150–151, 165, 170–171, 175, 180, 247, 264–265, 291–292, 295, 302–303
Bagasse, 238
Baking powder, 226–227
Bananas, 253–254, 280
Barley, 213–214, 232
Beef, 282
Beet sugar, 240–241
Beets, sugar content, 237
Behenic acid, 198
Benzoyl peroxide, 180
Beriberi, 92
Beverages, alcoholic, 262–263
  nonalcoholic, 259–261
Bidwell-Sterling apparatus, 84–85
Biotin, 100–101
Biuret reaction, 65
Blanching, 160
Boneblack, 132, 239
Botulism, 165
Bread, baking powder, 226
  staling, 225
  standards, 224–225
  yeast, 223–225
Breakfast foods, 266–267
Browning, enzymatic, 159–161
  nonenzymatic, 161–162
Buckwheat, 214
Butter, 24, 27, 114, 130, 142, 174–176
  colors, 275
  consumption, 191
  refrigeration, 150–151
  standard, 175, 176
Butylated hydroxyanisole (BHA), 31, 192
Butylated hydroxytoluene (BHT), 31, 192
Butyric acid, 9, 10, 11, 128

Cacao beans, 246–248
Cacao products, 245–250
Caffeine, 136